National-Level Planning in Democratic Countries

Town Planning Review (*TPR*) Special Studies

Edited at the Department of Civic Design, University of Liverpool, by PETER BATEY, DAVID MASSEY and DAVE SHAW

The scope and thematic coverage of this series reflect the interests of **TPR**, focusing on all aspects of town and regional planning and development in countries with advanced industrial economies and in newly emergent industrial states. From the broad fields of theory, policy, practice, implementation and methodology, among the planning interests of the series are: urban regeneration; environmental planning and management; strategic and regional planning; sustainable urban development; rural planning and development; coastal and estuary management; local planning; local government and planning; transport planning; planning history; and urban design. The following volumes are published:

Planning for Cities and Regions in Japan, edited by Philip Shapira, Ian Masser and David W. Edgington
Vol. 1, 1994, 213pp., ISBN 0–85323–248–2

Design Guidelines in American Cities: A Review of Design Policies and Guidance in Five West-Coast Cities by John Punter
Vol. 2, 1998, 224pp., ISBN 0–85323–893–6

Rebuilding Mostar: Urban Reconstruction in a War Zone by John Yarwood and others
Vol. 3, 1998, 160pp., ISBN 0–85323–903–7

Proposals for future volumes in the series are welcome and should be sent to *The Editors, Town Planning Review, Department of Civic Design, The University of Liverpool, Abercromby Square, Liverpool, L69 3BX, UK*

National-Level Planning in Democratic Countries

An International Comparison of City and Regional Policy-Making

Edited by RACHELLE ALTERMAN

LIVERPOOL UNIVERSITY PRESS

First published 2001 by
LIVERPOOL UNIVERSITY PRESS
4 Cambridge Street
Liverpool L69 7ZU

British Library Cataloguing-in-Publication Data
A British Library CIP record is available

ISBN 0-85323-845-6

Typeset in 10½/12½ Plantin by
Wilmaset Limited, Birkenhead, Wirral
Printed in the European Union by the Alden Press, Oxford

To Doron, my partner in all,
and to our joint ventures—Edan and Nora.

TABLE OF CONTENTS

LIST OF FIGURES

LIST OF TABLES

PREFACE

This book would not have come about were it not for a particular event—one might say, a coincidence—which brought to my attention the need to study national-level planning. A major crisis in Israel brought to the front burner the need for multi-sectoral long-range planning at the national level.

The crisis arose from external events. During the last months of the Soviet Union, international conditions changed, and mass emigration of Jews and family members was allowed. The estimate in 1990 was that within three to five years, Israel, with a population of 4.5 million at that time, should expect to take in 1.5 to two million people. Understandably, a feeling of crisis overtook government bureaux which were concerned about the impact of such an avalanche on housing, land use, the environment, economic development, and many more issues.

Although national-level planning institutions and powers were—and still are—ample in Israel, by 1990, long-range multi-sectoral planning at the national level, which had its heyday in the 1950s and 1960s, had become a distant memory, gradually withering away. A team of planners and academics decided to take the initiative and show government the way. The team was headed by Adam Mazor, a leading planner-architect and professor at the Technion-Israel Institute of Technology, where the country's major planning school resides. At first regarded as an academic-professional project outside government, the 'Israel 2020' project was later adopted by a consortium of government bureaux, while still maintaining its out-of-government status.

The 'Israel 2020' project (described in detail in Chapter 11) took it upon itself to create a new style of planning at the national level: no longer a blueprint land-use plan and command-style implementation system that Israel was still carrying from the 1950s, but a new style that would take Israel into the twenty-first century. We were attempting to create an integrated policy covering land use, infrastructure, economics, environment, water, agriculture, and social policy.

New modes of plan-making would not be enough; there should also be new ways of institutionalising national-level planning so as to fit better with the trends of deregulation, privatisation and changes in governance styles that Israel, like most other advanced-economy countries, was undergoing. We were therefore seeking to know more about alternative modes of national-level planning as carried out in other democratic countries, and it was my role in the team to find out.

I quickly discovered that the literature on the subject was scarce. I therefore proposed that we create our own knowledge base by studying how national-level planning operates in a sample of ten democratic countries with advanced economies, representing a variety of sizes, geographic locations and governance systems. Using my prior familiarity with 'planning systems' through my comparative research on land-use planning issues, I was able to locate and contact a highly knowledgeable researcher from each of the ten countries. From the response of each of the prospective authors, I learned that for them, as for me, the challenge of sorting out what planning was being carried out at the national level was a new one. None of us recalled having ever debated this topic in an academic or professional conference. With a set of common guidelines that I developed, the authors were requested to write up their description and assessment of national-level planning in their own country.

We convened at the Technion for a comparative seminar, where we shared what we had found with each other and with the other members of the 'Israel 2020' team. Having discovered the dearth of published research on national-level planning, we assumed that planners, decision-makers and researchers from other countries would also be interested in our findings. The draft volume issued by the 'Israel 2020' project was submitted for review to the Liverpool University Press editors. The anonymous reviewers' useful comments were used as guidelines for the format of this book. After a process of my own editing and, where necessary, rewriting and re-editing, the chapters were updated by each of the authors so as to be accurate to late 2000.

This book represents a unique linkage of research with practice. It was born of the needs of planning practice, albeit a very special and ambitious specimen of planning. I would hazard to guess that there are few cases in which a planning project generates research at such a scale. This unique link should be credited to Adam Mazor. His unsurpassed vision and professionalism have produced not only one of the most ambitious national planning enterprises anywhere in the West, but have also spun off many layers of knowledge. This is one of them. It is therefore a pleasure to thank Adam Mazor and the 'Israel 2020' project for stimulating my curiosity and for supplying the infrastructure that has made this book possible.

My thanks go also to Guy Kav-Venaki, at the time a graduate student of planning and my superb research assistant, who organised the logistics of the joint seminar and the draft volume with the greatest skill imaginable. Special thanks to the editors of the *Town Planning Review* special series and to the three anonymous reviewers, whose comments and guidance were priceless. And not least, I am very grateful to all thirteen contributors to this book, who have been not only most knowledgeable and insightful, but also extremely cooperative and patient, having tolerated my repeated queries and been willing to do last-minute updates.

RACHELLE ALTERMAN, *February 2001*

ABOUT THE CONTRIBUTORS

Rachelle Alterman Holding the David Azrieli chairs in town planning at the Technion-Israel Institute of Technology, Professor Alterman has degrees in urban and regional planning, social science and law from Israeli and Canadian universities. Her authoritative research on planning theory, comparative planning law and institutions, and comparative land policy has been published internationally in leading journals, and she has authored and edited several books.

Michael J. Bannon Professor Bannon, a prominent Irish analyst of urban and regional planning policy, is Dean of the Department of Regional and Urban Planning, University College, Dublin, Ireland and one of the foremost authorities on planning in his country. He is the author of the chapter on Ireland in the European compendium on land-use planning systems.

Göran Cars Associate Professor and Deputy Head at the Department of Infrastructure and Planning, Royal Institute of Technology, Stockholm, Sweden, Göran Cars' professional interests include housing, infrastructure provision, urban regeneration and development. He has also focused on planning processes and negotiations as a method for decision-making in urban planning.

David W. Edgington David Edgington is Associate Professor in the Department of Geography, University of British Columbia. His research centres on Japanese urban and regional restructuring, Japan's trade and overseas investments in the Pacific Rim, and other economic-geography topics. Having published widely on economic change in Japanese cities, he is author or editor of several books, including *Planning for Cities and Regions in Japan* (*Town Planning Review* Special Studies, No. 1).

Stig Enemark Associate Professor and Reader in business management at the Department of Development Planning, Aalborg University, Denmark, Dr Enemark is regarded as a leading authority on the Danish national land-use planning system, and is the author of the chapter on Denmark in the European compendium on national land-use planning systems.

Malcolm Grant Professor and Head of the Department of Law and Land Economy, Cambridge University, UK, and previously Professor of Law at the University College, London, Malcolm Grant has authored several major books and he is nationally and internationally recognised as a leading authority on planning and land-development law and policy in the UK. He has also served on many parliamentary and government commissions.

Bjorn Hårsman Director of a private-sector Institute of Regional Research and formerly Director of the Stockholm Region Government Planning Agency, Bjorn Hårsman is also a Professor with the Department of Regional Planning at the Royal Institute of Technology, Stockholm, and is one of the leading analysts on Swedish housing and regional planning, about which he has published extensively.

Ib Jorgensen Ib Jorgensen is Associate Professor at the Department of Development Planning, Aalborg University, Denmark. He joined academia after many years in professional practice in the private sector. He is a well-known Danish critic on the social aspects of urban and regional planning and on decision-making in planning, including public participation.

Jerold S. Kayden Jerold Kayden is Associate Professor of Urban Planning at the Graduate School of Design, Harvard University. Trained in law and in city planning, he has published widely on topics that bridge planning and law, and take an international perspective. In a short time, he has joined the small group of leading American thinkers on land-use planning law and policy in the USA and internationally.

Gérard Marcou Gérard Marcou is Professor of Public Law at the University of Paris I Panthéon-Sorbonne. With degrees in law, economics and public policy, he is one of the leading French experts on planning and land-development law and policy, also holding a prominent office at the Institut Français des Sciences Administratives.

Hans J. M. Mastop Professor and Dean of the Faculty of Policy Sciences, Department of Spatial Planning, University of Nijmegen, Professor Mastop is a highly respected authority on decision-making, implementation, and innovation in Dutch urban and regional planning, both in the Netherlands and in the European Union.

Paula Russell Paula Russell is Assistant Lecturer in the Department of Regional and Urban Planning, University College, Dublin, Ireland. Since receiving her Master's degree in Regional and Urban Planning in 1993, she has worked as research assistant and project coordinator with the Department and other university divisions. She is currently pursuing a PhD on the social impacts of urban renewal.

Gerd Schmidt-Eichstaedt Professor and Dean of the Department of Environment and Social Sciences at the Technical University of Berlin, Gerd Schmidt-Eichstaedt is internationally known as one of the prominent analysts of German planning and land-development law and policy, and is the editor of an international comparative book on land-use planning systems.

Paul-Hidehiko Tanimura Paul-Hidehiko Tanimura is Professor and Chair of the Institute of Policy and Planning Sciences; former Director of the Office of Planning at the University of Tsukuba, Japan; and former Dean of the College of Socio-Economic Planning. He has studied planning in Manitoba, Canada, and at the University of Tokyo, and has served as a senior planner in Tokyo. Professor Tanimura is one of the best-known Japanese authorities on regional planning and planning methods.

ONE

NATIONAL-LEVEL PLANNING IN DEMOCRATIC COUNTRIES: A COMPARATIVE PERSPECTIVE

Rachelle Alterman

National-level planning in democratic countries has been almost all but ignored by researchers in urban and regional planning since the reconstruction years following the Second World War. Having become identified in many people's eyes with communist regimes and coercive government practices, national-level planning fell into some disrepute. Yet, this book will show that planning is carried out on the national level to some degree in each and every one of the ten countries studied, even though the goals, degree of comprehensiveness, subjects, institutions, format, powers and effectiveness differ widely from country to country. There are even modest trends whereby, on the threshold of the twenty-first century, national-level planning is growing in importance in democratic, advanced-economy countries. These trends point to the need to revisit planning theory.

Why study national-level planning?

Little attention has been given to the study of national-level planning in Western countries for many decades. The attention of planning theorists in recent years, as expressed in the majority of topics for empirical research and the themes of normative debate, has tended to focus on decision-making modes relevant more to the local and individual levels than to the national one. The three compendiums of planning theory published in the 1990s (Campbell and Fainstein, 1996; Mandelbaum et al., 1996; Stein, 1995) do not include even a single chapter devoted to the types of issues, institutions and modes of decision-making typical of national-level planning.

This book was born of necessity. It is not the result of a library search for lacunae in knowledge, but of a real-life need for knowledge about how different countries handle their land-use (or 'spatial') planning issues at the national level. The need was Israel's—a country that ostensibly already has a high degree of national-level planning, but where a group of planners and academics involved in the ambitious 'Israel 2020' planning team[1] was seeking to know more about alternative modes of national-level planning. I began to search the literature for ideas. Is national-level planning still relevant in the era of deregulation, privatisation, globalisation of markets and communication, political federations, and deep changes in governance styles and social trends (Hall, 1993)? What is the range of ways in which national planning is prepared, adopted and implemented in democratic advanced-economy countries? What are the emerging issues and trends in modes of national-level planning? Cross-national learning, we assumed, would be an effective way of getting some answers to these questions. This book shares what we found with planners and decision-makers in other countries for whom many of these questions are no less relevant.

Since very little published analysis of national-level planning in democratic countries was found, we opted for the 'home made' approach. We invited leading researchers from ten democratic, advanced-economy countries to do the research for us. We purposely resisted what might have been considered a natural tendency to look only for countries widely reputed *a priori* to have 'good' or 'exemplary' national-level planning (such as the Netherlands or Japan). On the contrary, we were looking for as broad a sample of advanced-economy countries as possible, located in different parts of the world, with different types of constitutional structures, different population and area sizes, and different types of needs and constraints. From among the 18 OECD member countries with the highest Human Development Index scores (see Table 1) I selected nine: the United States, Ireland, the United Kingdom, Sweden, Denmark, the Netherlands, France, Germany and—the only Far East country in that group—Japan. Together with Israel, these ten countries represent a good proportion of the world's advanced economies. Such a broad range of countries should represent the gamut of degrees, modes, and approaches to national-level planning. For each of the countries selected, we invited a leading scholar in urban and regional planning policy or planning law to write a critical account of national-level planning in his or her country. The group was convened in a joint seminar for mutual presentations and discussions.

Each one of the researchers related that this was the first time he or she had been asked to think in particular about what occurs in planning at the national level. Indeed, none of us could recall any national or international academic or professional conference where the focus of analysis and debate was on the desirability of national-level spatial planning. The absence of a body of systematic knowledge on national-level planning could also be discerned from the fact that few of our authors cited any literature on national planning. The analysis was new and challenging for us all.

In this chapter, I shall first define 'national-level planning' and then present some background geographic and demographic data on the sample countries. As a starter for the task of comparison, I begin with an attempt to classify and rank the group of countries by the degree to which national-level planning has been institutionalised in each. This classification is correlated with population density. I then identify the major trends in national-level planning, as they emerge from a comparative reading of the ten chapters. These trends pertain to the variety of reasons for the inception, perseverance or demise of national-level planning, to its political-ideological contexts, to the emerging modes and styles whereby it is carried out, and to the tools and problems of implementation. I conclude by pointing out some of the challenges that the findings hold for planning theory.

Defining 'national-level planning'

I prefer the term 'national-level planning' to 'national planning' not so much because the latter may for some be tainted by the history of coercive planning, but because in this book we are not looking necessarily for *the* national, comprehensive planning enterprise. Rather, we would like to focus on any spatial planning carried out at the national level. We are also not dealing with what are known in Euro-English as 'national planning *systems*', 'spatial planning systems', or 'statutory planning systems'.[2] These terms refer to the legal and institutional bases for enabling urban and regional planning and regulation in a particular country (also known as 'town and country planning' in the UK, and 'land-use planning' in the USA). The vast majority of countries in the world do have some such 'system', and certainly all advanced-economy countries do. While the laws and regulations are usually enacted for the country as a whole, and in that sense are 'national', the major part of the legislators' attention is usually focused on enabling land-use planning and development regulation at the local and regional levels, down to the approval of a particular development. The fact that there is a national planning *system* established by parliamentary legislation does not necessarily mean that there is also planning or policy-making that is carried out at the national level.

Planning 'systems' have drawn considerable attention not only from legislators and planners, but also from researchers. There is extensive literature that analyses and evaluates how these systems operate and what they achieve. There is an even larger literature on particular aspects of spatial planning systems such as local plan-making, development permits, land-value implications of regulation, procedures for public participation, negotiation with developers, tools for farmland or historic preservation, financial aspects, etc.[3] The 1990s have seen the most rigorous and comprehensive, though not the first,[4] effort made to date to analyse planning systems in European countries. But *The EU Compendium of Spatial Planning Systems and Policies*, commissioned by the European Union (European Commission, 1997),[5] can devote only a few pages to the policies and instruments vested at the national levels.

The focus of this book is quite different. In my guidelines to the authors, I asked them figuratively to cut off the layer vested at national level and carried out by any government or quasi-government body at that level. The meaning of 'national-level planning' in this book is thus both narrower and broader than 'spatial planning systems'. It is narrower because it focuses only on those planning functions carried out at the national level. Planning carried out at the regional and local levels is not our focus, and is mentioned only for the purpose of explaining the division of labour between the national and the other levels, or as the context for implementation of national-level plans or policies. Our meaning is wider because we cover not only statutory 'land-use' planning in the traditional, regulatory sense, but also planning and policy-making carried out by national-level agencies outside the statutory 'system'. Thus, the authors were also requested to analyse *sectoral planning and policy-making*, such as transportation, environment, housing, economic development, parks, agriculture—whatever are spatial policy areas determined at the national level.

A set of common guidelines was given to each of the invited authors. The guidelines requested that they describe and analyse the goals, subjects, functions, instruments, and modes of implementation of national-level planning. The authors were also challenged to evaluate critically the successes and failures of national-level planning as they see them. We purposely avoided a rigid check-list format, preferring that each chapter draw a complete picture as viewed by the author, based on our general guidelines. Given the pioneering nature of this enterprise and the lack of an *a priori* shared definition of what comes under the umbrella of national-level planning, there inevitably remain some differences in the interpretations given by the various authors to the span of topics to be covered, especially regarding sectoral planning.[6]

In this chapter I will try to weave together a picture of national-level planning in democratic, advanced-economy countries as it emerges from the reports of the ten countries included in our sample. My purpose is to draw out both the shared and the differing elements, and to identify any common trends that may have emerged. But first, a few comparative indicators to highlight the wide range of geographic, demographic and economic contexts that characterise our sample.

Comparative background variables

Some argue that there is no justification for studying national-level planning because countries differ a great deal in size. A small country, they argue, can be regarded as equivalent in planning terms to a single region in a large country. This argument is unacceptable since nations are legal-institutional entities that, like people, come in various sizes, but have similar limbs. Indeed, there are important similarities among many countries, regardless of their size, in the basic hierarchy of planning functions on the sub-urban, urban, and often also the regional levels. In this book we ask to what extent there are also similarities and differences in the allocation of planning functions to national-level institutions.

Table 1 provides some background data to characterise our sample countries. Since we are dealing only with advanced-economy countries, the differences in Gross National Product per capita are relatively small compared with most countries of the world. However, if one compares the ten countries among themselves, there are considerable differences in affluence: Israel is the least affluent,[7] while the USA is the most affluent.

Table 1 *Selected physical, demographic and economic indicators in the sample countries, 2000*

Country	*Population*	*Births per woman*	*Annual immigration as percentage of total population***	*Population annual growth rate (%)*	*GNP per capita (US$) adjusted for purchase power parity* (1999 estimates)	*Human Development Index* (rank in the world 1999)***	*Surface area (sq. km)*	*Population density (residents per sq. km)*
USA	265,179,000	2.06	0.35%	0.91 %	33,900	3	9,629,091	28
France	59,329,700	1.75	–	0.38 %	23,300	12	547,030	108
UK	59,511,700	1.74	0.42%	0.25 %	21,800	10	244,820	243
Germany	82,797,400	1.38	1.34%	0.29 %	22,700	14	357,021	232
Netherlands	15,892,200	1.64	0.77%	0.57 %	23,100	8	41,532	383
Denmark	5,336,400	1.73	–	0.31 %	23,800	15	43,094	124
Sweden	8,873,100	1.53	0.02%	0.02 %	20,700	6	449,964	20
Ireland	3,797,300	1.91	–	1.16 %	20,300	18	70,283	54
Japan	126,550,000	1.41	–	0.18 %	23.400	9	377,835	335
Israel	6,100,000*	2.60	1.00%	1.67 %	18,300	23	20,770	294

Sources
US Government—*World Factbook* 2000 (estimates for mid-2000, unless stated otherwise).
* *Israel Statistical Yearbook 2000*
** *The Europa World Yearbook 1997*, Europa Publications Limited, London
*** *U.N.D.P. Human Development Report*, 2000.

Population sizes differ immensely. The largest populations are found in the USA and Japan. The smaller populations are in Ireland, Denmark, Israel, Sweden, and the Netherlands, in ascending order, with population sizes between 1.5 and eight per cent of the USA's. The medium-size populations are found in France, the UK, and Germany. Surface areas also differ greatly, with Israel the smallest—about two per cent of the area of the USA. Other small countries are the Netherlands, Denmark, and Ireland, and the rest are medium-sized. Population densities also vary a great deal among the countries: Sweden, the USA and Ireland have lower densities; the Netherlands, Japan and Israel are at the high end (Israel, with the highest population growth rate, can be expected to surpass the other countries in the sample). Less variation exists in demographic attributes. Most advanced-economy countries have an almost-zero natural growth rate. Israel is an exception, with a relatively high growth rate. Germany and Israel are the only countries with significant immigrant-intake (Israel's current rate of one per cent is much lower than the four per cent rate in 1990–92 during the mass influx of immigrants from the former Soviet Union).

We did not pre-judge countries on the basis of their constitutional structure. Whether federal or unitary, nations could, in theory, elect to have or not have national-level planning in some degree and form. We made sure to include within our sample two federal countries—the USA and Germany—and have indeed found that they differ immensely from each other on the point of national planning. Other legal-institutional variables, such as the number and sizes of local authorities, degree of local government power, or central government structure, have also been raised as reasons why national-level planning should not be studied comparatively. The countries in our sample differ widely on many such variables, but these are not necessarily related to national-level planning, as we shall see shortly.

Degrees and formats of national-level planning

National-level planning comes in many shapes and forms. This emerges vividly from reading the ten country chapters. There are, at the same time, also some shared formats. I shall not attempt a systematic comparison of all the similarities and differences. The information provided by each of the country chapters is too rich for that. Most of this chapter is therefore devoted not to a mechanistic comparison, but to an attempt to point out selected aspects, issues or trends that, in my judgement, are of special interest. I shall begin with an attempt to summarise a few key aspects of the formats and powers of *comprehensive* spatial planning (excluding sectoral planning). Based on this comparison, I will offer a rough classification of the sample of countries by 'degree of institutionalisation' of national-level spatial planning.

THE FORMATS OF NATIONAL-LEVEL PLANNING

There is no consensus among the countries on the format for national-level planning, neither comprehensive, nor sectoral. The ten countries vary not only in the 'software' of goals and values to be furthered (to be discussed in a later section), but also in most aspects of 'hardware'—institutions, instruments and procedures.

National-level spatial planning can be classified into two schematic types: comprehensive and sectoral. Comprehensive planning (called 'facet planning' by Mastop in the Dutch chapter) seeks to take as integrated and multi-sectoral a view as is feasible, both geographically and subject-wise, and seeks to guide the use of land for all types of needs in a coordinated fashion. By this I do not mean, of course, a fully comprehensive view of all aspects of national policy as called for by the now defunct *rational-comprehensive model* in planning. That model is rightly regarded as impossible in most real-life formats of public policy-making and administration. I mean a planning or policy-making perspective that uses the integrative potential of a spatial, land-use view, to develop policies that go beyond the areas of responsibility of any particular sectoral government ministry. Full comprehensiveness is, of course, hard to achieve, so most 'comprehensive planning' would be less comprehensive than the 'ideal type'.

At the other extreme, sectoral planning deals with a single sector such as transportation, housing, agriculture, parks or health. Usually, sectoral planning is carried out by a specific agency in charge of regulating or initiating projects in that particular area. Since public planning is never done in a vacuum, sectoral planning obviously also needs to coordinate with other sectors, to some degree (Alexander, 1998). Much day-to-day planning at any level, and certainly at the national level, should be classified somewhere in between comprehensive and sectoral planning. The relationship between comprehensive and sectoral planning at the national level will engage us several times more in this chapter.

Table 2 focuses on *comprehensive* national-level spatial planning. It summarises the answers to five of the questions that the authors were requested to answer:

- Is there a formal national-level plan or set of policies, and if not, are there any informal substitutes?
- Which national-level institutions are responsible for such planning or policy-making?
- What are the procedures for preparing such policies and for their approval?
- What is the legal status of such policies?
- What are the modes and means of implementation?

In addition, the authors were asked to answer the following questions about *sectoral* planning:

- In what subject areas is sectoral planning carried out at the national level and who are the agencies in charge?

Table 2 *The formats of national-level (comprehensive) spatial planning in ten democratic countries*

Country	*Is there a national spatial plan or some substitute?*	*Institutions in charge*	*Procedures for plan/ policies preparation*	*Legal status of the plan/ policies*	*Modes of implementation and coordination*
USA	No national plan or comprehensive spatial policy. Partial substitutes—federal policies (laws) in selected sectoral or multi-sectoral areas.	The Congress and Senate. The Federal Government's specific Offices, as legislated in each policy case.	Almost every federal action regarding land-use policy requires special authorising legislation. Usually accompanied by wide-scale public and professional debate.	No legal authority for national spatial planning. In a few areas, the constitution permits direct federal control. In most others, the Fed relies on special sectoral or multi-sectoral legislation.	Planning powers reside in local governments. Most states have weak authority. The Fed relies on incentives to states and locals to stimulate local or regional multi-sectoral planning.
Ireland	No national plan or comprehensive policy; no regional plans. EU-stimulated 'National Development Plan' but this is just project proposals. Intention to prepare national plan/ guidelines.	The Department of Environment and Local Government. Planning National Appeals Board (which is required to update itself regarding multi-departmental policies).	The National Development Plan was prepared by the Department of the Environment according to EU funding conditions only (Community Support Framework).	No national-level plans under Irish planning law. The 'National Development Plan' has only contractual status (regarding the EU).	The 'Plan' obtains EU funding. EU monitors the projects. Implementation through public and public-private partnerships. Minister has oversight authority over local plans—rarely used.
Sweden	No national plan. 'Vision 2009' is a research document with little impact. Occasional national planning directives to implement projects of national import.	The Ministry of the Environment. The National Board of Housing, Building and Planning. The County Administrative Boards.	Vision 2009 prepared by a professional, not a political body. Extensive public consultations, compiled for review by Interior.	The Planning and Building Act makes no mention of a national plan. Some legislated criteria allow intervention in local plans. Central government has administrative authority to make policies.	Regarding Vision 2009—no clear modes stated. National criteria and projects of national interest maintained through oversight powers over local planning by the county boards.

UK	No national plan. Highly effective substitute—set of (30) Planning Policy Guidelines (formerly called Circulars) without maps; Regional Planning Guidelines (with maps).	The Secretary of State for the Environment, Transport and the Regions (a 1997 integration of DOE and DOT).	Until recent years, national guidelines were prepared with consultation only within government. Since the 1990s, greater public exposure; hearings and public report expected.	No legislated basis for the PPGs or RPGs! Local plans must relate to 'relevant considerations'. Parliament's Select Committees becoming involved in oversight.	The Secretary of State has extensive oversight and intervention powers through objections in hearings, the inspectors, call-in power; hears appeals.
The Federal Republic of Germany	Federal Spatial Planning Report (advisory). Guidelines for Regional Planning—deal with inter-regional equality, development priorities, environment, transportation, European status.	The *Bund*. Ministry of Spatial Planning, Building and Urban Development. Spatial Planning Advisory Council advises the Minister. Conference of *Bund* and *Länder* planning Ministers; research Agency for Construction and Spatial Planning.	The Federal Report is prepared by Ministry of Spatial Planning at regular intervals. Presented to *Bund*. Incorporates all decisions of the Advisory Council and the Conference of Ministers, who prepare Guidelines.	The *Bund* legislates framework for supra-local spatial planning. The Federal Spatial Planning Act mandates preparation of Federal Report—advisory only. Mandates coordination with the *Länder*. Implementation is exclusively in hands of the *Länder*—legal separation of powers.	Extensive and effective vertical and horizontal forums for coordination and mutual compliance. Consultation with the Conference and the Advisory Council. Coordination of *Länder* targets with the Federal Report and Guidelines and local and regulatory plans. Monitored by Agency.
Denmark	A National Planning Report presents goals and policies, coordinates with EU policies. The planning law divides the country into three land uses. National directives used to implement specific projects or policies.	Minister of Energy and the Environment; the National Planning Department. The Parliament has a specific planning role.	The Report is prepared and submitted to Parliament after each national election; by the National Planning Department in cooperation with other agencies. Broad public debate through hearings.	The Report is mandatory to prepare but is not legally binding—'persuasive'. Directives are binding. Revision of regional plans based on Report. Local and regional plans mandatory to prepare and revise and cover whole country.	Plans at lower level must not contradict higher levels ('framework control'). Minister has authority to order counties and municipalities to ensure national policies; can veto or call in a plan. Usually negotiated, not imposed.

Table 2 *Continued*

Country	*Is there a national spatial plan or some substitute?*	*Institutions in charge*	*Procedures for plan/policies preparation*	*Legal status of the plan/policies*	*Modes of implementation and coordination*
France	In 1995 Act—*Schéma national d'aménagement et de développement du Territoire* (draft, 1997). New Bill will likely recast as 'strategic choices' plan. New powers to issue *directive territoriale d'aménagement*.	Ministère de l'aménagement du territoire; Conseil national (CNADT); CIAT—Comité interministeriel—DATAR Délégation à l'aménagement.	Prepared by the Ministry. Approval by the Conseil. Adopted by Parliament. Reappraisal every five years. Consultations with all government levels. Sectoral schemes and state-region plan conventions approved by CIAT.	Mandatory to prepare National Scheme and sectoral policies—compliance by government bodies only, now more flexible. Directives are binding on all lower government bodies. Conventions binding in contract law.	Through financial planning and budgeting. Since 1998, recast as 'collective service schemes'. Scheme to guide long-term sectoral policies, regional schemes, and the 'State-region plan conventions'.
Japan	Since 1945, five National Development Plans (spatial-economic—latest in 1998). Plus short-term economic plans (13). Separately—a national land-use plan that broadly designates five land uses.	Under PM are several planning agencies, including National Land Agency with six Bureaux, headed by a Minister of State. Advisory expert Land Development Council.	Development Plans prepared by Agency, debated in the Council to reflect popular opinion. Approved by Cabinet. Process of ongoing inter-agency and interest-group consultation.	Comprehensive Land Development Act (economic-spatial plans) not 'binding'—consensus. Land-use national plan authorised by National Land Use Act. Indicative, not binding (like economic plans). Separate legislation for local and regional planning.	Economic planning and budgeting are key tools to implement and coordinate among policy areas. Also personnel-sharing method of coordination. Local and regional urban planning is separate, not well integrated.

The Netherlands	Four national spatial strategic 'Notes', last one updated 1993 as 'Fourth Note Extra'. Fifth Note preparation begun 1998. Full national coverage by provincial strategic plans. First National Plan for the Environment Plus initiated, 1993.	Cabinet Council of Physical Planning and Environment Spatial Planning Advisory Council; Ministry of Housing, Physical Planning and Environment (VROM). National Spatial Planning Committee; National Spatial Planning Agency.	Minister prepares through Agency. Professional consultation with Spatial Planning Committee; interest group input through Advisory Council. Approval by Cabinet Council. May be debated in Parliament. Extensive public exposure and debate.	1965 Act on Spatial Planning authorises but does not mandate national planning; plans are non-binding, indicative. 1993 Act on Environmental Management mandates integrated national (and provincial) environmental plans. 1993 Act on infrastructure projection.	Higher government levels can intervene in local decisions (rare). Extensive coordination mechanisms among sectoral planning (e.g. National Spatial Planning Committee for compulsory interagency coordination). Key National Projects can override local.
Israel	National comprehensive land-use middle-range plan ('#31' 1993); new long-range plan under preparation ('#35'). Extra-governmental multi-sectoral indicative plan ('Israel 2020'). About 25 statutory sectoral land-use plans.	The Minister of the Interior and the Planning Administration. The National Planning and Building Board (mostly inter-ministerial). Six District Planning and Building Commissions. The National Lands Administration.	Comprehensive plans initiated by Ministry or by National Planning Board. Mandatory consultation only with District Commissions; informal public hearings a growing trend. Approval by Board and Cabinet. Sectoral plans—same (but different initiators).	National spatial plans authorised by 1965 Planning and Building Law. Not (effectively) mandatory to prepare or update, but once prepared strictly binding on all public and private interests. Same status to sectoral statutory plans.	Strict consistency required of district, local, and detail plans and permits. Minister has wide call-in powers. National and district planning bodies based on central-government representation for inter-ministerial coordination. National plans can impose specific projects.

- What are the interrelationships between comprehensive and spatial planning?
- How effective has the implementation process been, and how desirable the outcomes of both comprehensive and sectoral planning (in broad-brush terms)?

The table shows that there is no consensus on any of these questions: on whether there should be a document called a national 'plan' or 'policy' and what its format should be; on the identity of the ministry (or other bodies) charged with national-level policy-making; on the procedure for developing and approving such policies—who should prepare them and how they are to be approved; on the legal status of such policies; or on the types of implementation expected and the modes of carrying them out. Similarities frequently turn out to mask significant differences. For example, in the column titled 'institutions in charge', one can see that the ministry responsible often has the word 'environment' in its title, yet the actual span of responsibilities and the degree of coordination among component parts often differ significantly from country to country.

DEGREE OF INSTITUTIONALISATION OF NATIONAL-LEVEL PLANNING

Although some degree of national-level planning does exist in every one of the countries represented, the differences in degree are large. Figure 1 represents an attempt to classify the set of countries by degree of 'institutionalisation' of *comprehensive* national-level planning. The classification is relative, and is not deduced from a theoretical model. It is based on a comparative reading of the ten countries' reports and on consultation with the authors. My assessment is admittedly relative and judgemental, weighing together several of the variables and criteria summarised in Table 2, and adding the criterion of effectiveness not included in the table. The concept of 'institutionalisation' thus brings together several criteria related to comprehensive planning:

- Has there been an attempt at national-level spatial planning as expressed in a set of formal plans or policies? (These might be called 'plans', 'policy plans', or 'guidance statements'—the name does not necessarily correlate with the degree of importance or impact).[8]
- How comprehensive is the view taken by that set of plans or policies?
- To what degree are the procedures for making and approving such policies or plans well established in law, administrative procedure, or political action?
- How effective are the legal, administrative or financial powers available for the implementation of these policies?

I have called those countries with a low degree of institutionalisation of national-level planning, somewhat humorously, the 'have-nots'. This category includes the USA (not quite used to being classified among the 'have-nots') alongside Ireland

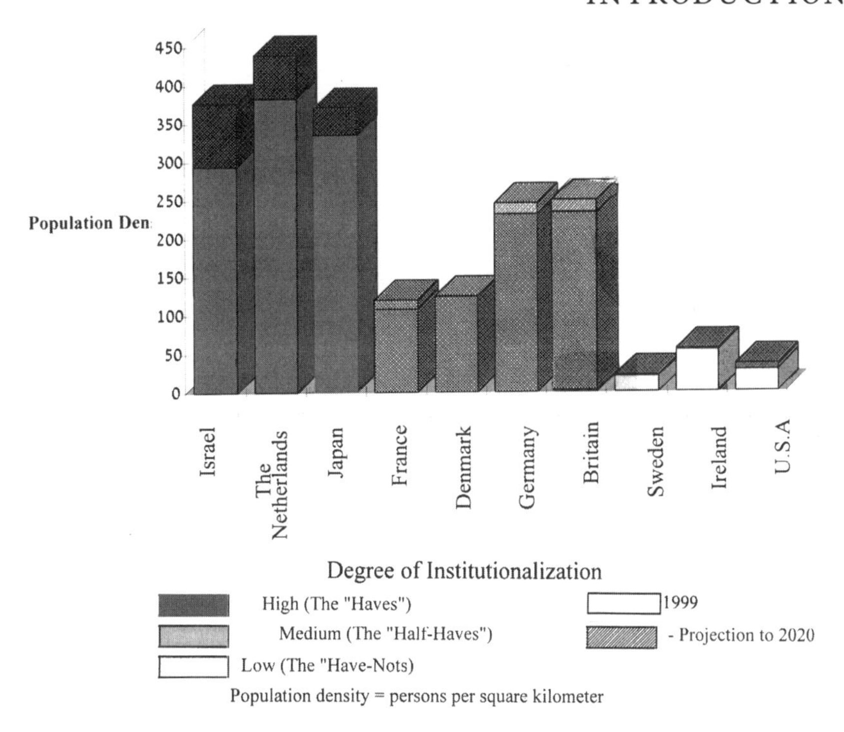

Figure 1. *Degree of institutionalisation of national planning in ten countries by population density*

and possibly Sweden (our two Swedish authors were at odds as to whether Sweden should today be classified in this category or the middle one). In these countries, comprehensive spatial planning exists at the local-government level, is relatively weak at the regional level, and is absent from the national level, except through sectoral planning. (The latter may be weakly or strongly institutionalised—that is not shown in the figure.)

At the other extreme are the countries with a relatively high degree of institutionalisation of planning—the 'haves'. This category includes the Netherlands, Israel and Japan. In these countries, comprehensive spatial planning at the national level is a distinct and visible task of specific government bodies, with a basis in law or administrative practice and with established procedures for approval and implementation. Yet implementation is by no means necessarily of the 'command and control' type. These three countries differ a great deal in the modes and formats of planning and implementation, as well as in degrees of effectiveness over time.

The middle level of the 'half-haves' includes France, as well as Denmark, Germany and Britain, more or less in that order. France could perhaps have been included with the 'haves' based on its ambitious 1995 legislation that sought to re-establish strong national-level spatial-economic planning. But in 1999, reflecting the change brought about by the 1997 national elections, that legislation was amended. I have thus classified France as falling within the medium-degree category. One can guess that most members of the European Union could be classified in this category.[9] The order of the chapters in this book has been arranged according to the order of the countries in Figure 1.

As may already be apparent, the classification by degree of institutionalisation is not technical nor simple. For example, it is not enough that a country has a document called a 'national plan' or 'national vision' in order to qualify for the higher category. For instance, Sweden, which we classified among the 'have-nots', does have such a national document, but our Swedish authors, Cars and Hårsman, do not attribute much weight to it.[10] A similar picture emerges from Ireland. A converse example comes from the UK, which has no 'national plan' or 'national vision', but which I classified in a higher category than Sweden and Ireland because its alternative mechanism—the Planning Policy Guidance—is much more effective in conveying and implementing central government's spatial policy than the Swedish or Irish tools. A third example relates to legal status. In Denmark, the preparation of the national report is mandatory in law, while in the Netherlands it is not. Yet the Netherlands, whose national-level planning is highly institutionalised in administrative practice and is very effective, is classified among the 'haves', whereas Denmark is classified among the 'half-haves'.

CAN THE DEGREE OF NATIONAL-LEVEL PLANNING BE CORRELATED WITH OTHER VARIABLES?

The classification in Figure 1 probably brings to readers' minds many hypotheses to try to 'explain' the differences, using, for example, some of the variables in Table 1. But as the rest of this comparative chapter and this book will show, the number of variables along which countries can be compared is almost endless. The state of knowledge regarding comparative planning in general, and national planning in particular, is so rudimentary that one should not take too seriously any attempt to test ostensibly hypotheses linking some such variable with degree of planning as a 'dependent variable'.

Having given this 'waiver', I did undertake one such exercise. Population density is a variable that can be hypothesised as possibly indicating an objective 'need' for spatial planning on the national level. In countries with lower densities, decision-makers and the electorate may not understand comprehensive national-level planning as an urgent requisite for the management of land resources. Indeed, the author of the US chapter, Kayden, offers this as one explanation of why national-level spatial planning has never taken root in his country. So, in Figure 1 I placed the classification of degrees of institutionalisation against the current and projected population densities.

A correlation does seem to emerge (one should not go so far with this analysis as to undertake statistical tests). The three countries with a high degree of institutionalisation of national planning—Israel, the Netherlands and Japan—also have the highest population densities, expected to reach 350–450 persons per square kilometre by 2020.[11] The countries with medium population densities ranging from approximately 100 to the mid-200s—France, Denmark, Germany and Britain—are classified as 'half-haves'. Whereas the three countries with the lowest population density levels ranging from approximately 50 down to 20—the USA, Ireland, and possibly Sweden—are classified with the 'have-nots'.

However, even this seeming correlation is less than convincing. Under the current state of knowledge we can by no means predict the emergence of national-level planning based on high population densities. Thus, although Sweden has the lowest density among the ten countries, until the early 1990s, while the 'Swedish model' of state intervention ruled high, Sweden had a much higher degree of national planning than it does today. And although France has the third lowest population density, it too was in the past considered a model of strong national-level planning, both in traditional land-use planning and control, and in the special type of spatial-economic planning which is France's *fortesse*. France attempted a revised 'come-back' to this model through its 1995 legislation, thus possibly requalifying for classification among the 'haves'.

The conclusion is that the decision to develop spatial planning on the national level is not predetermined by some specific set of variables that indicate a 'need' for planning. The decision to adopt national-level planning is a question of choice by decision-makers and voters. In other words, Davidoff and Reiner's (1962) classic point about the Choice Theory of Planning holds for national-level planning too.

Differing reasons for the introduction of national-level planning and its goals

The ten reports show that countries do indeed differ widely in the reasons that may have prompted decision-makers to introduce national-level planning or to phase it out at a particular time.

A TOOL FOR NATION-BUILDING

The three countries classified above as the 'haves' historically shared the view that national-level planning is an essential tool for nation-building. The Netherlands, with a long history of planning and regarded by many as the world's flagship of national spatial planning, has had a historic 'head start'. Mastop attributes this to its existential need for concerted public management of water in the lowlands and the need for a national effort for land reclamation. In recent decades, although these essential infrastructure and environmental policies have been routinised and these needs more or less assured, national-level spatial planning has not been phased out. Rather, it has shifted its focus to the challenges of meeting the internationalisation of the economy, of competing better within the European Union, of conserving the environment, and of meeting housing needs.

Japan and Israel, so distant geographically, socially and historically, are the closest in terms of the major role assigned to national-level planning as a tool for nation-building. One can guess that decision-makers among Europe's newly formed nations or its ethnic quasi-autonomous units may today also be considering the possibility of using strong national spatial planning as a tool for nation-building.[12] Japan and Israel also share the story of national-level planning, in danger of decline, being resurrected to serve new national goals. As Taminura and Edgington report, Japan used strong national planning in the post-Second World War years as a tool for repairing massive destruction and, even more, as a means for creating the astounding transformation of a traditional Asian society with a weak economy into a Western-style economy. Additional goals were added through the years to address Japan's scarce land resources and environmental sensitivity. It would be interesting to follow the story of Japan during its current economic recession. Will national planning be called upon to serve as a major tool for economic resuscitation?

Israel's story includes many parallels to both the Netherlands and Japan. The need for long-term planning to assure the life-sustaining water supply in the dry Middle East is reminiscent of the Netherlands' effort at managing the dangers of excess water. The social goal of nation-building after war and the challenge of creating an advanced economy from a developing country are reminiscent of Japan's goals. In contrast to the Netherlands and Japan, however, Israel is the only country within our sample and with few counterparts among advanced-economy countries worldwide where national-level planning is also assigned a distinct military-security (or perceived security) role. My account of Israeli national spatial planning, however, shows that by the latter 1970s, once the essential nation-building tasks were more or less fulfilled, national-level planning began a steep decline, which would probably have led to its eventual weakening from within. A major national crisis brought about by mass immigration—regarded as a nation-building goal in Israel—helped to 'save' national planning. It 'opened the eyes' of politicians and government officials to the need to redirect the old nation-building goals towards the environment and infrastructure, as is more typical of other Western countries.

REDUCING INTER-REGIONAL DISPARITIES WITHIN THE COUNTRY

The desire to reduce inter-regional inequalities within the country is a shared goal that has helped trigger national-level planning (independently of EU policies to be discussed later) in several of our sample countries both within Europe and outside it. In France this was the distinctive goal in the 1960s and 1970s, the heyday of national planning. As Marcou aptly explains, 'planning' in the sense of *aménagement du territoire* carries a very special meaning in France, referring not to traditional land-use planning but to the spatial aspects of socio-economic development. This same goal was also largely behind the enactment of the 1995 spatial planning act, but the focus there was on ensuring equal access to public services and utilities in all parts of the country. In Denmark and Sweden, reducing regional inequalities was a strong goal in the past, but is no longer so, since it has been more or less achieved. Alleviating regional disparities has also been an important, though not overriding, goal in Germany after unification. It has also been an important goal in Japan, where regional inequalities are still quite large. In Israel, the parallel policy is called 'population distribution' with the intent that economic incentives in the peripheries would not only help to equalise socio-economic levels but would also help to attract more population so as to contribute towards nation-building.

THE EU'S SPATIAL POLICIES AS A TRIGGER FOR NATIONAL-LEVEL PLANNING

During the 1990s, the European Union adopted the alleviation of inter-regional disparities as a key goal, as is clear from its spatial planning documents *Europe 2000* and *Europe 2000+* (European Commission, 1994) and the Spatial Development Perspectives approved in 1999. It has also been the focus of some of the EU's major implementation tools such as the structure funds and the community support policy (Faludi, 1997; Giannakourou, 1996; Kunzmann, 1995; Prodi, 1993; Shaw et al., 1996; and Marcou here). EU policies stipulate that eligibility for various types of incentives, such as the structure funds, is conditional upon proof that an adequate level of planning on a broader-than-project scale has accompanied the set of priorities submitted by each member country.

All seven reports from the European countries in this book (and one could have added reports from some other European countries)[13] mention how recent EU spatial planning policies and funds have either served as a trigger for the introduction of some degree of national-level planning or have reinforced the utility of existing practices. As Mastop (1998) puts it, 'for western Europe, the development of the new politico-institutional super-structure of the EU triggers important changes. Europe is building its own "nation" and as Europe is becoming a "Europe of regions", the various regions are building their "nations" too.' But the responses to the EU policies have differed among member countries, sometimes to an extreme degree. The story of Ireland is striking in this respect in a 'sweet and sour' way. Bannon and Russell's analysis shows how the promise of EU funding and its linkage with specific requirements for proof that some planning has been done beyond the project-by-project level became the trigger—indeed, the *only* real trigger—for initiating national-level planning in Ireland in the 1990s. However, in Ireland, such planning is not yet much more than a routine for obtaining funding. However, the authors note that in the future, national planning may become entrenched in its own right. Cars and Hårsman also mention the influence of EU policies as a trigger for a 'vision' statement for Sweden, but are very sceptical about the document's utility. Ireland and Sweden also happen to be classified as belonging to the low-level category of degrees of national-level planning.

Other European authors, including Grant for the UK, Enemark and Jorgensen for Denmark, Marcou for France, and Schmidt-Eichstaedt for Germany, are less sceptical about the impact of EU policies on planning in their countries. We learn, for example, how European policies have added extra flare to France's national policies about regional disparity alleviation and how Denmark was able to leave that goal entirely to the EU level and to make room in its national planning policy for a new focus on environmental and infrastructure goals. They also assess their country's existing national-level planning modes and procedures in terms of their adequacy for meeting EU requirements, and come to positive conclusions. Note that in my scale of degree of institutionalisation of national planning, the UK, Denmark, France and Germany are classified in the mid-level. Mastop for the Netherlands, which holds the highest degree of institutionalisation of national planning within the EU, does not even raise the question of planning adequacy. Marcou adds that competition among nations and regions at the European level today also includes the quality of the *planning function* itself and that French decision-makers are aware of this. We further learn from Faludi (1997) that the German decision-making structure for coordinating national-level planning with the *Länder* (see Schmidt-Eichstaedt's chapter) has become a model for the EU's coordinative Intergovernmental Conference of Ministers in charge of planning in their respective countries.

ENVIRONMENTAL, INFRASTRUCTURE, AND NIMBY-MITIGATION GOALS

In several countries, the rising public support for environmental values and the search for environmentally friendly infrastructure policies have replaced the goals of the previous decades and, one can surmise, have thus helped to maintain national-level planning. As Enemark and Jorgensen tell us, environmental goals and the improvement of international transportation links are today the key motivations for national-level planning in Denmark. Infrastructure planning and the prevention of NIMBYs ('not in my back yard') are key goals of national-level planning in Sweden, but unlike Denmark, in Sweden these areas represent exceptions to the general trend to reduce national-level planning and intervention. Environmental considerations, and especially a growing concern about the need to overcome resistance from local constituents to essential national projects, are today major goals and *raisons d'être* for national planning in the countries with the higher population densities—the Netherlands, Japan, Israel and the UK.

The USA is an example of a country where environmental considerations have played a double role. It is a probable reason why comprehensive national planning has not taken root, but at the same time it has been the stimulus for most of the (modest) examples of (sectoral) national-level planning that do exist. On the one hand, Kayden notes that the country's generous land and environmental expanses have been perceived as allowing 'room for mistakes', making the need for national planning less urgent. On the other hand, a quick glance at the US paper shows that many, if not most, of the sectoral areas where national-level planning and intervention did take root are environmental areas or, more recently, environment-plus-infrastructure areas. Kayden also notes that in Hawaii, Vermont and Oregon, the pioneering states where state-level planning first took root, the goal of protecting agricultural land—another environmental goal—was a prime motivation. Some of the other states that joined later (state-level planning exists, as yet, only in a small minority of states) were motivated by additional environmental goals such as reducing pollution, managing congestion and making more efficient use of infrastructure.

ACCOMMODATING GROWTH PRESSURES AND DEVELOPMENT NEEDS

For some countries, management of a high level of growth pressures has been a major goal for national-level planning. Growth accommodation has been a prime goal in Israel both in its initial years and during the 'resurrection' of national-level planning in the 1990s. It has also figured high in Japan, not so much as a result of demographic growth, but as a result of steep economic growth. In the Netherlands, although the demographic rate of growth is not high and is similar to other European countries, national-level planning has continuously seen growth accommodation as a prime goal. This seeming contradiction is solved if one notes that the Netherlands' long-standing ideological commitment to 'a home for every household' has exposed planning decision-makers to the commitment of meeting the growing demand for housing due to decreasing household sizes.

In Germany, Schmidt-Eichstaedt tells us, although demographic growth has been low, there has been a growing need to upgrade the infrastructure, housing and public services in the former East German *Länder*. In Ireland, Bannon and Russell hypothesise that the current relatively high rate of growth in household formation might in the future lead to a higher degree of national-level planning than currently. In the USA, those Federal programmes oriented to the management of sprawl and transportation congestion are a reflection of the need for accommodating growth—although in the USA there are great differences from region to region and therefore the programmes are also different. Indeed, I would add that, in the USA, 'growth management'—a term widely used as an American euphemism for planning—is the umbrella goal for much of state and local-level planning.

By contrast, the absence of growth does not seem to be related to the phasing out of national-level planning, reconfirming the applicability of the Choice Theory to national-level planning. The authors of the French, UK and Danish reports tell us that their countries have in recent years had a very low level of demographic growth and an almost stagnant development rate, and that there is an excess supply of most types of built-up space. Yet in none of these countries has national-level planning been phased out; on the contrary, it shows signs of strengthening.

CRISES OR THEIR ABSENCE

Finally, the emergence—or non-emergence—of national-level planning in some countries is related to crises. While such a linkage must have existed in the post-war years in many countries, most authors did not cite the existence of a crisis as an important stimulus for planning in recent decades. The story of the mass immigration crisis in Israel in the 1990s is an exception, and is indeed the most distinctive story of the linkage between crises and planning to be found anywhere among advanced-economy countries since the Second World War (Alterman, forthcoming). A possible runner-up is the story of German unification, but from Schmidt-Eichstaedt's chapter we do not learn of any major impact that this change has had on the degree and modes of national-level planning. Kayden hypothesises a link between the perceptions of the existence of a crisis to explain why, on three occasions, the USA came closer to adopting some form—albeit very mild—of national-level planning. The absence of a real crisis situation, surmises Kayden, may have contributed to the fact that land-use planning at the national level has not been reintroduced.

Trends in national-level planning

Despite the great differences among the ten countries in the formats and goals for national-level planning, there are also some trends that are shared by several of our sample countries and that can probably be generalised to some other countries not represented here. I have identified 16 trends that seem to me to reflect directions that should engage debate and further study. These will be classified into five groups:

- Trends related to national political and administrative contexts;
- Trends in national comprehensive planning;
- Trends in national sectoral planning;
- Emerging styles of planning and implementation; and
- Shared problems and dilemmas.

TRENDS RELATED TO NATIONAL POLITICAL AND ADMINISTRATIVE CONTEXTS

The state of planning at any level has always been linked to the state of government, and changes in ideological views regarding government intervention and styles of governing have always influenced planning. This linkage is even sharper where national-level planning is concerned. Two important trends emerge from the country reports that herald the promise of a new relationship between attitudes to government intervention and to national-level planning.

The release of national planning from the shackles of political ideology

The ideological link between attitudes to government intervention and to national planning may be almost dead. This is one of the more important—and surprising—findings of this comparative study. Only in one country within our sample—Sweden—do we still today find a linkage between changing ideologies about government intervention and degree of national planning. Sweden's is a story of the rise and subsequent demise of national-level planning. In the post-war years, rapid industrialisation, urbanisation, and a concern for the over-exploitation of natural resources led to the introduction of considerable central government planning and intervention in many spheres of life. The ideologically espoused and world-renowned 'Swedish model' of government intervention came into being. Improvement of social and housing conditions and equalisation throughout the country were the leading goals. However, as Cars and Hårsman tell us, by the 1990s, excessive government spending and intervention in the market had led to a steep economic decline. So, an overdose of national planning—not necessarily in spatial policy—can be 'credited' with the recent reluctance of Swedish leaders to adopt a degree of national-level spatial planning commensurate with most other European countries in our sample, such as neighbouring Denmark.

The finding of the 'death of ideology' emerges from the reports of all the countries except Sweden. In chapter after chapter, we learn that the former debates between the party-ideological views on left and right, once so vehemently pro or con planning, are no longer relevant to the attitudes towards national-level planning. Some of the authors make this point overtly, but probably without guessing that this is a shared picture; others make it indirectly, by keeping silent about any significant changes in the status of national planning as a result of a swing of power from left to right, or from right to left. There are even two countries—France and the UK—in which a reverse direction of linkage could be observed in the 1990s.

The UK is a country where controversies about town and country planning powers have, during the post-war years, occupied a central place in the well-documented debates between left and right. As Grant aptly puts it:

> [A] powerful consideration in both national and local politics has been ideology. British politics in relation to land policy was, between 1945 and 1985, bedevilled by a simplistic dichotomy between socialism and conservatism ... That is not true of the Labour Government elected in May 1997...

Grant and others (see also Thornley, 1991) point out that Margaret Thatcher's highly conservative government, while ostensibly seeking to weaken planning and government intervention, in effect led to greater centralisation of planning powers and greater involvement of the national government in local and county planning. Grant notes that, during the Tories' last years in office, Parliament began to be involved in specific national planning policies. Furthermore, public support for planning (recast as 'sustainable development') actually increased. The Labour Government that took over in 1997 did not find it necessary to make too many changes in the powers of national-level planning. So, national-level planning has become oblivious even to a rather dramatic party swing!

The French story is an even stronger case in point. The 1995 Guidance Act on Spatial Planning and Development (*aménagement du territoire*), which called for extensive and ambitious national spatial-economic planning, was in fact enacted by a right-leaning government. And ironically, in 1997, the incoming left-wing government was the one to recall the Act for review and made it more moderate in national-level powers.

From Kayden's account one learns that even in the USA the relationship between political ideology and planning is at times obscure, or paradoxical. The only attempt to enact a law introducing land-use planning powers at the federal level was defeated in the 1970s, and such intervention is regarded by both parties as anathema and has not been proposed again. During the post-war years, the Democrats usually supported rather more (sectoral) policy-making at the federal level than did the

Republicans. But in recent years even these differences have been fading. Paradoxically, in a country where the ideology supporting private enterprise and property has always been held at the centre, in recent decades one can also find policies—proposed and implemented by all administrations—that permit extreme intervention in the right to develop land, without paying compensation, for purposes such as wildlife protection, wetlands preservation, or open coastline access. Such freedom to intervene is, I may add, at times more excessive than in some other countries in our sample, although they may have a much higher degree of institutionalisation of national-level planning.[14] The recent legislative proposals in some states to establish the right to compensation for some of the above-mentioned controls take no account of whether the Democrats or the Republicans are ruling in Washington.

In Ireland, there seems to be another ideological paradox. The minimalist degree of comprehensive national-level spatial planning is attributed by Bannon and Russell to the very strong ideological commitment in Ireland to the freedom to use private property. At the same time, we learn that Ireland is a highly centralised country, with a large dose of sectoral national-level planning and intervention in various areas.

The authors of the two remaining countries in the 'half-haves' category, Germany and Denmark, make little mention of political debates over planning. But the most telling test should be regarding the countries in the 'haves' category—the Netherlands, Japan and Israel. Even in these countries, the desirable degree of national planning is no longer a prominent issue of party debate. But this fact is apparently not mutually known. Assuming (as I too had done before undertaking this analysis) that party-political debate about national planning still occupies a major position in other countries, Tanimura and Edgington present Japan as an exception, noting that its extensive degree of national planning has been supported by all governments, and has never become a party-political issue. We now know that Japan is no exception.

So, in all the countries in this book except Sweden, national-level planning has in recent years become almost immune to the vicissitudes of party-political changes.

Trends of decentralisation and the general weakening of governments

Two tendencies emerge that appear to contradict each other: on the one hand, governments are being weakened; on the other, national-level planning is probably gaining strength. The first trend is dealt with here, the other in the next section.

In the latter decades of the twentieth century, most advanced-economy democracies experienced a weakening of governmental powers in general and central government powers in particular. This occurred through decentralisation, privatisation, deregulation, fiscal problems, and the legitimation crisis in governments.[15] Among these trends, decentralisation deserves a special focus because it appears at first sight to be at odds with national-level planning.

Most of the authors report that the 1980s and the early 1990s have seen devolutionary trends in which some central government powers have been moved to the regional or local levels or to special agencies. The most striking case-in-point is France—a country that, until the early 1980s, was one of the most highly centralised in the West. Since then, France has taken a swing to the other extreme by devolving many central government powers, including land-use planning, to the local level. Considering the fact that France has thousands of tiny local authorities, such devolution was an extreme measure indeed.

In Denmark and Ireland, local governments have always had considerable powers. In Sweden, legislative changes that devolved some land-use planning powers occurred in 1987, ahead of the more general collapse of the Swedish Model in the early 1990s. Thereafter, devolution occurred in many spheres. In the Netherlands, although decentralisation trends in the 1990s have been soft and gradual, they continue to occupy a central position in debates about the proper role of national planning, as emerges vividly from Mastop's account. Even in federal Germany, where the constitution clearly stipulates the division of authority among levels of government, decentralisation trends have been occurring, sometimes in institutionally ingenious ways, as recounted by Schmidt-Eichstaedt. And in the USA—a highly decentralised country where one can hardly think of any further steps towards decentralisation—Kayden reports about a recent trend to decentralise some federal powers in sectoral planning to the state level, or to a partnership of the state and private bodies. This has gained momentum especially in the environmental area.

The UK and Israel were the exceptions to this trend—but may have now joined it through some devolutionary steps taken during the last years of the 1990s. Grant reports that the Labour Government has re-established the Greater London Council and in 1999 devolved extensive powers, including planning, to Scotland and Wales. And I report about the first ever change in the planning law in Israel which devolves to the local planning authorities a modicum of independent power to approve some types of amendment plans where these make only minor changes in land use. However, on the *de facto* rather than *de jure* level, significant decentralisation trends have been occurring in Israel as well since the early 1980s.

And finally, in Japan the situation has been rather paradoxical. Tanimura and Edgington tell us of the extensive powers of central government in economic planning and certain topics of land-use planning that are salient to central government. But I would like to add from my own observations of Japanese planning practice that, alongside these powers, Japanese local authorities are not as highly encumbered by central government in their local land-use planning and development control decisions as it seems on paper, and their independence is probably on the rise.[16] This has led to a rather low quality of the built environment in many Japanese towns and cities outside the major urban centres.

Perceived excesses of decentralisation have, in some countries, led to recentralisation of selected powers. Thus in France, as Marcou reports, excessive decentralisation of land-use planning powers to thousands of small local authorities has caused problems in coordinating spatial planning at the regional level and in implementing essential national or regional projects. A few corrective measures were taken in the 1990s, without changing the basic decentralised structure. A similar trend occurred in Sweden, as reported by Cars and Hårsman. Central government powers were strengthened where necessary to override local opposition to a project deemed essential. In the Netherlands, as Mastop reports, in 1994 a so-called 'NIMBY Act' was legislated for a similar purpose. In the UK, where central government and the national assemblies already have considerable powers to call in local plans for the Minister's approval, Grant reports that there may be an emerging tendency for Parliament to intervene as well through a 'public bill'. It has done so in special cases where controversial national projects were involved (such as the Channel Tunnel and King's Cross station).

TRENDS IN COMPREHENSIVE PLANNING

Comprehensive planning is the more ambitious form of national planning. I have noted that the ten countries may be classified into three levels by the degree of institutionalisation of such planning. Has comprehensive planning been on the increase or decrease in recent years? Comprehensive planning at the national level would benefit if a comprehensive view were also adopted at the regional level or at the supra-national level. Among our sample countries, three trends can be observed concerning the linkage between the national, regional and supra-national levels of comprehensive planning.

The strengthening of comprehensive national-level planning

Despite the global trends weakening governments, in several of the sample countries, national-level planning shows signs of strengthening. Except for the Swedish case, one cannot point out a correlation between the trends towards decentralisation noted above, and a weakening of national-level planning. Only in Sweden did these trends bring about a marked decline in national planning in sectors such as housing and economic development and a reluctance to adopt comprehensive spatial planning. In all the other countries in our sample, national-level planning or particular aspects of it have become more highly institutionalised, sometimes concomitantly with decentralisation and deregulation.

This is true even for the three countries with the highest degree of institutionalisation of national planning. In Japan, where economic-industrial planning did show signs of weakening in the 1990s, the same cannot be said of land-use planning. The current economic crisis and the worsening environmental problems may even lead to its further strengthening. In the Netherlands, despite public debates about the proper role of national spatial planning and the steps taken towards decentralisation and deregulation, national planning is likely to grow in strength. This can be credited to the 1993 legislation that mandates a national *environmental* plan.[17]

In France, the renewed importance of national-level spatial planning is striking. In 1995, the ambitious national planning act returned to some of the legacy of the 1970s, when French national-level planning was at its height; but, as Marcou stresses, the emphasis was no longer on pure economic planning, but rather on spatial socio-economic planning. And although the implementation of this act was partial and was subsequently changed, it will probably prove to have had some effect on rejuvenating national-level planning.

National-level planning in the UK also shows signs of becoming stronger. Since 1988, the former government circulars have become public documents (Planning Policy Guidance, or PPGs). They now increasingly resemble national planning policies and are not too distant from what in other countries are regarded as components of a national spatial policies plan. Since 1992, local authorities in the UK have been required to adhere to local plans (except if there are 'material considerations' that justify otherwise); and because such plans come under the guidance of the PPGs, central government and the two national assemblies now have potentially stronger implementation tools than in the past.

In Israel, the 1990s saw the impressive re-emergence of national-level planning to a degree of comprehensiveness it had not enjoyed since the 1950s, and with a higher degree of public exposure than ever before (previously public exposure was very low in cross-national comparative terms). In Ireland, national-level planning had its modest advent in the 1990s and a new attempt to issue national planning directives showed promising mementum in 2000. Finally, in Denmark, and even in federal Germany where the constitution bars the federal government from carrying out direct land-use planning, one can point out some signs, albeit more subtle than the other examples, that national planning is growing stronger.

The increasing importance of regions and the regional level

A regional view has always been one of the hallmarks of a comprehensive spatial-planning approach. All the European reports note the increasing importance that national governments attribute to regional planning, regional institutions and regional alignments (compared with the past focus on the local level). This growing prominence finds expression in both the contents and procedures of national-level planning policies. In Germany, France and Denmark, the goal of furthering regional perspectives finds formal expression in the constitution or in national planning legislation. The attention on regions probably reflects three trends: growing environmental awareness and therefore greater attention to the environmental distinctiveness of different regions; social changes that encourage greater attention to ethnic diversity and regional pride; and, of course, the EU policies targeted at the regional level. The EU has adopted a set of policies that emphasise regional planning within member countries in order to reach greater equalisation of socio-economic levels in the EU as a whole. The EU regional planning policies offer lucrative funding for that purpose. The impact of EU policies on strengthening regional planning is already apparent in France, Germany, Ireland and Denmark, and in the future may also become apparent in Sweden, the Netherlands and the UK.

A stronger supra-national perspective

Alongside the inwards regional perspective, there is also a trend in various parts of the world towards spatial planning at the supra-national level. This trend is most notable within the European Union, which publishes its own spatial policies plans for the entire area—*Europe 2000+* (European Commission, 1994) and *European Spatial Development Perspectives* (European Commission, 1999). Supra-national planning also occurs among neighbouring countries, such as the Benelux strategic plan mentioned in Mastop's report. This trend has increasingly engaged European researchers in planning and related areas. There are dedicated research funds and specialised research institutes and journals[18] that are helping to create a knowledge base.

This trend reflects not only EU goals and procedures, but also deep worldwide changes—globalisation of markets, information, education and culture. Where international regional agreements enable cross-border planning, such as between the USA, Canada and Mexico, one can expect increasing supra-national planning in sectoral areas. The peace process in the Middle East has also brought about attempts at cross-border planning between Israel and its neighbours, Jordan, the Palestinian Authority and Egypt.

TRENDS IN SECTORAL PLANNING

Sectoral planning and comprehensive planning are often in tension with each other, competing for the attention of the public, politicians, government officials and professionals. This tension is especially strong at the national level because there, comprehensiveness is difficult to achieve. Three trends can be gleaned from the country reports and these offer some fresh perspectives on the relationship between comprehensive and sectoral planning.

The subjects for sectoral planning—some shared, others differing

One of the hypotheses with which this book began was that the topics assigned to sectoral planning at the national level would differ from country to country. There may also be a difference in whether urban and regional planners recognise that sectoral planning areas constitute part of the patchwork whereby spatial policies are made.[19] Some of the areas subject to sectoral planning at the national level are shared by many of the countries in this book and have usually existed for many decades. These one could call the more 'traditional' areas—highways and railways, parks and forests, national heritage sites, sensitive resources, bodies of water or coastlines. To these one could add the newer environmental areas such as air quality and toxic waste disposal. Other traditional areas, albeit somewhat less consensual, include housing, urban renewal and agriculture.

There are also significant differences among the countries in the areas covered by national-level sectoral planning. Some of these differences may reflect an objective need; others are probably a question of choice reflecting values and goals. In smaller and more densely populated countries, there may be a tendency to vest, at the national level, more sectoral NIMBY and LULU[20] policies than in larger countries, where they would be left to the local or regional levels. In Israel, for example, inter-town roads, railways, power plants, airports, toxic waste sites, mining and excavation sites, gas pipes, telecommunication lines, and even cemeteries and non-toxic waste disposal sites, are all handled at the national level by means of statutory sectoral plans. However, there is no consensus even among the more densely populated countries. In the UK, some of the areas that in Israel are subject to national sectoral plans, such as airports, are regarded as appropriate for the local and regional levels. Whenever the Minister deems it necessary, he or she can promote national interests through the activation of the plan call-in powers.[21]

Other differences probably reflect the importance assigned to particular policy areas. In some countries, additional, non-traditional areas are subject to national-level planning, such as higher education, science and health services. New environmental planning areas, such as wildlife preservation, have also been added. In some other countries these have not (yet?) benefited from national sectoral planning. In Japan and France, where there is a tradition of strong economic development planning, sectoral planning for industrial development has been prominent. In Israel, on the other hand, there is a separate national statutory plan for tourism, but not one for industry (the latter is handled through a national policy of differential financial incentives).

Some strengthening of sectoral planning

The existence of sectoral planning is a confirmation of the 'disjointed incrementalism' theory of public policy and planning (Braybrooke and Lindblom, 1963). That theory recognises the *realpolitik* dynamics of public bodies and stakeholders who wish to control their own interests. The advantage of disjointed incrementalism and of sectoral planning is a smoother planning process and, because of the closer correlation between the planning and implementation bodies, a hoped-for smoother implementation process. The price is, of course, poor coordination with other sectors and, at times, unexpected implementation problems (Christensen, 1999). I suspect that there are some trade-offs between comprehensive and sectoral planning.

In parallel with the trend noted above regarding the growing prominence of comprehensive national-level spatial planning in some countries, one can also (hesitantly) point out a trend in several countries whereby national-level sectoral planning in particular areas may also be growing in strength—with or without a concomitant enhancement of comprehensive planning. The expansion of sectoral planning at the national level may be an expression of a wider recognition of the need for more national planning, which finds expression in sectoral planning rather than in the more administratively and politically difficult comprehensive spatial planning. This trend is sometimes triggered by a political commitment to a particular policy area, by the availability of outside funding in particular areas, such as through the EU, or by international political or trade agreements. For example, in the USA, sectoral areas promoted in the 1990s included several policy areas related to environmental and transportation issues that have high political visibility and electoral support.

Not all countries share this trend. In a series of policy areas in the UK, the national government has in fact reduced its involvement. Grant reports that agriculture is the only sectoral ministry left with national-level planning powers (and these, too, have been relaxed).[22] Other areas previously vested at the national level—water, roads and rail—have either been privatised or assigned to a special agency.

New paradigms for broadening and coordinating sectoral planning

There are signs of a new tendency to soften the raw edge of disjointed action. New paradigms may be emerging that seek to build into sectoral planning better mechanisms of integration with other sectors than is possible through regular interministerial coordination. The five paradigms enumerated below represent mid-way compromises between fully comprehensive spatial planning, and planning that relies entirely on disparate sectors.

The first such model calls for sectoral planning that weaves together two or three selected areas into a single programme, without yet taking on the comprehensive spatial-planning view. Under this model, sectoral planning takes a 'look over the shoulder', so to speak, towards adjacent sectoral areas.

An interesting example comes from the UK. Grant tells us of a 1998 programme that seeks to encourage the integration of transportation with land-use planning—two areas that have suffered from poor coordination between them. But the clearest examples of this tendency come from the USA. Kayden analyses a long list of federal environmental policies that have been expanded to include what he aptly calls 'a slice of land-use planning' (coastal zones, wetlands, wildlife, transportation, air quality). A US federal programme—ISTEA[23]—encodes the multi-sectoral view into its emblem. This programme is tailored to encourage an integrated view of land-use, transportation and air-quality considerations, and this integrative view has already won itself the acronym LUTRAQ. As Kayden puts it:

> ISTEA introduced cutting-edge change at the national level. Since transportation investment decisions obviously have a major impact on land use and environmental quality, this legally imposed relationship between plans and institutions represented an important break with past planning practices.

ISTEA sets up a special institutional implementation tool—the Metropolitan Planning Organisations (MPOs)—which are major innovations in the US tradition of local government. They are responsible for preparing long-range transportation plans in conjunction with the states, which are required to prepare their own long-range plans for areas outside the jurisdiction of the MPOs.

Israel provides a second paradigm for interrelationships between sectoral and comprehensive planning. From 1965 until the early 1990s, national statutory planning took the form of sectoral planning only, rather than of comprehensive land-use planning. Ironically, sectoral plans were actually based on the legal basis intended for a comprehensive national plan.[24] Under this legal umbrella, some 25 national sectoral plans have been prepared to date. The legislators' intention that a comprehensive view be taken was met only indirectly—through the composition of the body in charge of approving statutory plans, which includes representatives of all relevant government ministries. But despite this institutional structure, the accumulation of sectoral plans, often uncoordinated and even contradictory, never became a substitute for an integrated view. Only in 1991 was the first initiative taken to prepare a comprehensive statutory plan. This dramatic 'comprehensive turn' may also be credited with spinning off more sectoral plans in new subject areas and enhancing their quality. Thanks to the new comprehensive plans, the sectoral plans are now cast within broader coordinative mechanisms than before. A similar dynamic is described by Mastop for the Netherlands, where the national plan titled 'Fourth Report+' is credited with stimulating considerable sectoral planning in areas such as infrastructure, environment, agriculture and economic development.

A third type of interrelationship between comprehensive and sectoral planning comes from Japan, as described by Tanimura and Edgington. In Japan, economic planning is the strongest among our sample countries, and it is institutionally distinct from national land-use planning. The comprehensive economic plan is used as an effective coordinating mechanism for most other sectors, including the spatial planning sectors. The tool is the requirement that a five-year plan be prepared by all sectors under the guidance of the economic plan. Another ingenious Japanese coordinative mechanism is based on the personnel element: the national economic planning agency seconds officials from several other agencies, who work as part of that agency on a rotational basis, thus bringing in their 'mother agency's' perspective and improving the mutual coordination of decisions.

A fourth paradigm of the relationship between sectoral and comprehensive planning comes from the Netherlands. Indeed, Mastop's account of national-level planning there can be seen as the story of the perennial search for the right balance between comprehensive (called 'facet') and sectoral planning. He vividly describes the various concepts and experiments that have been attempted, decade after decade. The Achilles' heel of sectoral planning in the Netherlands is the notoriously poor coordination among land-use, environmental and water-management plans (and this despite institutional location of the first two topics in a single ministry). Mastop reports on recent provincial-level experimental programmes for integrating these three sectors into a single planning document. If these are successful, government has promised that the national plan currently in preparation—the Fifth Note—will integrate a 'note' about environmental planning too. Thus we see a new paradigm emerging, whereby expanded sectoral planning is subsequently integrated into a comprehensive planning view.

And finally, an even more fascinating paradigm may be emerging from the Netherlands, with a twist. Mastop (1998) and Priemus (1998) tell us of the alignment for the preparation of the new 2030 plan. Behold: the four major sectoral ministries, each of which has traditionally prepared its own sectoral plan, are on the verge of espousing a new role. The four ministries are Economic Affairs; Transport and Public Works; Water Management and Agriculture, Nature Management and Fisheries; and Housing, Spatial Planning and Environment. The latter is the ministry legally and traditionally in charge of land-use planning, and in all previous national plans it has taken the lead but coordinated poorly with the others. This time, each of the other three ministries has prepared a policy document which, in effect, is a comprehensive alternative conception to the national plan! One can therefore anticipate that the debates about the 2030 plan will be based on new grounds and with new dynamics. Instead of a debate between a lead agency in charge of comprehensive planning and sectoral agencies that in fact act as lobbies for their interests, the debate will now be between alternative comprehensive conceptions, from different loci. Thus, the notoriety of poor coordination among sectoral plans may have led the sectoral agencies to realise that the best way to effect comprehensive planning is by preparing it themselves!

EMERGING STYLES OF PLANNING AND IMPLEMENTATION

For many, national planning carries the image of a command-and-control approach to planning and implementation. This image may reflect the fact that national-level planning often takes a longer view than local planning, deals with subjects that might be anathema to local communities, and is carried out by central government. National planning might be viewed as more distant from the consumers of planning than is local planning. Surprisingly, the picture that emerges from the ten chapters is a far cry from this image. The styles whereby planning and implementation are carried out at the national level turn out to be not too dissimilar from the styles of local and regional planning and public policy-making that have been evolving in democratic countries in recent decades.

The use of incentives to improve local and regional planning and comply with national policies

Increasingly, national governments are using incentives rather than command-and-control styles to encourage comprehensive planning at the local and regional levels, and thereby encourage compliance with national policies. This is carried out through a type of quasi-voluntary 'deal'. If the local level undertakes a comprehensive planning process, which also implements national policies, it will be granted more financial resources, or more legal powers. An example of this is the 1983 French planning law which authorised central government to devolve full land-use and building-control powers to the local authorities, provided that a local plan had been prepared and approved. Conditioning the transfer of funds upon the improvement of planning has been the prime technique adopted by the federal government in the USA since the 1950s and, more recently, by the German federal government.

In the USA, a whole pageant of policy areas have been implemented in this manner. As Kayden reports, some have been more prominent in the past, such as housing, urban renewal and neighbourhood regeneration, while others are the current stars, such as integrated transportation modes and various areas of environmental quality. Through this approach, a national government can gain three things. It can improve local and regional planning (in the EU case, as we saw above, it is also used to improve national-level planning); it can gain control over the timing of planning; and it can further the implementation of national-level planning goals by assuring that these goals are absorbed into local or regional plans.

From 'blueprint'-style plans to flexible policy documents

Theorists and legal scholars concerned with local land-use planning have been engaged in the decades-long debate about the advantages and disadvantages of flexible broad-brush policy plans versus 'blueprint'-style plans (Alterman, 1981; Alterman and Hill, 1978; Faludi, 1970; 1987; Haar, 1956). Where *local* plans are concerned, the law and practice in this regard does vary significantly.[25] Indeed, in some of the chapters in this book, the authors complain that overly rigid local plans have become a national problem that inhibits or slows down implementation of national policy. But at the national level, the debate has clearly been decided in favour of flexible policy plans.

In all but one of the countries where there is a document equivalent to a national plan of some form ('plan', 'report', 'note', 'guidelines', '*schéma*'), that document is described as a flexible set of policies or guidelines (see Table 2). Such is the case for the Netherlands, Japan, France, Denmark, Germany, the UK and Sweden. In France, the 1995 law which anticipated a relatively rigid type of plan was amended so as to turn the *schéma national* into a document that presents a set of *choices*.

Most of the planning documents make use of maps to convey the spatial aspects of policies, but maps are no longer the *sine qua non* for national spatial planning. In the UK, maps have purposely been avoided in the PPGs. Written policies are the major form of communication in national planning documents in most countries.

The only exception to this trend is Israel, where the tradition of blueprint-style plans that specify rules and regulations has survived longer than in the other countries in our sample. However, even in Israel, a change of course may have begun. The 'Israel 2020' plan, completed in 1997, was the first policies plan, but it was not a statutory plan, nor even a government-initiated one. So the real test regarding the emerging style of plans will only occur when the final version of the comprehensive National Outline Plan 35 is submitted for approval. Will the planners be able to withstand the pressures from both government bureaux and environmental groups to be more and more specific? I have my doubts.

From a binding to an advisory status in implementation

The offshoot of the flexible policy style of plans is that in the implementation arena most of the documents play an advisory, guiding function rather than a legally binding one, as can be seen in Table 2. This book shows that the legal status regarding plan *preparation* has largely been disengaged from the legal status regarding plan *implementation*. Although many of the national planning documents among the countries in our sample are mandatory to prepare (Germany, Denmark, France, Japan), that does not mean that they are binding on development-control decisions (or, in American planning parlance, full 'consistency' is not required). Lower-echelon government units may be expected, by legal or administrative requirements, to relate to national plans (the British term to 'take regard' is appropriate here). Central government may use various methods of oversight, incentives or sanctions to encourage lower-echelon agencies to adhere to national planning policy, but it cannot bind private and non-governmental bodies to it. In other words, the command-and-control style of plans at the national level is not the prevalent implementation mode—except for Israel, where national statutory plans are strictly binding on all parties (but are not mandatory for government to prepare).

The correlation between the plan's legal status and its degree of effectiveness is not very strong. For example, in the UK, the 'National Policy Guidelines' are not mandatory to prepare; indeed, they have no direct grounding in any legislation, yet national government has rendered them rather binding on local authorities through administrative practice. In the Netherlands too, it is not mandatory to prepare national Notes, yet these policy statements have to date been rather effective—not through a command-and-control process, but an advisory, guiding one.

There is an exception to this trend. Several of our authors report a 'backlash' where projects of salient national interest are concerned and where there is a fear that local government or private interests might block or delay an important project. In those cases, central government issues a very specific plan or directive that requires full and detailed compliance. Such a trend is reported from Sweden and France (greater use of national directives), the UK (where a 'private bill' technique is of late being used), and the Netherlands (where a 1994 amendment to the planning act is nicknamed the 'NIMBY Law'). In Israel, many of the statutory national plans are in fact project-specific NIMBY-bypass plans.

More public participation alongside a growing NIMBY problem

The fact that the national planning reported in this book is carried out in democratic countries would lead one to expect that public authorities would allow opportunities for public participation. The question is, on which rung of Arnstein's (1969) famous 'ladder of public participation' do they stand? Do the authorities just ensure provision of information to allow objection by injured parties, or do they expand participation towards a more proactive process?

The distinctive finding from most of the countries is that public participation modes in the 1990s have in general gone 'up the ladder', regardless of where they were stationed before. Denmark, France, Germany, the Netherlands and the UK are countries that have had a relatively high level of participation for many decades, relative to many other countries (the UK's temporary regression in the 1980s notwithstanding). The authors of these chapters point out that there has been a distinctive expansion in participation modes in the 1990s. The resort to conflict resolution in a conscious manner (as distinct from just an outcome of regular planning administration) is an emerging theme in some of the chapters, such as the Netherlands, Denmark and Germany.[26] While the Irish and American authors do not refer to such a trend, that is probably because in these countries extensive public participation is already taken for granted.

The trend of expanding participation holds even for countries where government administrative culture has traditionally been less open. Thus, in France, where the prefects still have significant authority and can implement national planning directives (DTAs[27]), they have been resorting to public participation more broadly than required by law. A similar picture emerges from Israel, where the National

Planning and Building Board has been increasingly willing to hear stakeholders even though the law exempts the National Board from that requirement. In Japan, 'participation' is traditionally practised in different ways than in the other countries—it focuses on collaboration with specific interest groups, rather than on expanding access to the public at large.

The enhancement of public participation does not always hold. A reverse tendency has also been emerging as part of the NIMBY-bypass trend noted above. This, however, applies only to selected cases where the national government perceives that a particular policy or project is an essential one, sees time is pressing, and fears that local government and private interests will halt the project. We learn from the Dutch, Swedish and French reports that, during the 1990s, central government has enacted new laws or adopted procedures that are specifically designed to override local opposition by consciously limiting the opportunities for public participation and litigation in the courts. In Israel too, the early 1990s saw a temporary reduction in participation rights when government saw an impending crisis as justifying such procedures and enacted a temporary law, which has since been terminated (Alterman, 2000). In none of the cases, however, have public participation rights been totally curtailed.

From a command-and-control planning process to a negotiated, collaborative one

One of the most consistent themes emerging from the reports is that, at the national level, the command-and-control mode of plan-making is in demise, whereas the negotiated and contractual mode is on the rise. From most of the reports we learn that central governments are shifting their implementation energies from the more traditional 'instruments' of implementation that had engaged analysts for decades (Alterman, 1982) to negotiated contracts. These are carried out on a wide span of issues and with a growing range of government bureaux, other agencies and private bodies.

We learn of many types and layers of negotiated agreements. They are being used to create joint policy between *hierarchical* agencies, such as central government, regions and local authorities (here the French state-region plan conventions and the German federal state and local conferences are notable examples). They are also used between horizontal agencies, such as among sectoral agencies at the central government level (here Japanese practice is a case in point) and among local authorities (the French and more recent UK practices provide examples). Negotiations are also being used to develop and coordinate policies among *hierarchical and horizontal* agencies at once—what Mastop calls 'diagonal' negotiations (here Dutch and Danish practices are distinctive). And finally, negotiated contracts are employed between government agencies, a variety of non-governmental organisations (NGOs), and private corporate bodies. This mode—the change from *government* to *governance*—is taking root in several European countries in our sample, and has also been adopted in the 1990s by experimental US federal environmental programmes. It is also a veteran Japanese practice.

Indeed, in some of the chapters—especially Germany, France and Denmark—

the focus on negotiation is so prominent that the accounts no longer resemble the traditional accounts of planning 'systems' where hierarchical structures and rules of compliance used to occupy centre stage. The implications of this trend are captured by Grant in his report on the UK:

> [T]he language of partnership is having a powerful effect on current public policy management.... Partnership is at one level a sign of weak government, and an indication of the dependency of national and local governments upon others to secure public policy objectives. But it is also a potent tool for consensus building in generating those objectives and delivering the policies necessary to achieve them.

The finding about the growing prominence of negotiated, collaborative planning modes at the national level is among the most important in this book. This finding would not have been a surprise had it applied to the local and regional levels. After all, the shift to negotiated modes accords perfectly with currently prominent theories of urban and regional planning, such as Healey's 'collaborative planning', Innes' 'communicative planning' or Forester's 'deliberative planning' (Healey, 1997; Innes, 1995; 1996; 1998; Forester, 1999). These theories, however, focus in their examples mostly on the neighbourhood and local levels—the 'shaping of places', as Healey calls the collaborative process. Our findings show that national-level planning, despite its greater distance from consumers and often its longer time range, is adopting modes that are not dissimilar to local planning.

SHARED PROBLEMS IN IMPLEMENTATION

The emerging trends regarding national-level planning may end up painting a picture that is too positive. Despite the signs that such planning may be gaining strength in democratic countries, and despite the adoption of more open, participatory, flexible and implementation-oriented modes, national planning is not immune to problems. Each chapter outlines the specific difficulties encountered in that country. I have found that some types of problems are shared by several countries. Here are three of them.

Continuing problems of inter-agency coordination

Because of its geographic and organisational span, national-level planning, whether sectoral or comprehensive, must deal with many agencies and must find ways of coordinating among them. After all, government structure is always *disjointed*, to borrow Braybrooke and Lindblom's (1963) term once again. Even in this summary chapter we saw evidence of the great efforts that planners and decision-makers are making to structure into the planning and implementation process improved methods of coordination. In some chapters—most distinctively the Netherlands and Germany—the search for effective ways of coordination is almost the *leitmotif*. Yet only one author—Schmidt-Eichstaedt for Germany—expresses satisfaction with the degree of inter-agency coordination achieved. Enemark and Jorgensen for Denmark also voice few complaints.

All other authors persist in reports about the continuous search for better mechanisms of coordination because existing ones are not satisfactory. For example, in the Netherlands a new format of coordination is tried out every few years, testifying to the decision-makers' perpetual dissatisfaction with past performance. Inter-agency coordination of national-level planning may indeed be a 'wicked problem' (Rittel and Webber, 1973).

Lengthy procedures and growing litigiousness

Many of the authors in this book voice the same twin complaints. On the one hand, planning procedures at the local and regional levels are increasingly perceived as taking too long, and national authorities are increasingly being called in to help. Grant calls this the 'procrastination culture'. On the other hand, the tendency of stakeholders to resort to legal procedures against planning authorities is steeply on the rise. These two trends are, of course, linked in a paradoxical way. When procedures take a long time, various stakeholders—those interested in promoting development and those interested in halting it—have greater reason or opportunity to go to court. But resorting to litigation is a sure recipe for delaying procedures further—often taking the control of timing totally out of the hands of the planning authorities.

The longer timescales involved and greater likelihood of litigiousness probably reflect several shared trends: the increase in environmental awareness, the availability and wide dispersion of information about spatial planning and its impacts, the generally higher education levels of the public, and greater affluence that allows both public and private parties to the disputes to invest in further actions and in litigation.

Complaints about lengthy procedures and frequent litigation are expressed in the German, Dutch, British, Swedish and Israeli chapters. Kayden mentions that in various states of the USA, legislation has increased landowners' compensation rights, thus, I would add, in effect lowering the litigation threshold. He does not, however, complain about lengthy procedures and delays since these are probably less problematic in the USA, where planning procedures usually have fewer 'clearance points' than in the various European, Israeli or Japanese systems. Recent legislation in the Netherlands has given citizens the right to demand compensation for a growing range of public decisions. In Israel, the litigation threshold has been lowered in several ways: through the establishment of new, highly accessible quasi-judicial and judicial planning bodies, a lowered threshold of compensation, and greater access to information about planning procedures. In Japan, the problem is not litigiousness (that is not typical of Japanese society) but the reverse—bypassing legal procedures through rampant bribery, as noted by Tanimura and Edgington.

Some of the authors report new attempts to reduce either litigiousness or procrastination, or both. We learn that in Germany new legislation stipulates that appeals about certain types of planning initiatives in the new *Länder* can no longer stall the approval of such plans. In Sweden, 'projects of national impact' have special speedy procedures. In the Netherlands, national 'key projects' such as airports also have special speedy procedures, whereby appeals and litigation are in principle restricted to the initial stage of the approval process. During the second, detailed stage, no more appeals from local government or private parties can be heard. In France, special directives can be used to streamline procedures. In the UK, the Minister has powers to speed up procedures where necessary, but in some prominently problematic cases in the late 1990s, Parliament has had to intervene. Finally, in Israel a paradoxical process has been occurring. On the one hand government has initiated amendments to the planning law that were intended to reduce quasi-judicial and judicial litigation. But on the other hand, the Knesset (parliament) dissipated these goals and achieved the converse: access to quasi-judicial and judicial litigation has actually been increased. No type of national-level planning, no matter how urgent it may be, is today immune to full scrutiny by the courts.

These problems are not innocuous. They are the symptom, not the malady. The maladies arise from built-in contradictions in the trends noted above. Flexible plans do not enable specificity and predictability. Stakeholders want to know what will happen on a particular site. Approval of flexible policy only delays conflict, it does not make it go away. Public participation, open information, flexible plans and negotiated contracts are indeed the desirable ways of conducting planning in democratic countries and are the paths along which planning theorists have been leading planning students and practitioners. But national-level planning deals with complex administrative and interest group structures, and with the middle or long term. In this context, no prescribed ways apparently hold the recipe of enduring consensus.

Conclusions

Despite the 13 positive trends concerning national-level planning noted above, and the evidence that such planning is on the rise in democratic countries, the reports in this book by no means indicate a full consensus about its desirability. National planning must justify itself even in those countries that already have a high or a medium level of it. The authors of the chapters on the Netherlands, the UK and Israel anticipate that the debate about the desirability of the existing level of national planning will intensify in the coming years. And at the other end of the scale, the authors of the chapters on Ireland, the USA and Sweden report continuing resistance and doubt about the utility of national-level planning.

For some, national planning still carries the image of a command-and-control approach to planning and implementation. This image may reflect the intrinsic characteristics of national-level planning in terms of time-span and distance from the consumers. Furthermore, national planning sometimes deals with subjects that might be anathema to local communities. However, despite the shared problems, the emerging trends of national-level planning do indicate that 'something is going right'.

On the threshold of the twenty-first century, the practice of national-level planning is becoming increasingly attuned to changes in governance style, in society and in the economy. Such planning, as this book shows, is becoming adaptable as a useful tool for shaping government policy in democratic countries. The styles whereby planning and implementation are carried out at the national level turn out to be not too dissimilar to the styles of local and regional planning and public policy-making that have been evolving in democratic countries in recent decades. Although national-level planning has been virtually ignored by planning theorists and by empirical researchers, those theorists who have argued for the 'communicative turn' in planning and for planning as collaboration should be very pleased. The general trends in planning that they predicted seem to be occurring even on the national level!

At the beginning of the twenty-first century, the picture of national-level planning that emerges from this book is very different from what it was in earlier decades. The twentieth century that saw the rise of some of the most coercive misuse of national planning by non-democratic regimes, and witnessed their demise, has sidestepped successfully the danger that any and all national planning would be viewed negatively. Surprisingly, in most countries represented in this book, national planning (or its absence) seems to have become immune to the churning of political-ideological debates. If this trend perseveres, we may be seeing growing debate in democratic countries about the pros and cons of enhancing national planning, but these debates will no longer cut across the traditional party-ideological lines of left and right. They will no longer focus on whether planning *per se* is desirable, but rather on the values and goals that are to be pursued and on the degree of effectiveness of alternative modes and formats of planning. National planning would then be largely seen as an instrumental rather than an ideological issue. Such planning will increasingly be viewed as a useful tool for democratic, advanced-economy countries that are seeking to improve their standing in a world of increasing economic and demographic competition, and are eager to answer voters' demands for better global and local environmental management.

The ten accounts of national-level planning provided in this book will, we hope, provide a varied set of models from which to learn and a rich information base on which to conduct future debate.

REFERENCES

ALBRECHTS, LOUIS (1998), 'The Flemish Diamond: Precious Gem and Virgin Area', *European Planning Studies*, **6**, 411–24.

ALEXANDER, ERNEST R. (1998), 'Planning and Implementation: Coordinative Planning in Practice', *International Planning Studies*, **3**, 303–20.

ALTERMAN, RACHELLE (1981), 'The Planning and Building Law and Local Plans: Rigid Regulations or a Flexible Framework', *Mishpatim (Laws)* (Law Journal of the Faculty of Law, Hebrew University, Jerusalem), **11**, 179-220 (in Hebrew).

ALTERMAN, RACHELLE (1982), 'Implementation Analysis in Urban and Regional Planning: Toward a Research Agenda' in Patsy Healey, Glen McDougall and Michael Thomas (eds), *Planning Theory: Prospects for the 1980s*, Planning Series, Vol. 23 pages 225–45, Oxford, Pergamon.

ALTERMAN, RACHELLE (1988), *Private Supply of Public Services: Evaluation of Real Estate Exactions, Linkage and Alternative Land Policies*, New York University Press, New York, paperback edition 1990.

ALTERMAN, RACHELLE (1990), *Comparison of Statutory Planning Systems in Selected Countries* (monograph), Haifa, Klutznick Center for Urban and Regional Studies, The Technion–Israel Institute of Technology (in Hebrew).

ALTERMAN, RACHELLE (1997), 'The Challenge of Farmland Preservation: Lessons from a Six-country Comparison', *Journal of the American Planning Association*, **63**, 220-43.

ALTERMAN, RACHELLE (2000), 'Land-use Law in the Face of a Rapid-Growth Crisis: The Case of the Mass-Immigration to Israel in the 1990s', *Washington University Journal of Law and Policy*, **3**, 773–840.

ALTERMAN, RACHELLE (in preparation), *Planning in the Face of Crisis: Land and Housing Policy in Israel.*

ALTERMAN, RACHELLE and MORRIS HILL (1978), 'Implementation of Urban Land Use Plans', *Journal of the American Institute of Planners*, July, 274–85.

ARNSTEIN, SHERRY R. (1969), 'A Ladder of Citizen Participation', *Journal of the American Institute of Planners*, July, 216–24.

AVE, GASTON (1996), *Urban Land and Property Markets in Italy*, London, University College of London Press.

BRAYBROOKE, D. and C. E. LINDBLOM (1963), *A Strategy of Decision: Policy Evaluation as a Social Process*, Glencoe, IL, Free Press.

CAMPBELL, SCOTT and SUSAN FAINSTEIN (eds) (1996), *Planning Theory*, Madelen, MA and Oxford, Blackwell.

CHRISTENSEN, KAREN STROMME (1999), *Cities and Complexity: Making Intergovernmental Decisions*, London, Sage Publications.

DAVIDOFF, PAUL and THOMAS REINER (1962), 'A Choice Theory of Planning', *Journal of the American Institute of Planners*, **28**, 103–15.

DIETERICH, HARTMUT, RICHARD H. WILLIAMS and BARRY D. WOOD (eds) (1993–1996), *European Urban Land and Property Markets* (in series), London, University College of London Press.

DIETERICH, HARTMUT, EGBERT DRANSFELD and WINRICH VO (1993), *Urban Land and Property Markets in Germany*, London, University College of London Press.

EUROPEAN COMMISSION (1994), *Europe 2000+—Cooperation for European Territorial Development*, Brussels and Luxembourg, Office for Official Publications of the European Commission.

EUROPEAN UNION, EUROPEAN COMMISSION (1997, 2000), *The EU Compendium of Spatial Planning Systems and Policies*, Luxembourg, Office for Official Publications of the European Commission.

FALUDI, ANDREAS (1970), 'The Planning Environment and the Meaning of Planning', *Regional Studies*, **41**, 1–9.

FALUDI, ANDREAS (1987), *A Decision Centered View of Environmental Planning*, Oxford, Pergamon.

FALUDI, ANDREAS (1997), 'European Spatial Development Policy in Maastricht II?', *European Planning Studies*, **5**, 535–43.

FORESTER, JOHN (1980), 'Critical Theory and Planning Practice', *Journal of the American Planning Association*, **46**, 275–86.

FORESTER, JOHN (1989), *Planning in the Face of Power*, Berkeley, CA, University of California Press.

FORESTER, JOHN (1999), *The Deliberative Practitioner: Encouraging Participatory Planning Processes*, Cambridge, MA, MIT Press.

GIANNAKOUROU, GEORGIA (1996), 'Towards a European Spatial Planning Policy: Theoretical Dilemmas and Institutional Implications, *European Planning Studies*, **4**, 595–613.

HAAR, CHARLES (1956), 'The Contents of the General Plan: a Glance at History', *Journal of the American Institute of Planners*, **21**, 66–70.

HABERMAS, JÜRGEN (1975), *Legitimation Crisis*, Boston, Beacon.

HALL, PETER (1993), 'Planning in the 1990s: An International Agenda', *European Planning Studies*, **1**, 3–12.

HEALEY, PATSY (ed.) (1994), *Trends in Development Plan Making in European Planning Systems* (Working Paper No. 42), Newcastle upon Tyne, Centre for Research in European Urban Environments, Department of Town and Country Planning, University of Newcastle upon Tyne.

HEALEY, PATSY (1997), *Collaborative Planning: Shaping Places in Fragmented Societies*, London, Macmillan.

INNES, JUDITH E. (1995), 'Planning Theory's Emerging Paradigm: Communicative Action and Interactive Practice', *Journal of Planning Education and Research*, **14**, 183–89.

INNES, JUDITH E. (1996), 'Planning through Consensus Building: A New View of the Comprehensive Planning Ideal', *Journal of the American Planning Association*, **62**.

INNES, JUDITH E. (1998), 'Information in Communicative Planning', *Journal of the American Planning Association*, **64**, 52–63.

KALBRO, THOMAS and HANS MATTSSON (1995), *Urban Land and Property Markets in Sweden*, London, University College of London Press.

KUNZMANN, KLAUS R. (1995), *Defending the National Territory: Spatial Development Policies in Europe in the 90s*, Dortmund, Fakultat Raumplanung, Fachgebiet Europaische Raumplanung, Universitat Dortmund, Reprint 07/95 (from F. Knipping [ed.] [1994], *Federal Conceptions in EU Member States: Traditions and Perspectives*, Baden Baden, Schriftenreihe des Europaischen Zentrums für Forderalismus-Forschung, Band1, Nomos Verlagsgesellschaft.

LLOYD, GREG (1999), 'The Scottish Parliament and the Planning System: Addressing the Strategic Deficit through Spatial Planning', in J. McCarthy and D. Newlands (eds), *Governing Scotland: Problems and Policies*, Aldershot, Avebury, 108–20.

LLOYD, M. G. and J. McCARTHY (1998), *Constitutional Reform and National Planning Options in Scotland* (paper presented at the twelfth annual conference of AESOP [the Association of European Schools of Planning], Aveiro, Portugal, 22–25 July.

MANDELBAUM, SEYMOUR J., LUIGGI MAZZA and ROBERT W. BURCHELL (eds) (1996), *Explorations in Planning Theory*, New Brunswick, NJ, Rutgers University Press.

MASTOP, J. M. (HANS) (1998), *National Planning: New Institutions for Integration* (paper presented at the twelfth annual conference of AESOP), Aveiro, Portugal, 22–25 July.

NEEDHAM, BARRIE, PATRICK KOENDERS and BERT KRUIJT (1993), *Urban Land and Property Markets in The Netherlands*, London, University College of London Press.

ÖSTERREICHISCHE RAUMORDNUNGSKONFERENZ (OROK) (1996), *Austria within the Framework of Spatial Development Policy in Europe*, Vienna, OROK.
PRATS, YVES (1990), 'Land Use Control' in Vincent Renard and Joseph Comby (eds), *Land Policy in France*, Paris, ADEF, 27–37.
PRIEMUS, HUGO (1998), *Four Ministries, Four Spatial Planning Systems? On the Future of Dutch National Planning* (paper presented at the twelfth annual conference of AESOP), Aveiro, Portugal, 22–25 July.
PRODI, ROMANO (1993), 'The Single European Market: Institutions and Economic Policies', *European Planning Studies*, **1**, 13-23.
RITTEL, H. and M. WEBBER (1973), 'Dilemmas in a General Theory of Planning', *Policy Sciences*, **4**, 155–69.
SCHMIDT-EICHSTAEDT, GERD (1995), *Land Use Planning and Building Permission in the European Union*, Germany, Deutscher Gemeindeverlag Verlag W. Kohlhammer (in English and German).
SHAW, DAVID, VINCENT NADIN and TIM WESTLAKE (1996), 'Towards a Supranational Spatial Development Perspective: Experience in Europe', *Journal of Planning Education and Research*, **15**, 135–42.
STEIN, JAY M. (1995), *Classic Readings in Urban Planning: An Introduction*, McGraw-Hill.
SUSSKIND, L., S. McKEARNON and J. THOMAS-LARNER (eds) (1999), *The Consensus Building Handbook: A Comprehensive Guide to Reaching Agreement*, Corwin.
SUSSKIND, L. and J. CRUIKSHANK (1987), *Breaking the Impasse: Consensual Approaches to Resolving Public Disputes*, New York, Basic Books.
THORNLEY, ANDY (1991), *Urban Planning under Thatcherism: The Challenge of the Market*, London and New York, Routledge.
WILLIAMS, RICHARD H. and BARRY WOOD (1994), *Urban Land and Property Markets in the U.K.*, London, University College of London Press.

NOTES

1 The 'Israel 2020' project was initiated and headed from 1990 to 1997 by Adam Mazor, Israel's leading architect-planner practitioner and professor at the Technion–Israel Institute of Technology. More is said about this initiative in the chapter on Israel and in the Preface. This book is the outcome of the exceptional link of practice and academic knowledge production that was enabled by Mazor's leadership.

2 In Britain an equivalent term is 'town and country planning system', while in American English there is no fully equivalent term, though a close approximation would be 'land-use planning' or 'urban and regional planning' (without the word 'system' attached).

3 Even if we restrict ourselves to the English language, the literature on each of these topics is too extensive to be cited here. A few of these topics have also benefited from cross-national comparative analysis, such as plan-making styles in European countries (Healey, 1994), building permits (Schmidt-Eichstaedt, 1995), agricultural land preservation tools (Alterman, 1997), and the division of labour between planning authorities and developers in financing public services (Alterman, 1988).

4 Previous compendiums include Alterman (1990) and Schmidt-Eichstaedt (1995).

5 The authors are Vincent Nadin, Peter Hawkes, Sheila Cooper, David Shaw and Tim

Westlake. In 1997 the EU published the summary volume under the name of the Commission, and in 2000 it published most of the 15 country reports.

6 One of the chapters—that on Ireland—does not cover sectoral planning. The others do, but the span and depths vary somewhat.

7 In view of the resumption of hostilities in the Middle East, the gap between Israel and the other countries in this book is expected to increase.

8 The *EU Compendium of Spatial Planning Systems and Policies* (1997), which classifies all EU countries in terms of the planning instruments used, identifies two types of instruments of national-level comprehensive planning, which it calls either a 'national perspective or plan', or a 'general policy guidance'. These two categories are probably not fully parallel to the difference between our 'haves' and 'half-haves' categories (see text below) because we are looking not just for the existence of the instrument, but also for its legal powers and degree of effectiveness.

9 Most European countries not included in this book appear in one or other of the categories of national-level instruments mentioned in the *EU Compendium* (see above note). In the 'national perspectives or plans' category are included Austria (with both instruments), Finland and Greece. In the category of 'general policy guidance' are included Italy and Luxembourg. As noted in the previous note, these two categories are probably not fully parallel to the difference between our 'haves' and 'half-haves' categories. Two countries—Spain and Portugal—are classified as having only sectoral policies at the national level, and only one country, Belgium, is classified as having no national-level planning instruments.

10 Indeed, in the first two versions of their chapter, the Swedish authors hardly mentioned the national Vision document, until I as editor requested a specific discussion of it.

11 While Israel's population density is currently not much different from the UK's and considerably lower than the Netherlands or Japan, Israel does have a large uninhabitable region, a desert, which the Netherlands and the UK do not (Japan has high mountains). Israel is expected to surpass all the other countries in density, not only effectively (discounting the desert), but also nominally, due to its higher demographic growth rate.

12 Scotland and Flanders are two such examples. See the detailed proposal made by Lloyd (1999) for instituting a distinctive layer of national spatial comprehensive planning in Scotland on the eve of the creation of the Scottish Parliament in 1999, and see the description of the Flanders plan by Albrechts (1998) and by Mastop (1998).

13 See, for example, Österreichische Raumordnungskonferenz (1996).

14 This point does not emerge from the reports in this book, but is part of ongoing research by this author on compensation rights.

15 The concept of the 'legitimation crisis' was introduced by Habermas (1975) as part of the 'critical theory' school. He has had great impact not only through the theorising of social scientists and planners (such as through the writings of Forester [1980; 1989]), but has also impacted on governance styles, such as through the concept of 'reinventing government'.

16 I am basing these observations on my two-month study visit to Japan in January and February of 1994.

17 Some might argue that in the Netherlands the new planning process associated with the 2030 plan now in progress shows some signs of a weakening of support for national-level planning. But given the high level of institutionalisation and effectiveness of Dutch planning, I would hazard a guess that these are not signs of a weakening, but of institutional

realignment and changes in formats and styles of planning, a conclusion hinted at by Mastop (cf. Mastop, 1998; Priemus, 1998).

18 Such as *European Planning Studies*.

19 Kayden notes that in the USA planners generally do not regard sectoral planning as part of land-use planning. I should hope that this view is not shared by planners in some of the other countries discussed here.

20 Locally unwanted land uses.

21 At the time of publication, the High Court declared the involvement of ministers in planning decisions to be incompatible with the European Convention on Human Rights. An appeal is pending against that decision. See note 5 in Malcolm Grant's chapter.

22 See Alterman (1997).

23 ISTEA stands for Intermodal Surface Transportation Efficiency Act of 1992 (see Kayden's chapter).

24 The Israel 1965 Planning and Building Law talks of the preparation of a 'national outline plan', whereby the legislators probably meant a comprehensive plan. However, the law also allows for the preparation of 'partial outline plans' if necessary.

25 Although in many of our sample countries the law and practice regarding local plans have been changed to call for strategic policy plans, this has probably not yet reached consensus status. In other parts of the world, blueprint-style local plans are still the majority rule (see Schmidt-Eichstaedt, 1995; European Commission, 1997). My own comparative research in several countries also supports this statement.

26 The literature on conflict-resolution methods as applied to public decision-making has its origin in the USA, but is gaining presence in Europe and elsewhere, as is indirectly visible from some of the chapters in this book. The literature on conflict resolution or Alternative Dispute Resolution is today very extensive. Two key items are Susskind et al. (1999) and Susskind and Cruikshank (1987).

27 *Directive territoriale d'aménagement.*

TWO

NATIONAL LAND-USE PLANNING AND REGULATION IN THE UNITED STATES: UNDERSTANDING ITS FUNDAMENTAL IMPORTANCE

Jerold S. Kayden

Listed in order of importance, three levels of government in the United States exercise legal authority over land-use planning and regulation: the local level, the state level, and the national level. This article addresses the least important, the national level, concluding that the national government fundamentally does not practise *de jure* or *de facto* national land-use planning and regulation as that term would be specified and understood internationally. Instead, local governments and, increasingly, state governments play the dominant role in making and administering plans and regulations affecting the use of land.

Although not the generator of an orchestrated comprehensive national approach, the national government nonetheless has a substantial impact on the use and development of private and public land through an uncoordinated patchwork of national laws, institutions and actions. The five noteworthy patches cover the following areas: first, environmental regulation; second, management of nationally owned land; third, transportation policy and finance; fourth, housing and economic development subsidies; and fifth, anti-land-use planning and regulation. Singly or together, these patches do not assemble themselves into a coherent and comprehensive approach worthy of the sobriquet 'national land-use planning'.

Terminology

Terminology is too frequently a hobgoblin of international comparative analysis (Alterman and Kayden, 1988, 22–23), so it is wise to specify at the outset the underlying definitions of key words used in this article. In the United States, 'land-use planning' and 'land-use regulation' are terms regularly employed to describe governmental efforts to guide and control the use and development of land.[1] Unlike usage elsewhere in the world, however, the term 'land-use planning' does not subsume the term 'land-use regulation'. As employed in the United States and in this article, 'land-use planning' is a process conducted by the public secctor to analyse and recommend in a comprehensive manner, from social, economic, environmental, infrastructure capacity, aesthetic, and all other relevant aspects, the best present and future uses of geographically specified territories. The usual product of public-sector land-use planning is a 'land-use plan', consisting of text and maps, covering a defined area such as a neighbourhood, municipality, or other area. The plan itself does not directly control the use of land, although it may indirectly control use by influencing or directing the very regulation, such as zoning and subdivision, that expressly control such use.[2] 'National' land-use planning is land-use planning conducted by national government agencies.

'Land-use regulation' refers to laws such as zoning and subdivision that directly control the use and development of privately owned land.[3] Land-use regulations may also apply to publicly owned land, although the government bodies that enact such laws frequently exempt themselves from their application, and higher-level government bodies do not normally have to follow regulations enacted by their lower-level counterparts. 'National' land-use regulation is a land-use law enacted and administered by national government agencies.

Other adjectival words, such as territorial, spatial and positive planning, are not a usual part of the American legal and professional planning vocabulary, although they may be found in academic literature, especially that written for international audiences. Sectoral planning such as economic or social welfare planning, and planning for infrastructure projects, public land, public housing and economic development, among others, are not understood by American planners to be part of the conventional land-use planning practice.

Explaining the relative absence of national land-use planning and regulation

Although it has been assumed, at least until a recent series of interpretive US Supreme Court opinions, that the national government has the constitutional authority to enact national laws enabling, mandating or guiding national land-use planning and regulation for every square metre of America,[4] national legislative and executive branches have consciously eschewed such exercise. Most significantly, there is no national land-use planning or regulatory law denominated as such, nor any other differently named national law that might be construed as a functional equivalent.[5] Furthermore, there is no articulated or commonly accepted national land-use policy emanating from executive or legislative branches that, short of a law, still guides national executive or legislative actions. Members of the professional, academic and lay communities involved in land-use planning and regulation would not characterise the actual national role as primary when it comes to the planning and regulation of land in America.[6]

The lone attempt to enact legislatively a national land-use policy law,[7] albeit a toothless one, foundered politically over 25 years ago. Introduced by then Senator

Henry M. Jackson on 29 January 1970, Senate Bill S. 3354, the 'National Land Use Policy Act of 1970', proposed to establish:

> a national policy to encourage and assist the several states to more effectively exercise their constitutional responsibilities for the planning, management, and administration of the Nation's land resources through the development and implementation of comprehensive 'Statewide Environmental, Recreational and Industrial Land Use Plans,' (hereinafter referred to as Statewide Land Use Plans) and management programs designed to achieve an ecologically and environmentally sound use of the nation's land resources.[8]

The law did not authorise the national government to plan, let alone regulate, the use and development of land or the location of infrastructure. Instead, it merely intended to engage national, state and local levels in a process of consultation and information exchange on matters of land use, a step that would nonetheless represent a substantial expansion of the extant national role. The original Jackson bill never came to a vote in the Senate, but subsequent versions, watered down from the original and receiving only lukewarm support from the Senator himself, eventually garnered majority support in the Senate, but failed in the House of Representatives (Daly, 1996). Simply put, even the thin gruel of national land-use policy represented by S. 3354 and successor bills proved politically inedible.

THE FEDERAL GOVERNMENTAL STRUCTURE AND CONSTITUTIONAL ALLOCATIONS OF AUTHORITY

What are the possible explanations for this national absence? To begin with, in a federal system whose central government emerged after, rather than before, its constituent states, it is not altogether surprising that the allocation of governmental roles and responsibilities, in land use as in other areas, would not necessarily radiate from the 'top down'. Born in 1787 and 1788 when the ratifying states and the people adopted a constitution establishing a national government of expressly delegated powers while retaining undelegated powers for the states, the 'United States of America' remains fundamentally layered in its legal and institutional approaches to most problems.

As a constitutional matter and for purposes of generalisation here, the national government enjoys supreme authority to enact laws that govern matters affecting the interests of more than one state, even if individual states or citizens disagree.[9] For example, Article I, Section 8, grants to Congress the power to regulate commerce among the several states (the so-called 'Commerce Clause').[10] It may easily be argued that, because the use of land by one party inevitably affects others, and because such 'others' may reside across state lines from the initial user, the Constitution axiomatically grants the national government the power to control local land use. Recent United States Supreme Court decisions have chipped away, however, at the automatic assumption of sweeping national authority, casting a sceptical eye at national efforts to regulate in such areas as gun control and, most recently, certain types of wetlands.[11] Since the Constitution does not explicitly grant Congress[12] or the President[13] the right to plan or control the use of land, it is no longer as senseless as it once was to question the assertion of federal power in this area.

THE HISTORICAL RECORD

In addition to the layered authority inherent in the United States federal system and recent interpretations about the allocation of constitutional authority, it has historically been the case that legislative and administrative efforts in land-use planning and regulation have predominantly occurred at the state and local levels. The fifty states have unchallenged authority under their so-called 'police power'[14] to plan and regulate the use and development of land. Under that authority, they have fully legislated the structure within which land-use planning and regulation take place, enacting laws that enable, mandate and guide local governments in their adoption of local land-use plans and regulations (Meck, 1996, 1–17). Ironically, the states initially relied on the national government, not for legal approval, but for encouragement and inspiration, copying language from model acts on zoning and city planning drafted by *ad hoc* national advisory committees appointed by then Secretary of Commerce, later President, Herbert Hoover and designed to jump-start local zoning and planning (US Department of Commerce, 1924; 1927).

Even as states have legislated the basic structure of planning and regulation, however, they have played only a supporting role to that assumed by a level of government not even constitutionally recognised: the local government. The *realpolitik* of American land-use planning and regulation today, as for most of the twentieth century, is that local governments—those tens of thousands of cities, towns and villages scattered across the nation—exercise the greatest *de jure* and *de facto* control over the use and development of the majority of land holdings in the United States.

Virtually all local governments have adopted local zoning ordinances to control land use within their political boundaries, and such laws have enjoyed a near mythic status as untouchable local government prerogatives, even when such regulations stand in the way of higher-level goals and even though the legal authority for such local actions paradoxically devolves from higher-level laws. Planning is similarly local in practice. The most common land-use plan, prepared separately by individual municipalities, covers the whole of that municipality and is known as the comprehensive, general or master plan. Land-use plans spanning a region larger than one municipality, let alone a state, are still the exception rather than the rule, although there is some movement up the ladder over the past 25 years as part of a 'quiet revolution' in land-use administration (Callies, 1996, 19–26).

THE CONTEXTUAL EXPLANATIONS

The American context helps to illuminate why American society, politics and policy favour local, rather than national, planning and regulation of land use and development. Contextual explanations derive from aspects of American history, culture, laws, institutional capacities, political structures, economic systems, demography, land utilisation and ownership patterns, stages of nationwide development and the like. Measuring these factors in the American context, as against the context of another country, reveals answers explaining the absence of a national role, an absence unusual in comparison with countries around the world.

One of the most important, although not dispositive, factors is the size of the United States, and the resulting greater possibility of topographical, economic, social and cultural variations that intrinsically complicate efforts to plan and regulate centrally. The idea of an official based in Washington, DC, or even a regional representative, attempting to plan and control land-use development for each and every American city and town is on the face of it an operational nightmare.

Furthermore, the United States does not face overwhelming imperatives, such as grave natural resource deficiencies, neighbouring military threats, or the challenges of early stages of economic development, that would render national land-use planning and regulation a matter of national survival rather than one of arguably greater efficiency and equity. Indeed, it has been at times of great national crisis that the American national government has come closest to introducing far-reaching initiatives involving aspects of land-use planning. For example, when faced with the development of the midwestern and western portions of the country in the 1800s (Wolf, 1981, 36–66), the Depression of the 1930s,[15] and the urban riots of the late 1960s,[16] the national government sponsored initiatives involving some degree of national planning approaches. Without the constant presence of such concerns, however, the imperative to institutionalise national planning and regulation never bore fruit.

The primacy of private property and private market ideology provides another explanation for the meek national role. In a country where the percentage of privately owned land is estimated to be roughly 60 per cent of the total (Wolf, 1981, 443), it is understandable that the national government may have less of a claim or interest in what should happen on that land than in countries where land is or has historically been owned predominantly by the state.[17] Americans especially value private property in land, and that sentiment has reduced the ability of all levels of government to plan and regulate in ways that unduly upset property owner expectations.[18] Overall, the fundamental role of the private sector in the American economy fosters a strong counterweight to government authority, and this reality touches land-use regulation as well as other public–private interactions.

Another explanation for the absence of national land-use planning and regulation is the preference many individuals appear to express for local control over their lives. Some observers suggest that smaller, i.e. local, governments are inherently more responsive to their citizens than larger far-away governments (Frug, 1980, 1057ff). To the extent that this is a correct evaluation, then owners and neighbours may prefer to find their land locally planned and regulated by persons they know rather than by persons more remote by measures of distance, knowledge, and susceptibility to influence.

Examining the limited role of the national government in land-use planning and regulation

What role, then, does the national government play in terms of land-use planning and regulation? The national role takes its shape from a patchwork of substantive laws, institutions and actions that singly and together exert an uneven, albeit substantial, influence over the use and development of land. This patchwork has arisen in response to specific land-use problems that suggest national, rather than local, solutions, and is composed of five principal patches:

- Environmental regulation;
- Management of nationally owned land;
- Transportation policy and finance;
- Housing and economic development subsidies; and
- Anti-land-use planning and regulation.[19]

ENVIRONMENTAL REGULATION

Since 1970, the national government has enacted a bevy of environmental laws that make it, rather than the states and local governments, the leader in environmental protection. These laws include the following:

- National Environmental Policy Act (1970);[20]
- Clean Air Act (1970);[21]
- Clean Water Act (1972);[22]
- Coastal Zone Management Act (1972);[23]
- Endangered Species Act (1973);[24]
- Safe Drinking Water Act (1974);[25]
- Toxic Substances Control Act (known as ToSCA) (1976);[26]
- Resource Conservation and Recovery Act (known as ReCRA) (1976);[27]
- Surface Mining Control and Reclamation Act;[28] and
- Comprehensive Environmental Response, Compensation, and Liability Act (known as CERCLA or 'Superfund') (1980).[29]

This national environmental regime is composed of one law-making institution (Congress), two hybrid law-making, law-administering institutions (national administrative agencies and states), one law-reviewing institution (judiciary), and two actively participating non-public parties (non-governmental organisations and regulated parties). Congress drafts and adopts the laws, setting fundamental goals and policies within each piece of legislation while at the same time delegating enormous discretion to national administrative agencies. The national administrative agencies implement the laws on an ongoing basis. They draft and adopt regulations announcing environmental standards or mandating technology, grant or deny permits to industry, local governments, and sometimes private developers, approve state and local environmental plans that satisfy the requirements of the national law and administrative regulations, and enforce conduct through administrative and judicial action. The Environmental Protection Agency, the national government's central environmental institution, has primary administrative responsibility for the key environmental laws (Clean Air, Clean Water, CERCLA, ReCRA, etc.). Other national executive branch agencies, especially the Departments of Interior, Agriculture, and Commerce, and the Army Corps of Engineers, also have formal roles under some of these laws.

The states now increasingly work in partnership with, or may even supplant, the national government. This was not always the case. The initial years of the national environmental intervention relied heavily on a top-down, centralised 'one size fits all' administrative approach. The rationale for centralisation at the highest jurisdictional level was surely plausible. Pollution does not recognise political boundaries. The acidic air emitted by coal-burning midwestern power plants results in acid rain wreaking havoc on the forests and lakes of New England. Depending on which way the wind blows and water flows, individual states, let alone cities, could not be assumed to act in the best interests of their neighbours when they acted in their own self-interest.

Devolution of environmental authority to the state level, however, has grown dramatically in recent years. States now prepare their own plans and programmes that, once certified by the responsible national administrative agency as satisfying the letter of the national law and implementing regulations, allow the states themselves to administer the environmental plan and programme and grant or deny permits to local parties.[30] In addition, states may adopt even more stringent standards than those required under the national law.

The reviewing institution, the judiciary, considers challenges from private and public actors to a given law or administrative action (Wald, 1992, 519–46). Non-governmental organisations and regulated parties participate in the administrative rule-making process and are also able to bring lawsuits against agencies. They will assert, for example, that the environmental administrator has failed to adopt an environmental standard, has acted arbitrarily and capriciously unsupported by substantial evidence in the record, or has otherwise improperly interpreted the law under which it operates.[31]

Land-use planning and regulatory content of environmental laws and institutions

Taken alone or together, the national environmental laws and institutions cannot be characterised as creating a regime of national land-use planning or regulation, and they do not pretend to do so. They are expressly single-issue laws focused on one environmental medium or issue. For example, the Clean Air Act and Clean Water Act may employ 'command-and-control' strategies dictating 'end of pipe' or 'end of smokestack' pollution results, mandate use of specific control technologies, or impress efficiency-oriented 'market-based' regulatory solutions that allow companies to trade 'pollution permits' (Stewart, 1992, 547–55). What they intentionally do not do, however, is provide the opportunity to balance comprehensively the range of other concerns, from economic development and social equity to infrastructure capacity and quality of life, that undergird boilerplate land-use planning and regulation. They do not mandate preparation of national, state or local land-use plans, nor do they determine across the typical land-use palette a set of uses, densities and shapes for development. Indeed, a 1990 amendment to the Clean Air Act expressly cautioned: 'Nothing in this [Act] constitutes an infringement on the existing authority of counties and cities to plan or control land use, and nothing in this [Act] provides or transfers authority over such land use'.[32]

At the same time, lawmakers, law administrators and judges have recognised the axiomatic reality that land uses and their location contribute to pollution, and that one of the obvious ways to control pollution may involve some attention to land-use planning and regulation. Indeed, in one extreme case, a judicial interpretation of the Clean Water Act triggered a court-ordered moratorium on all commercial development within the Boston metropolitan area until the state of Massachusetts developed a meaningful programme for the treatment of sewage being dumped in Boston Harbour (Haar and Wolf, 1989, 773). Not surprisingly, such extreme measures helped prod the legislature to act expeditiously to address the problem.

Although national environmental laws and institutions do not authorise or conduct national land-use planning and regulation as such, they nevertheless introduce slices of land-use planning and regulation as part of their overall regimes. These slices fall into five categories:

- Environmental plans with land-use elements;
- Collaborative planning among different levels of government;
- Nationally granted land-use permits;
- Environmental impact review; and
- Financial grants for local environmental clean-up.

First, environmental laws may require or encourage preparation of state or territorially-based environmental plans that may include land-use elements. For example, under the Clean Air Act, states prepare a 'State Implementation Plan' that establishes 'enforceable emission limitations and other control measures' demonstrating how the state will reach national primary and secondary ambient air quality standards.[33] In preparing this plan, a state may choose that mix of control measures deemed best suited to its particular situation, including 'Transportation Control Measures' that may include state and local land-use controls such as site plan review, mixed-use zoning, and transit-oriented design standards, encouraging mass transit and discouraging single-occupancy vehicle travel (Netter and Wickersham, 1993a, 145, 149–50). However, as described earlier, efforts by the Environmental Protection Agency to require local governments to include such controls in the arsenal of air pollution control weapons have met with Congressional disapproval.

Federal environmental laws may provide financial incentives for the preparation of plans and additional measures that affect the use and development of land. Under the Coastal Zone Management Act, for example, states have been eligible to receive financial grants from the national government to assist in preparing and managing a programme for the preservation of coastal areas.[34] Ongoing administrative grants are available to states that have obtained approval from the national government for their coastal management programmme. Such approval is contingent, among other things, on a 'definition of what shall constitute permissible land uses and water uses within the coastal zone which have a direct and significant impact on the coastal waters',[35] and upon demonstration that the state has assumed the authority 'to administer land use and water use regulations to control development to ensure compliance with the management program, and to resolve conflicts among competing uses.'[36]

Second, environmental laws may encourage collaborative decision-making between public and private sectors involved in land development and regulation to create plans that take into account interests beyond the environmental objective. Under the Endangered Species Act, for example, land that is designated critical habitat for endangered species is normally difficult, if not impossible, to develop.[37] However, in the face of political unpopularity over such draconian national intrusions, the Department of the Interior has experimented under one section of the Act to allow parties to work together on the preparation of so-called habitat 'conservation plans' that give play to non-environmental, as well as environmental, interests.[38] Pioneering efforts in California and Texas have tested successfully different institutional arrangements to oversee selection, planning and monitoring of such habitats, to allow development while ensuring achievement of biological diversity (Wheeler, 1996, 155–72).

Third, environmental laws may authorise national agencies to grant or deny permission to develop land. Under Section 404 of the Clean Water Act, for example, landowners may not dredge or fill, let alone develop, any wetland[39] in the United States without first obtaining a permit from the Army Corps of Engineers, a national agency concerned principally with the construction of dams and the maintenance of navigable waterways.[40] Because wetlands located in and near metropolitan areas are increasingly attractive development sites, the Section 404 review has played a significant and notorious role in a number of land development projects.[41]

Fourth, environmental laws may require national agencies to conduct environmental impact reviews for national actions having a significant effect on the environment. The National Environmental Policy Act requires all national agencies to 'utilize a systematic, interdisciplinary approach which will insure the integrated use of the natural and social sciences and the environmental design arts in planning and in decision-making which may have an impact on man's environment'.[42] Specifically, the Act requires all agencies of the federal government to:

> include in every recommendation or report on proposals for legislation and other major Federal actions significantly affecting the quality of the human environment, a detailed statement by the responsible official on (i) the environmental impact of the proposed action, (ii) any adverse environmental effects which cannot be avoided should the proposal be implemented, (iii) alternatives to the proposed action. . . .[43]

Private activities on land become subject to this review when they require some federal approval or grant of money constituting a 'major Federal action'.

In practice, the environmental impact statement requirement has had a profound impact on private and public development projects across the country. Although it is only advisory, in the sense that it does not by its own terms veto a project even if the review reveals such a project to be environmentally harmful, the transparency of information compiled for the statement, coupled with the procedural requirements for the review, have generated public opposition and lawsuits sufficient to cancel or modify proposed projects.[44] Additionally, NEPA has spawned state copycat legislation (so-called 'little NEPAs') mandating state and local environmental impact reviews that in many cases pack a more substantive land-use review wallop than that required by the federal NEPA itself.[45] Finally, NEPA has spurred similar provisions in other national laws requiring assessments to determine the impact of federal actions on historic properties (Section 106 reviews)[46] and parkland and historic sites (Section 4(f) reviews).[47]

Fifth, environmental laws may provide financial grants for local environmental clean-up efforts that make development or redevelopment of land possible. The Clean Water Act, for example, has provided sewage treatment facilities grants to state and local governments that provide infrastructure capacity to accommodate growth,[48] while the Comprehensive Environmental Response, Compensation, and Liability Act establishes a 'Hazardous Substance Superfund' to assist in the clean-up of hazardous waste sites, known as 'brownfield' sites, often located in or near the central city and often a barrier to inner-city redevelopment.[49]

Continuing disjunction between environmental and land-use policy

Even with the substantial reliance on land-use approaches by the national environmental regime, environmental policy and land-use policy in the United States still remain, to a disturbing extent, separate and distinct fields, created and implemented by different levels of governments and studied by different sets of academics and professionals. The average environmental policy-maker is as unfamiliar with zoning as the average land-use policy-maker is with the Clean Air Act. Yet zoning's predilection for single-family one-acre development patterns in many suburban areas may engender a substantial dependence on the automobile, whose polluting side effects are addressed after the fact by laws from Congress and implementing regulations from the Environmental Protection Agency.

The reasons for this disjunction are situational and functional. Land-use controls emerged locally in the early twentieth century to address inevitable conflicts between land uses. Such incompatibilities were not limited to health and safety concerns—the ambit of environmental laws—but instead extended to the gamut of issues—noise, odours, visual blight, and other offences to the senses—that traditional Anglo-American nuisance law had heretofore mediated. Zoning, the principal regulatory tool, codified nuisance law and helped stabilise residential property values by separating incompatible land uses into districts.[50] Zoning would tell landowners what they could do with their property through its trio of use, height and bulk restrictions.

State government, let alone national government, could hardly from such a distance tell residents where residential, commercial and industrial uses should go.

In contrast, national environmental policy emerged some 50 years later in the wake of growing concern about links between chemical pesticides and health woes (Carson, 1962, 1ff), links not specifically tied to local land-use concerns. Indeed, ecologists highlighted the holistic character of the environment, captured by the metaphor of 'Spaceship Earth', that inherently militated against localised approaches to environmental degradation.[51] The national environmental movement ultimately seized upon pollution standards for air and water, as opposed to land, thereby emphasising the fluidity of elements whose very nature transcends local boundaries and anchored geography.[52]

As the national government has become more interested in land-based pollution, such as wetland development and solid and hazardous waste disposal, and as local governments have become more concerned with water tables and conservation of open space, the intersection of environmental and land-use policy is becoming increasingly apparent. Furthermore, as more states require environmental impact assessments of private projects that effectively replicate land-use regulatory reviews, the desire for efficient government administration will continue to suggest closer scrutiny of duplicative efforts, with an eye towards merging aspects of the two regimes.

MANAGEMENT OF NATIONALLY OWNED LAND

The national government owns and manages roughly 30 per cent of all land in the United States (Wolf, 1981, 36–66). Although such an ownership stake might be thought to be a wedge for the national government in articulating a national land-use plan or policy, it has never been perceived or used as such. One reason is that most of this land is located in Alaska and a few western states, far from population centres and relatively inaccessible except to those interested in natural resource exploitation or tourism. Even a clearly articulated planned vision for this land would leave most Americans unaware and untouched on a daily basis. Furthermore, national goals and objectives for nationally owned land have run foul of locally perceived interests even in those few states where national ownership is an issue.[53]

The Federal Land Policy and Management Act of 1976 provides a base framework for the planning, management and use of nationally owned land.[54] Institutionally, the Department of the Interior (through its Bureau of Land Management, its Fish and Wildlife Service, and its National Park Service), and the Department of Agriculture (through its Forest Service), have prime responsibility for nationally owned land. Depending on the type of land, the legislative mandate set forth in the several relevant laws may prescribe a single use or multiple uses. For example, national parkland managed by the National Park Service is designated for the singular purpose of providing national parks for the enjoyment of present and future generations of Americans.[55] The Bureau of Land Management and the Forest Service have more contested mandates of allowing multiple land uses, with recreation and open space protection purposes competing head-to-head with logging, mining and grazing activities.[56] Here, the debate focuses as much on such issues as the fee structure for use of public lands by private licensees, including cattle owners, timber companies and mining interests, as on the nature of appropriate uses for the land.

TRANSPORTATION POLICY AND FINANCE

For the past 45 years, the national government has played a fundamental role in the planning and financing of much of the nation's transportation infrastructure, especially the coast-to-coast network of interstate highways. During most of this road-building period, the national government, through its Federal Highway Administration and Department of Transportation, provided 90 per cent funding, with states providing the remaining 10 per cent. Although this federal programme probably had greater impact on the land-use patterns of metropolitan America than any other national government-sponsored activity,[57] by lubricating suburbanisation trends, its legal framework only recently took strongly conscious note of the 'land-use' aspects of highways and other transportation projects.

In 1991, Congress enacted the Intermodal Surface Transportation Efficiency Act of 1992 (ISTEA),[58] setting forth a comprehensive intergovernmental planning approach that relied much more than ever before on federal, state and regional involvement and an affirmative obligation to prepare long-range transportation plans for urbanised areas. The Act did not dramatically alter previous responsibilities of the federal government, retaining its principal funding responsibilities for a wide variety of transportation projects. At the same time, the Act strengthened and expanded the notion of planning, including land-use planning, as part of delivering transportation projects. The key institutional players would be an existing or newly formed 'Metropolitan Planning Organization' (MPO),[59] responsible for preparing long-range transportation plans,[60] along with the states, which would be required to prepare their own long-range plan relying on the MPO-prepared plans as well as state-prepared plans for areas outside the jurisdiction of existing MPOs. The plans would consider not only direct transportation issues, but the impact of land-use and development patterns on transportation outcomes.[61]

MPOs and states also would prepare 'Transportation Improvement Programmes' (TIPs) that specify projects and programmes within the metropolitan area that would receive funding obtained in part from the national government.[62] The state must approve the locally prepared TIP, which must be consistent with the state's and MPO's own long-range transportation plan, and the state also would adopt its own TIP.[63] Significantly, ISTEA set aside three per cent of the project budget to pay for this level of planning, with two per cent going to the state and one per cent going to MPOs.[64] In a new gesture towards alternative transportation modes, ISTEA also set aside 10 per cent of its surface transportation block grant funding for so-called 'transportation enhancements' such as bicycle and pedestrian paths, scenic easements and other types of non-highway facilities.[65]

In addition to its intergovernmental consistency requirement, ISTEA established a conformity relationship outside itself with the Clean Air Act. Both the long-range transportation plan and the transportation improvement programme must be in conformity with the State Implementation Plan demanded under the Clean Air Act.[66] In this way, ISTEA introduced cutting-edge change at the national level by weaving together previously discrete actions into a land-use, transportation and air quality web. Since transportation investment decisions obviously have a major impact on land use and environmental quality, this legally imposed relationship between plans and institutions represented an important break with prior planning practices.[67]

HOUSING AND ECONOMIC DEVELOPMENT SUBSIDIES

Since the 1930s, the national government has distributed money directly to states, local governments, special purpose local agencies and private developers for housing construction and the redevelopment of urban and rural areas. Indeed, in the eyes of many, these distributions of funds have constituted America's national urban policy (*The President's National Urban Policy Report*, 1978, 11-2–11-3). At one time, traditional land-use planning itself received official recognition under the national urban policy development umbrella. Under the federal 'Section 701' programme of the 1950s, the national government distributed grants expressly for the preparation by local governmental units of urban land-use plans.[68] More recently, rather than receive money from the national government to pay for land-use planning, local governments have employed land-use planning to enhance their efforts to wrest available grant and loan money from the national government for urban redevelopment.

Specifically, the national Department of Housing and Urban Development (HUD), along with smaller government agencies (the Department of Commerce's Economic Development Administration and the Small Business Administration), makes intergovernmental transfers of money in two principal ways: block grants and categorical grants. Block grants are distributions of money as a matter of right to local governments for a broad range of urban development purposes. The size of the grant is determined according to a formula set forth in the authorising legislation. HUD ensures that the money is being spent in accordance with the requirements of the legislation, by reviewing reports which summarise what cities have done and will do with the money. This review, then, assures that local governments rely on aspects of land-use planning approaches to secure, and continue to secure, their block grant money.

Categorical grants are awarded competitively, with municipalities and developers applying to the national agencies to demonstrate why their proposed project or programme is superior to other proposals. Applications for such funds frequently demand some degree of planning in order to compete effectively pursuant to the stated requirements of the grant legislation and implementing agency regulations. Most recently, for example, under the 'empowerment zone' programme, HUD has reviewed applications from cities across the country for designation of an empowerment zone, a designation that brings financial rewards to that area.[69] The empowerment zone programme expressly requires a detailed and inclusive planning exercise by each city, bringing all stakeholders together to consider ways in which a poor area of the city might by regenerated.[70]

ANTI-LAND-USE PLANNING AND REGULATION

The most recent attempted incursion of the national government into land-use planning and regulation is, ironically, anti-land-use planning and regulation. The Congress has debated for several years a number of proposals that would provide legislative protection extending beyond what the Constitution provides for private property owners aggrieved by national government planning and regulatory actions. These congressional rumblings are a legislative response to increased activity in the courts. Under the national constitution's 'Just Compensation' Clause, which commands, 'nor shall private property be taken for public use, without just compensation', property owners have challenged local zoning regulations, coastal development controls, denial of wetlands permits and similar restrictions on property development.[71] Judges have struggled to define an acceptable balance between government regulation on behalf of the public's health, safety and general welfare, and the constitutional view that owners should not be forced to bear burdens more properly borne by society as a whole. By issuing important opinions establishing that, although owners are not entitled to the most profitable use of their property, they are entitled not to be denied all economically viable use,[72] the Supreme Court of the United States has raised the stakes in the government regulation versus private property game.

Interestingly, the congressional bills would essentially supplant the Court's constitutional jurisprudence, substituting a standard far less deferential to government regulation and concomitantly far more accommodating to the interests of owners: in a real sense, an anti-planning, anti-regulatory national government initiative. Under bills considered, but not enacted, by the Congress, monetary compensation would be paid to landowners every time a national government action reduced the value of an owner's property by more than 20 per cent or 33 per cent.[73] These ideas took their lead from a Presidential Executive Order (Executive Order 12, 630), issued in 1988, requiring all federal agencies to conduct a 'takings impact assessment' prior to the adoption of any regulation or action to determine whether or not the proposed regulation or action would unconstitutionally infringe upon the rights of property owners.[74] More recent proposed legislation in Congress attempts to make it faster and easier for property owners to gain relief in the federal courts.[75] Whether these types of bill will meet the same fate as Senator Jackson's S 3354 remains an intriguing question.

State and regional land-use planning and regulation

Although highly valued by Americans, the balkanised system of land-use planning and regulation, with every municipal tub on its own bottom, has generated sufficient problems to motivate some localities and citizens to yield to higher-level planning and regulatory solutions. Under the rubrics of 'state growth management' or 'state planning', an increasing number of states are introducing stronger regional and statewide planning and regulatory requirements. Indeed, these state roles begin to approximate to the roles played by national governments in many countries around the world. While this upward trend might theoretically suggest the possibility of a climb to the top, there is no evidence to suggest that such national views are emerging.

What specific problems are such higher-level approaches attempting to remedy? Lack of regional planning or control may create conditions suitable for sprawl development. Sprawl, it is said, needlessly consumes valuable open space and agricultural land, potentially increases costs of infrastructure by not taking advantage of existing infrastructure in developed areas, increases the cost of housing to the extent that higher densities are not allowed, and promotes reliance on the automobile that may produce greater congestion and air pollution.[76] Furthermore, the ability of suburban communities to wield zoning as an exclusionary measure may exacerbate the plight of some inner-city residents, making it economically difficult for them to obtain decent jobs and housing in outlying metropolitan areas (Haar, 1996, 4–8).

Not surprisingly, the states that have adopted laws going well beyond traditional authorisations for local zoning controls contain broad expanses of environmentally sensitive land, fear an onslaught of new residents moving from other states, or have experienced the problems of enormous growth. Vermont, Hawaii, Oregon and Florida—the first generation—were primarily concerned with the preservation of environmentally sensitive land and, in Florida's case, explosive growth and infrastructure impacts (Pelham, 1979, 1ff). The second generation, Washington, New Jersey, California, Rhode Island, Maine and Georgia, attempted to address a wider range of issues derived from their local context (Callies, 1996, 19–26). Although it is impossible to describe one single model that encapsulates all of these laws, it is possible to set forth key elements underlying typical state growth management laws.

- *State level planning and plans*: The state laws may call for preparation of a state planning document to guide the actions of state agencies. In the case of Hawaii, for example, the state actually prepared a land-use plan that geographically covered the state.[77] In other states, however, the plans tend to be strategic in nature, describing broad goals, objectives and policies (Stroud, 1996, 85–87).
- *Local plan preparation requirement*: The state laws may require local governments to prepare comprehensive plans and include sections within such plans dealing with specified issues such as affordable housing, open space and agricultural lands, transportation and so forth (Stroud, 1996, 86). The state laws may provide financial incentives to prepare the plans, and penalties for those communities that fail to prepare plans, including ineligibility for certain state funds or the removal of the local authority to regulate for certain purposes (Stroud, 1996, 86).
- *Consistency and review requirement*: The state laws may require horizontal and vertical consistency requirements for local plans and actions (Lincoln,

1996, 96–102). For example, all municipal actions must be in accordance with the municipally adopted comprehensive plan. The comprehensive plan must be consistent with state goals and objectives, or with the state plan itself. State actions within the municipal jurisdiction must be consistent with the local comprehensive plan. Finally, local plans may be reviewed, and even approved, by higher-level authorities.

- *Concurrency requirements*: The state laws may require that public facilities and services needed to support new development be available concurrently with the impacts of the new development. In Florida, for example, there have been concurrency requirements attached to six facilities and services: transportation, water, sewerage systems, solid waste, parks and recreation, and storm water management (DeGrove, 1992, 16–17). Interestingly, this 'pay as you grow' model may create unanticipated consequences, when development leapfrogs to areas of low traffic congestion where road infrastructure may be able to accommodate new growth, but where growth is otherwise not desirable.
- *Urban growth boundaries*: The state laws may authorise the drawing of a boundary line around a metropolitan area within which development is encouraged and outside which development is deterred. Oregon has adopted urban growth boundaries, and the city of Portland has operated within one for years.[78]
- *Areas of critical state concern*: The state laws may appoint or authorise a state agency to designate certain areas as sufficiently sensitive (almost always from an environmental point of view) to need special legal protection (Berry, 1996, 105–07).
- *Developments of regional impact*: The state laws may require developments of a certain size or impact to undergo special regional review above and beyond requirements imposed upon other development (Morris, 1996, 111–18).
- *Regional plans and planning agencies*: State laws may authorise the formation of a regional planning agency to prepare plans and regulate land uses for areas including more than one local government jurisdiction.[79]
- *Horizontal intergovernmental agreements*: State laws may authorise cooperative agreements between neighbouring local jurisdictions (Salkin, 1996, 39–40).
- *Fair share housing*: State laws may require municipalities to plan and regulate in ways that encourage development within the municipality of its 'fair share' of the region's present and prospective housing needs for all income classes (Moskowitz, 1996, 153–57).

Conclusion

This article has described the legal and institutional structure for land-use planning and regulation in the United States, concluding that the national government has not and does not practise what would constitute national land-use planning and regulation as that term is commonly understood in the United States and internationally. The reasons for this absence stem from a country-specific blend of constitutional, historical, cultural and economic ingredients that together favour local land-use planning and regulation over higher-level exercises. Recent experience indicates greater interest in and acceptance of state-level planning and regulation, especially in regions of the country facing high growth rates and threats to their environmentally sensitive lands. This 'rise up the ladder' to a higher level of government, however, shows no signs of topping out at the national level, and the conditions that have previously limited the national role show no signs of abating. Thus, while the national role will continue in patchwork fashion, a more comprehensive effort appears unlikely to emerge in the near future.

REFERENCES

ALTERMAN, RACHELLE and JEROLD S. KAYDEN (1988), 'Developer Provisions of Public Benefits: Toward a Consensus Vocabulary', in Rachelle Alterman (ed.), *Private Supply of Public Services: Evaluation of Real Estate Exactions, Linkage, and Alternative Land Policies*, New York, New York University Press.

BARNETT, JONATHAN (1995), *The Fractured Metropolis: Improving the New City, Restoring the Old City, Reshaping the Region*, New York, Harper Collins.

BERRY, JAMES F. (1996), 'Areas of Critical State Concern', in *Modernizing State Planning Statutes* (The Growing Smart Working Papers, Volume One), Chicago, American Planning Association.

CALLIES, DAVID (1996), 'The Quiet Revolution Revisited: A Quarter Century of Progress', in *Modernizing State Planning Statutes* (The Growing Smart Working Papers, Volume One), Chicago, American Planning Association.

CARSON, RACHEL (1962), *The Silent Spring*, New York, Fawcett Crest.

DALY, JAYNE E. (1996), 'A Glimpse of the Past—A Vision for the Future: Senator Henry M. Jackson and National Land Use Legislation', in Pace University Land Use Law Center Website, 4–10.

DeGROVE, JOHN (1992), *The New Frontier for Land Policy: Planning and Growth Management in the States*, Cambridge, MA, Lincoln Institute of Land Policy.

EISENBERG, EVAN (1998), *The Ecology of Eden*, New York, Alfred A. Knopf.

FRUG, GERALD (1980), 'The City As a Legal Concept', *Harvard Law Review*, **93**.

GORDON, PETER and HARRY W. RICHARDSON (1998), 'Prove It: The Costs and Benefits of Sprawl', *Brookings Review*, **16**.

HAAR, CHARLES M. (1955), 'In Accordance with a Comprehensive Plan', *Harvard Law Review*, **68**.

HAAR, CHARLES M. (1977), *Land-Use Planning* (3rd edn), Boston, Little Brown.

HAAR, CHARLES M. (1996), *Suburbs under Siege: Race, Space, and Audacious Judges*, Princeton, Princeton University Press.

HAAR, CHARLES M. and MICHAEL WOLF (1989), *Land-Use Planning* (4th edn), Boston, Little Brown.

INTERNATIONAL FINANCE CORPORATION (1995), *Land Privatization and Farm Reorganization in Russia*, Washington, DC, International Finance Corporation.

KAYDEN, JEROLD S. (1993), 'The Role of Government in Private Land Markets', in *Privatization of Land and Urban Development in Ukraine*, Kiev, United States Agency for International Development.

KAYDEN, JEROLD S., ALEX GAMOTH and VLADIMIR NOSIK (1995), *A Guide to Land Auctions in Ukraine*, Kiev, United States Agency for International Development.

LINCOLN, ROBERT (1996), 'Implementing the Consistency Doctrine', in *Modernizing State Planning Statutes* (The Growing Smart Working Papers, Volume One), Chicago, American Planning Association.

MANDELKER, DANIEL, JULES GERARD and E. THOMAS SULLIVAN (1986), *Federal Land Use Law*, New York, C. Boardman.

MECK, STUART (1996), 'Model Planning and Zoning Enabling Legislation: A Short History', in *Modernizing State Planning Statutes* (The Growing Smart Working Papers, Volume One), Chicago, American Planning Association.

MOE, RICHARD and CARTER WILKIE (1977), *Changing Places: Rebuilding Community in the Age of Sprawl*, New York, Henry Holt.

MORRIS, MARYA (1996), 'Approaches to Regulating Developments of Regional Impact', in *Modernizing State Planning Statutes* (The Growing Smart Working Papers, Volume One), Chicago, American Planning Association.

MOSKOWITZ, HARVEY (1996), 'State and Regional Fair-Share Housing Planning', in *Modernizing State Planning Statutes* (The Growing Smart Working Papers, Volume One), Chicago, American Planning Association.

NETTER, EDITH and JAY WICKERSHAM (1993a), 'Driving to Extremes: Planning to Minimize the Air Pollution Impacts of Cars and Trucks (Part I)', *Zoning and Planning Law Report*, **16**.

NETTER, EDITH and JAY WICKERSHAM (1993b), 'Driving to Extremes: Planning to Minimize the Air Pollution Impacts of Cars and Trucks (Part II)', *Zoning and Planning Law Report*, **16**.

PELHAM, THOMAS (1979), *State Land-Use Planning and Regulation*, Lexington, DC Heath.

SALKIN, PATRICIA (1996), 'Interlocal Approaches to Land-Use Decision Making', in *Modernizing State Planning Statutes* (The Growing Smart Working Papers, Volume One), Chicago, American Planning Association.

SCOTT, MEL (1971), *American City Planning*, Berkeley, University of California Press.

STEWART, RICHARD (1992), 'Models for Environmental Regulation: Central Planning versus Market-Based Approaches', *Boston College Environmental Affairs Law Review*, **19**, 547–55.

STROUD, NANCY (1996), 'State Review and Certification of Local Plans', in *Modernizing State Planning Statutes* (The Growing Smart Working Papers, Volume One), Chicago, American Planning Association.

The President's National Urban Policy Report (1978), Washington, DC, Government Printing Office.

US DEPARTMENT OF COMMERCE (1924), *A Standard State Zoning Enabling Act*, Washington, DC, Government Printing Office.

US DEPARTMENT OF COMMERCE (1927), *A Standard City Planning Enabling Act*, Washington, DC, Government Printing Office.

VON ECKARDT, WOLF (1979), 'Camelot Rises on Boston Bay', *Washington Post*, 14 October, H1.

WALD, PATRICIA (1992), 'The Role of the Judiciary in Environmental Protection', *Boston College Environmental Affairs Law Review*, **19**, 519–46.

WHEELER, DOUGLAS P. (1996), 'Ecosystem Management: An Organizing Principle for Land Use', in Henry Diamond and Patrick F. Noonan (eds), *Land Use in America*, Washington, DC, Island.

WOLF, PETER (1981), *Land in America: Its Value, Use and Control*, New York, Pantheon.

NOTES

1 Of course, the term 'land-use planning' also covers activities undertaken by non-government professionals on behalf of private development projects. 'Land-use regulation', on the other hand, is by definition a governmental activity.

2 Charles Haar's pathbreaking law review article first illuminated the different legal statuses for the comprehensive land-use plan, including the possibility that it might have no meaningful legal status vis-à-vis land-use regulation (see Haar, 1955, 1155–57). Today, the legal status of the 'comprehensive plan' varies and confuses, state to state (see Lincoln, 1996, 89–104).

3 Unlike its definition in some countries, the term 'land-use regulation' does not include taxation of land. In the United States, the exercise of the taxing power and the enactment and administration of tax laws are not described as 'regulating' economic activity.

4 Compare *Hodel v. Virginia Surface Mining & Reclamation Assn., Inc.*, 452 US 264, 288 (1981) (rejecting claims that national government is prohibited from regulating local land use of mining), with *Solid Waste Agency of Northern Cook County v. United States Army Corps of Engineers*, 531 US–, slip opinion at 13 (referring to possibility that claim of national jurisdiction over ponds and mud flats 'would result in a significant impingement of the states' traditional and primary powers over land and water use').

5 As is often true around the world, understanding a country's formal legal framework, as articulated by its declarative laws, does not necessarily provide adequate assistance in decoding the nation's true legal and institutional posture for a given area of law and policy. 'Black letter' laws and implementing institutions may mask, as much as reveal, empirical realities forged by history, culture, economics and politics. In the case of the United States, however, there is neither a *de jure* albeit non-functional national land-use law nor a *de facto* national land-use policy equivalent.

6 The 1977 edition of a leading land-use law casebook devoted only nine out of 1084 pages to something its index labelled 'national land use policy' (Haar, 1977, 1078). Even that slight entry disappeared in the casebook's 1989 edition (Haar and Wolf, 1989). Professor Daniel Mandelker's casebook, *Federal Land Use Law*, explores federal laws affecting the use of land, but never makes the ultimate claim that there is such a thing as a comprehensive

national land-use policy or comprehensive national land-use planning or regulation as those terms are employed in this article (Mandelker et al., 1986, 1ff).

7 S. 3354, 91st Cong., 2nd Sess. (1970).

8 Ibid., Section 402(a)

9 US Const. art. VI, sec. 2 ('This Constitution, and the Laws of the United States which shall be made in Pursuance thereof ... shall be the supreme Law of the Land').

10 US Const. art. I, sec. 8.

11 See, for example, *Solid Waste Agency of Northern Cook County v. United States Army Corps of Engineers*, cited in note 4; *Printz v. United States*, 521 US 898 (1997) (striking down parts of national gun control law as impinging upon residuary and inviolable state sovereignty guaranteed by provisions in the Constitution); *United States v. Lopez*, 514 US 549, 559–67 (1995) (striking down national gun control law as beyond the power of national government under Commerce Clause).

12 See US Const. art. I.

13 See US Const. art. II.

14 Although never mentioned in the national constitution, the 'police power' refers to the residual power of state government to enact laws that promote or protect the health, safety, morals and general welfare of its citizens.

15 The National Planning Board and the Tennessee Valley Authority are two examples of aggressive federal government responses to the Depression. See Scott (1971), 300–16.

16 President Lyndon Johnson's 'Great Society' programmes attempted to rejuvenate inner cities, in part through participatory planning efforts. See, for example, Demonstration Cities and Metropolitan Development Act of 1966, 42 USC Section 3301 (1996) (stressing constructive engagement of neighbourhood residents).

17 In the former Soviet Union, for example, states and republics attempting the difficult transition from state to mixed private-state ownership have begun to reconsider their national planning and regulatory legal framework to match the potential changes in ownership patterns. See Kayden et al. (1995), 43–82; International Finance Corporation (1995), 1–5; Kayden (1993), 55–69.

18 Even the Supreme Court of the United States has noted that private property in 'land', as distinct from private personal property, occupies a special historical place in American constitutional culture. See *Lucas v. South Carolina Coastal Council*, 505 US 1003, 1027 (1992).

19 A more inclusive list of patches might add the nation's agricultural and tax policies in so far as they affect the use and development of land.

20 National Environmental Policy Act of 1969, 42 USC Section 4321 et seq. (1970).

21 Clean Air Act, 42 USC Section 7401 et seq. (1970).

22 Federal Water Pollution Control Act, 33 USC Section 1251 et seq. (1972).

23 Coastal Zone Management Act of 1972, 16 USC 1451 et seq. (1972).

24 Endangered Species Act of 1973, 16 USC Section 1531 et seq. (1973).

25 Safe Drinking Water Act, 42 USC Section 300f et seq. (1974).

26 Toxic Substances Control Act, 15 USC Section 2601 et seq. (1976).

27 Resource Conservation and Recovery Act of 1976, 42 USC Section 6901 et seq. (1976) (also known as 'Solid Waste Disposal Act').

28 Surface Mining Control and Reclamation Act of 1977, 30 USC Section 1201 et seq. (1977).

29 Comprehensive Environmental Response, Compensation, and Liability Act of 1980, 42

USC Section 9601 et seq. (1980) (see also 1986 'Superfund Amendments and Reauthorization Act').

30 See, for example, Clean Water Act, 33 USC Section 1342 (b) (state plan).

31 See, for example, *United States v. Riverside Bayview Homes, Inc.*, 474 US 121, 126 (1985) (dealing with interpretation of regulations defining wetlands).

32 42 USC Section 7431.

33 42 USC Section 7410(a)(2)(A).

34 16 USC Sections 1454; 1455.

35 16 USC Section 1455(d)(2)(B).

36 16 USC Section 1455(d)(10)(A).

37 16 USC Section 1538.

38 16 USC Section 1539(a)(2)(A)&(B).

39 Administrative regulations prepared pursuant to the Clean Water Act provide the specific definition of a wetland.

40 33 USC Section 1344.

41 See, for example, *Florida Rock Industries, Inc. v. United States*, 18 F.3d 1560 (Fed. Cir. 1994), *cert. denied*, 513 US 1108 (1995) (wetlands designation preventing limestone extraction); *Loveladies Harbor, Inc. v. United States*, 28 F.3d 1171 (Fed. Cir. 1994) (wetlands designation preventing residential development).

42 42 USC Section 4332(A).

43 42 USC Section 4332(C)(i–iii).

44 See Von Eckardt (1979), describing community opposition to a proposed library in Cambridge, resulting in relocation to site in Boston.

45 See, for example, New York State Environmental Quality Review Act, NY Envtl. Conserv. Law Section 8-0101 et seq. (1998) (substantive little NEPA).

46 National Historic Preservation Act, 16 USC Section 470(f)(1998).

47 Department of Transportation Act, 49 USC Section 303(c)(1998).

48 33 USC Section 1281(g).

49 42 USC Section 9611.

50 See *Village of Euclid v. Ambler Realty Company*, 272 US 365, 387–89 (1926).

51 See, for example, Eisenberg (1998), 1ff., describing the history of recent ecological movements and aspects of environmental degradation.

52 Clean Air Act, 42 USC Section 7401 et seq. (1998); Federal Water Pollution Control Act, 33 USC Section 1251 et seq. (1998).

53 See, for example, 'Secretary Babbitt Predicted Strong Opposition to a Plan', *Inside Energy*, 4 May 1998, at 12 (describing local opposition from ranchers to land sale plan).

54 43 USC Section 1701 et seq. (1998).

55 National Park Service Act, 16 USC Section 1902 (1998).

56 See 43 USC Section 1712.

57 The national government has played a significant role in other large infrastructure projects, including navigable water, dam, energy and airport projects, but none has had as dramatic a land-use impact as the federal highway building programme.

58 23 USC Section 101 et seq.

59 23 USC Section 134(g).

60 23 USC Section 135(e).

61 23 USC Sections 134(f); 135(c).

62 23 USC Section 134(h).

63 23 USC Sections 134(h)(5); 135(f); 135(f)(12).

64 23 USC Sections 104(f)(1); 307(c)(1).

65 23 USC Section 133(d)(2).

66 42 USC Section 7410(a)(2)(A).

67 See Netter and Wickersham (1993b), 153, 155–58. At the time this article was being completed for publication, Congress had just reauthorized ISTEA as the Transportation Equity Act for the 21st Century (TEA-21), continuing many of its land-use planning innovations.

68 40 USC Section 461(c) (1982)

69 'Designation and Treatment of Empowerment Zones, Enterprise Communities, and Rural Development Investment Areas', 26 USC Section 1391 et seq. (1993).

70 Ibid., Section 1391(f)(2).

71 See, for example, *Lucas v. South Carolina Coastal Council*, 505 US 1003 (1992) (challenging coastal restrictions).

72 Ibid., at 1016, 1018, 1026, 1030.

73 HR 925, 'Job Creation and Wage Enhancement Act of 1995'; S 605, 'Omnibus Property Rights Act of 1995'.

74 Exec. Order No. 12, 630, 53 Fed. Reg. 8859 (1988).

75 See, for example, HR 1534 (1998).

76 See, for example, Moe and Wilkie (1997), 228–29, 245–49, 256–58; Barnett (1995), 25–26. Although some commentators are sceptical that sprawl is such a bad thing, see Gordon and Richardson (1998), 16, the perception that it is continued to inform many state actions.

77 Hawaii State Planning Act, Haw. Rev. State. Sections 226–1 et seq. (1978).

78 Oregon State Senate Bill 100 (1973).

79 See, for example, Cape Cod Commission Act, 1989 Mass. Laws 716 et seq. (1989).

THREE

STRUCTURES FOR POLICY-MAKING AND THE IMPLEMENTATION OF PLANNING IN THE REPUBLIC OF IRELAND

Michael J. Bannon and Paula Russell

Background on Ireland

With a land area of 70 282 square kilometres and a population of 3.62 million people, the Republic of Ireland is both one of the smallest and one of the least densely populated member states of the European Union. In European terms the country has been relatively poor and it constitutes a single Objective I region for assistance under EU structural funding programmes (i.e. less than 75 per cent of EU average per capita gross domestic product [GDP] in 1993).

Traditionally a predominantly rural society with a high dependence on agriculture, the country has undergone major economic and social transformations since the dawning of the 'modern era' in 1958. The population has increased from 2.8 million in 1961 to 3.62 million in 1996. Perhaps more significant has been the increased share of population in the younger age groups—approximately 43 per cent of the population is now under twenty-five years of age. The country is now experiencing rapid increases in both the rates of household formation and in the numbers entering the labour force (Bannon and Greer, 1998).

In terms of occupation and economic structures, the country has been radically transformed. Up to the 1950s, the Irish economy was inward looking and essentially closed. Post-war policies have redressed this situation, making Ireland one of the more open economies in the world as it seeks both to attract foreign investment and to market its products and services throughout the world. Exposure to free trade and to world competition has profoundly changed the composition of economic activity. Thus, the share of employment in agriculture has fallen from 42.9 per cent in 1949 to only 11 per cent in 1995. While the share of employment in industry has increased from 21.5 to 28.3 per cent over the same period, the hallmark of modern Ireland has been the growth of employment in service activities—increasing from 35.6 per cent of total employment in 1949 to 60.7 per cent in 1995. The growth of industry and industrial output reflects the role of direct foreign investment in the economy, particularly in the mid-1990s, while regional assistance has helped to spread such investment across the country. The growth of services is a reflection of the changing nature of production and the role of services in exports, both directly and indirectly. At the same time, the growth of service jobs is changing the nature of work and is also bringing an increasing number of women into the workforce.

But for planners, perhaps the greatest sign of change has been the growth of urbanisation. Between 1926 and 1996, the number of people residing in rural areas declined by over 500 000, while an additional 1.15 million people were recorded as living in urban areas of 1500 persons or more. Between 1961 and 1996, the percentage of the population living in towns increased from 46.4 to 58.1 per cent. With urbanisation have come major changes in lifestyles, social attitudes and behaviour patterns creating a gulf between urban and rural. This gulf was clearly seen in the voting patterns of the 1995 referendum on divorce legislation.

From a planning point of view, the rapid growth of urbanisation raises many issues. Urban growth has been largely confined to the five largest cities, and particularly to the capital city, Dublin, which has been described as 'an extraordinarily isolated giant' in the Irish urban landscape. The scale of recent urban development in relation to Ireland's resources and expertise has posed acute problems, as have the problems of coordination. There are numerous apparent contradictions evident in Irish policy and practice. Thus, alongside a strong commitment to private enterprise, there is a very high level of state and public sector involvement in many facets of the economy. Yet, despite this public sector involvement, there has been only a limited commitment to strategic policies and often a scepticism about long-range planning issues.

Perhaps the biggest challenge for modern planning lies in overcoming the psychological barriers and problems arising from the introduction of planning management, control and guidance in a society where many people have had rural roots. Land is 'sacred' and private property rights are enshrined in a written constitution. Attitudes to law, to regulation and to control are all influenced by religion, by culture and by the baggage of a post-colonial inheritance (Bannon, 1985; 1989).

Administrative structures for development

By and large, the Republic of Ireland may be viewed as a highly centralised state with a high level of administrative powers and control vested in the central government. The Irish administrative system consists of three levels:

1. *Central government*. This level is made up of two components:
   - The Civil Service and the fifteen Ministries of State, many of which have an involvement in or relation to planning matters.
   - Within the central government tier (see Fig. 1) we may also include over one hundred uni-purpose state sponsored bodies. These bodies, which employ over 75 000 persons, operate in fields such as utilities, transport, trade, development and promotion, research, etc. Many of

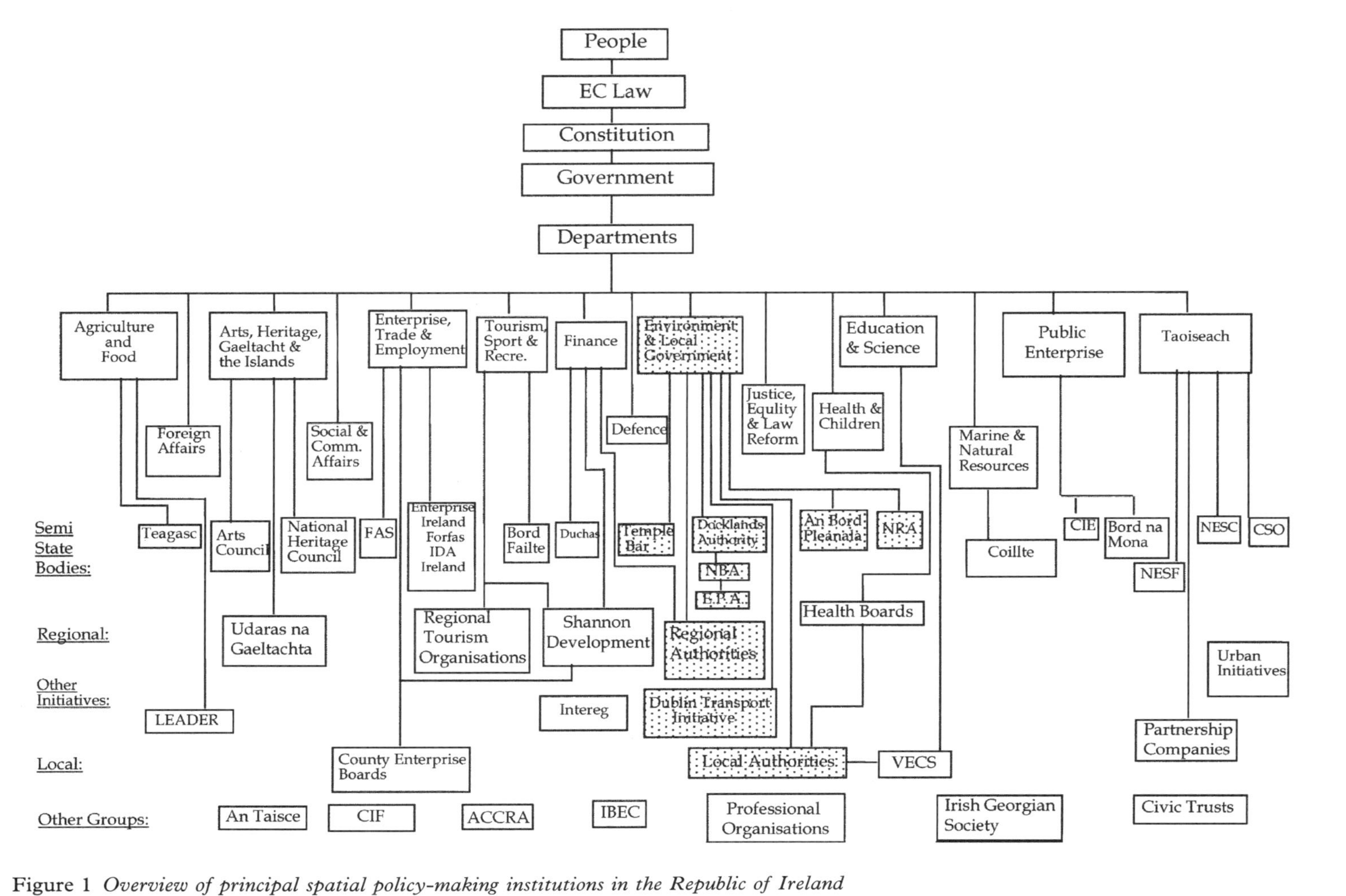

Figure 1 *Overview of principal spatial policy-making institutions in the Republic of Ireland*

these bodies were established to overcome the limitations of the Irish private sector and they often perform functions which elsewhere might have been devolved to local government.

2. *Regional authorities.* While many of the government departments and the state sponsored bodies have regionalised the implementation of their policies and programmes, the role of multi-purpose regional authorities is limited. In 1994, eight regional authorities were established to perform monitoring and coordinating roles which are still evolving. An exception is the Shannon Development Authority which functions as a regional development authority for its region, as does Udaras na Gaeltachta in the Irish-speaking parts of the country.
3. *Local authorities.* There are 114 local authorities of varying sizes and powers. They are essentially a unitary local government system and they function as the real second tier in the overall system of governance. Figure 2 sets out this local government system in a planning context. From the foundation of the state until the mid-1990s, it was evident that the scope and autonomy of local government was contracting relative to the central government. The structure and functions of local government are now under review, with serious efforts being made to enhance the powers and effectiveness of the major elements of the local government system (Department of the Environment, 1996).

This brief overview of the administrative system provides the context for a discussion of planning and planning powers (Chubb, 1992; Keane, 1982).

Modern planning procedures

Irish cities, towns and villages usually contain an important heritage of planned eighteenth-century streetscapes and buildings. Attempts were made to establish modern town planning movements in 1911–14 and 1939–45 (see Bannon, 1985; 1989). Modern planning, however, is linked to the enactment of the Local Government (Planning and Development) Act, 1963. While there have been six legislative amendments to this Act, it remains the basic bed-rock on which Irish planning has been built (see Table 2).

The 1963 Act established a highly decentralised system of mandatory land-use planning across the entire territory of the state. Under the Act, planning authorities were established (Fig. 2), and the operations and obligations of planning authorities were defined, as were procedures with respect to obtaining planning permissions, appeals against decisions and the establishment of enforcement procedures. The Act also dealt with necessary financial provisions for the payment of compensation and it conferred upon planning authorities the right to act as developers or to enter into joint ventures with private-sector or other operators. While subsequent amending legislation has strengthened the administration of the planning code, the 1963 Act

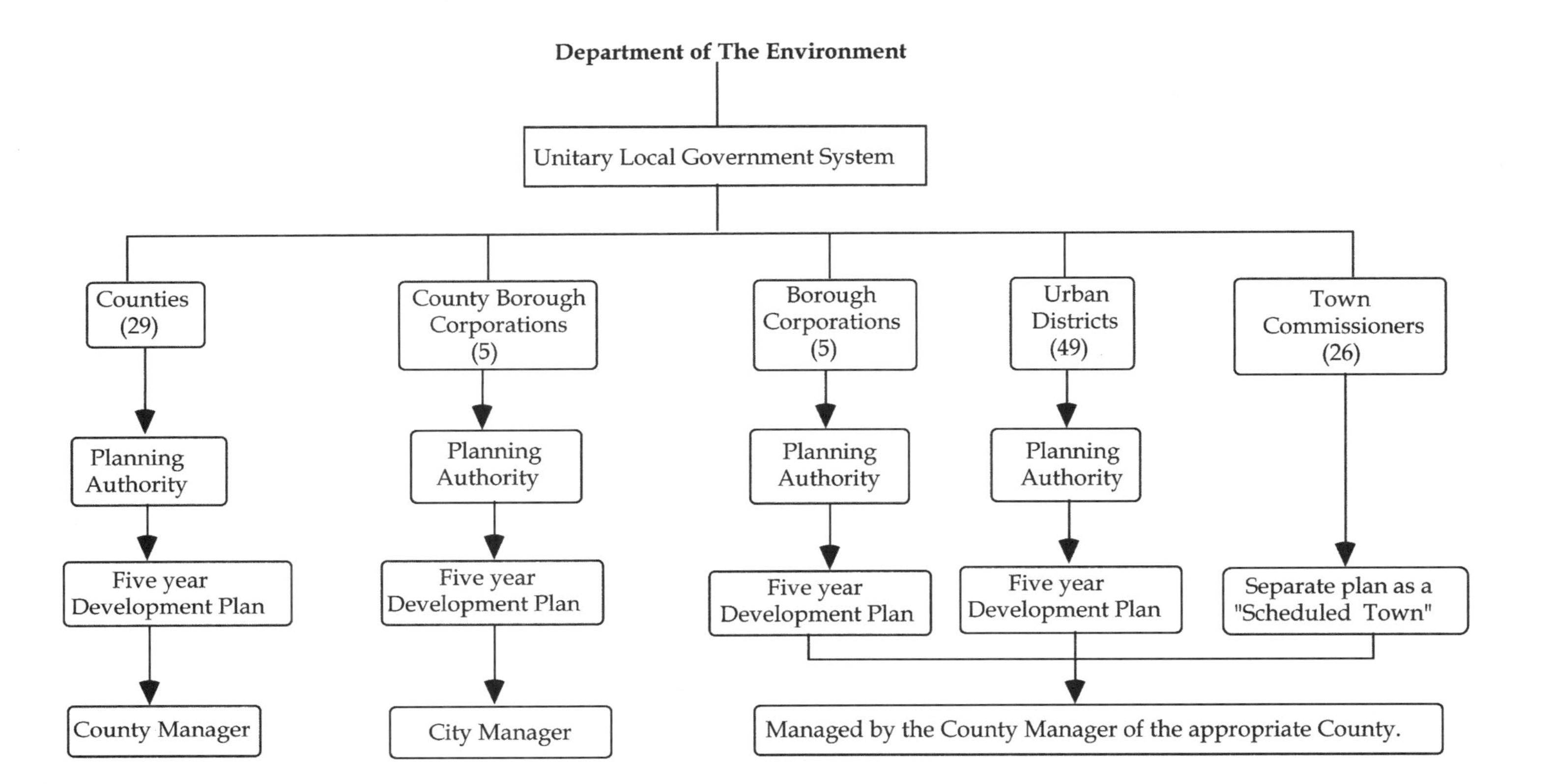

Figure 2 *The Irish local government system*

remains the basic and principal Act. Urban renewal programmes are implemented through separate legislation, as are most environmental matters.

Although Section 22 of the 1963 Planning Act made provision to enable the Minister 'to require the development plans of two or more planning authorities to be co-ordinated', there was no explicit reference to regions and no provision was made for regional planning. Although the Minister has extensive powers under the Act, there was no reference to national-level planning nor to how a 'national plan' might be formulated or implemented. Likewise, the regional authorities established in 1994 do not have a specific planning function.

Within this planning framework, different levels of administration play different roles. These are outlined briefly below.

The planning roles of each level of administration

This section briefly examines the critical functions of each level of Irish administration with respect to the operation and enhancement of the planning system.

CENTRAL GOVERNMENT: THE DEPARTMENT OF THE ENVIRONMENT AND LOCAL GOVERNMENT

The Department of the Environment and Local Government plays a central and leading role in Irish planning (Fig. 1). In the field of planning, the Department and its Minister are responsible for:

- Drafting and amending legislation and regulations on planning;
- Providing most of the funding for local (and planning) authorities;
- Ensuring that planning authorities prepare and review development plans and, where necessary, coordinate two or more planning authorities;
- The approval of various orders, including special amenity area orders;
- Making and issuing directives, guidelines and circulars; and
- Certifying environmental impact statements prepared by local authorities.

The Minister for the Environment is also answerable to Parliament for the operations of four state-sponsored bodies which have direct bearing upon the planning system and interact closely with planning authorities. These are:

- An Bord Pleanala—the Planning Appeals Board;
- The Environmental Protection Agency;
- The National Roads Authority; and
- The National Building Agency.

The Minister for the Environment is also responsible for the activities of two development authorities operating in inner Dublin—the Dublin Docklands Development Authority and Temple Bar Renewal/Temple Bar Properties Ltd (see Fig. 1).

In addition, the Department of the Environment and Local Government is one of the main implementing bodies with respect to the EU Community Support Framework and the various operational programmes and sub-programmes dealing with environmental services, aspects of transport, urban renewal, etc., which will be discussed further. This Department is also charged with implementing Ireland's national strategy for sustainable development (Department of the Environment, 1997).

As can be seen from Figure 1, a number of other government departments and semi-state bodies, notably the Department of Arts, Heritage, Gaeltacht and the Islands, have roles that impinge upon or interact with the planning system. It is worth noting that the Irish Planning Appeals Board is required to keep itself informed of the policies and objectives of all public bodies whose functions have, or might have, a bearing on the proper planning development of cities, towns or other areas, whether urban or rural.

REGIONAL AUTHORITIES AND DEVELOPMENT PLANNING

A regional tier of authorities came into effect on 1 January 1994. Their role is to perform 'such functions in relation to the co-ordination of the provision of public services in the region as are conferred on it by or under this section' (Local Government Act, 1991, S. 43[2]). Eight new authorities have been designated, and Table 1 outlines the counties, the population of the respective regions and the membership of each regional authority (see Fig. 3 for details).

Table 1 *Regional authorities*

Region	*Counties and county boroughs*	*Population 1991*	*Membership*
Border	Donegal, Leitrim, Louth, Cavan, Monaghan and Sligo	402 987	37
West	Galway City and County, Mayo and Roscommon	342 974	26
Mid-West	Clare, Limerick City and County, Tipperary North Riding	310 728	26
South-West	Cork City and County, Kerry	532 263	23
South-East	Carlow, Kilkenny, Waterford City and County, Tipperary South Riding, Wexford	383 188	35
Midlands	Laois, Longford, Offaly, Westmeath	202 984	23
Mid-East	Kildare, Meath, Wicklow	325 291	21
Dublin	Fingal, South Dublin, Dun Laoghaire/ Rathdown and Dublin City	1 025 304	29
Total		3 525 719	220

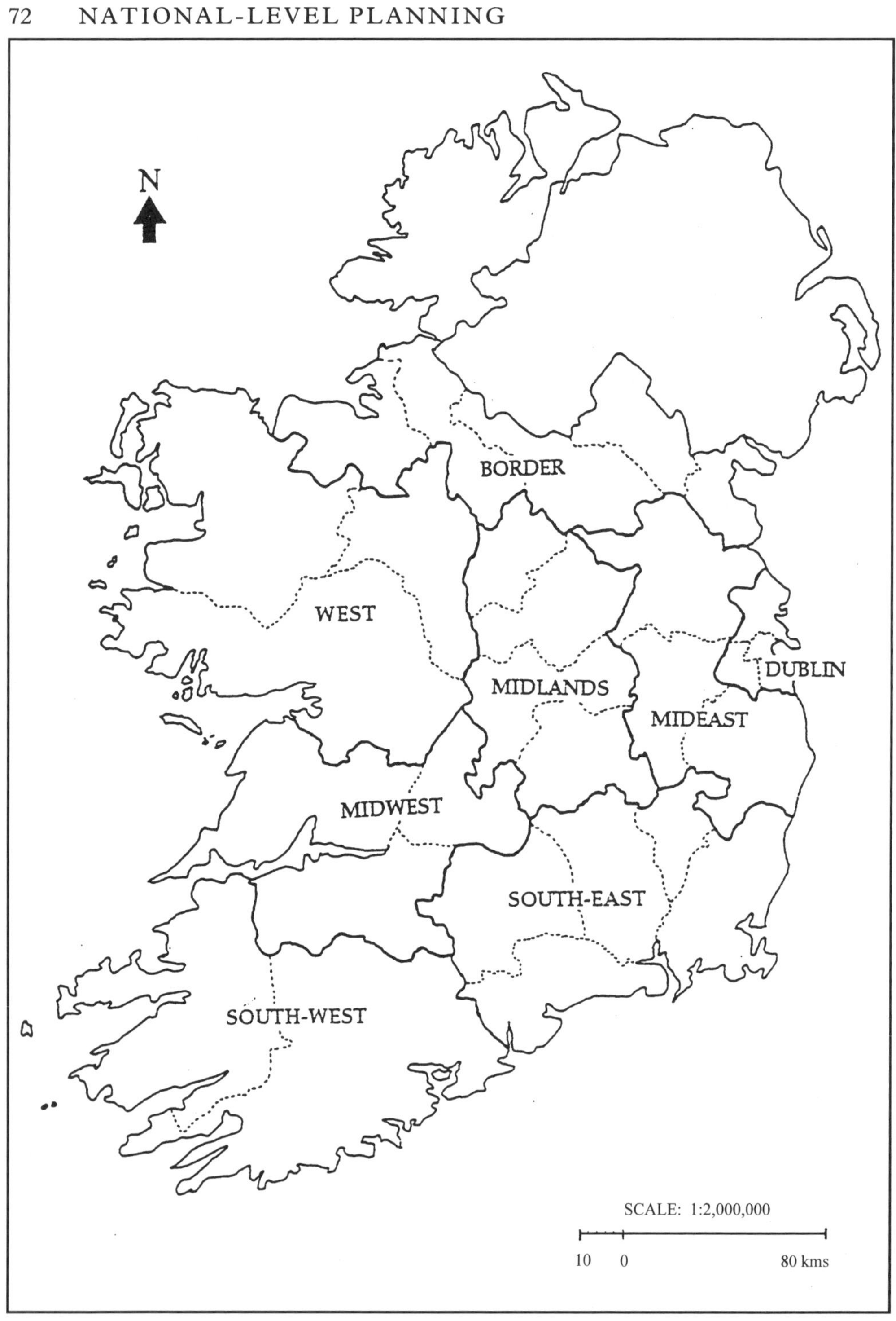

Figure 3 *Regional authority administrative areas*

The regional authorities have been allocated two principal responsibilities. First, they are to promote the coordination of public services within their respective regions and with contiguous regions. They are to develop joint actions by local authorities, ensuring that local actions and proposals are consistent with government objectives. They are to promote regional thinking and consciousness and review and coordinate local plans. Second, the regional authorities have a responsibility at the regional level to monitor and to review the implementation of the EU's structural fund programmes under the Community Support Framework, 1994–1999.

THE PLANNING FUNCTIONS OF LOCAL AUTHORITIES

As noted above (and shown in Fig. 2) there are some 114 units of local government in the Republic of Ireland, of which the twenty-nine county authorities and the five county borough corporations (cities) cover most of the territory and account for the major proportion of the population (see Fig. 4). The system of local government owes much to the Local Government, Ireland, Act of 1898, which consolidated the large number of authorities and created a democratically elected system. From 1922, the new Irish Government consolidated the various types of local authorities into a single system under the tight control of the appropriate government minister (now the Minister for the Environment and Local Government) (Grist, 1983).

Local authorities are run under a management system, whereby all functions are divided into 'executive' and 'reserved' functions. The overall control of the local authority is vested in an elected council, while day-to-day administration is in the hands of a manager and staff. The 'reserved' functions must be discharged by the elected members. Reserved functions include the adoption of estimates and the borrowing of money.

Under the Local Government (Planning and Development) Act, 1963, all local authorities except town commissioners became the planning authorities for their functional areas. Under s.2(2) of this Act, county councils, county or other boroughs and urban district councils were defined as planning authorities. There are, therefore, 88 separate planning authorities as of 1996.

The principal functions that are given to the planning authority are:

- To prepare and to revise development plans;
- To make recommendations and to enforce decisions on individual applications for planning permission;
- To have extensive powers relating to conservation and amenities; and
- To have the power to make orders controlling vandalism, litter and advertisements (Local Government [Planning and Development] Act, 1963 S. 52, 53 and 54).

A planning authority may also act as a developer in its own right. While matters such as the making or the review of a development plan or the making of an order that materially contravenes such a plan are 'reserved' functions to be discharged by the

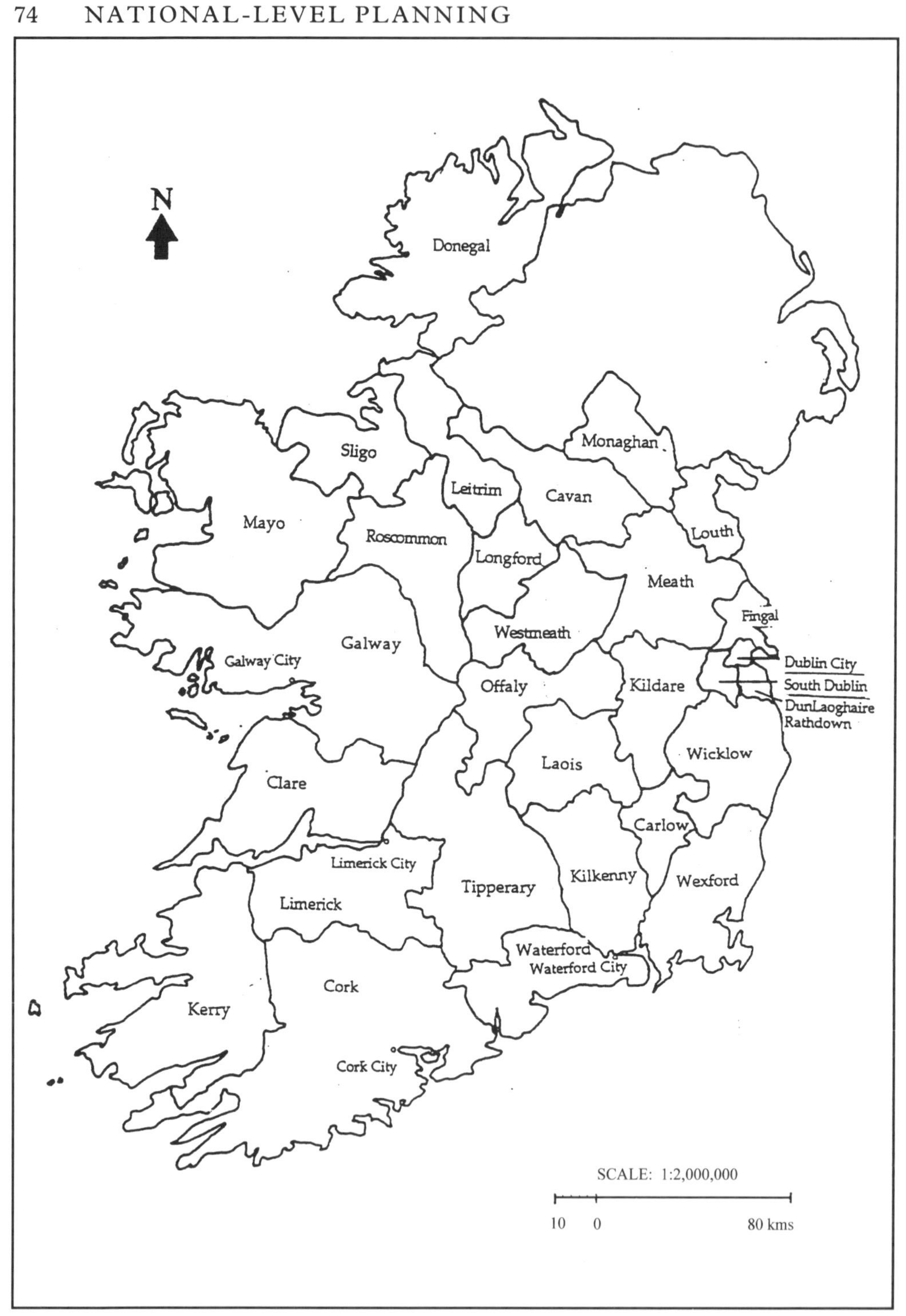

Figure 4 *Boundaries of administrative areas—county councils and county boroughs*

elected members of the local authority, decisions on planning applications are made by the manager, who is appointed. The manager bases these decisions on the recommendations of planning and other technical advisers and, where necessary, consults with the planning committee which is composed of elected representatives. Finally, the planning legislation confers extensive powers of land appropriation and access upon planning authorities to enable them to perform their functions under the Acts.

In its operation over the past thirty years, the planning system has done much to secure both ordered development and a degree of rational thinking in public policy. However, it can be argued that the effectiveness of planning has been constrained by reliance on a highly decentralised system operating without the benefit of effective regional coordination and in the absence of a strong proactive role by central government.

The search for a national development strategy

The absence of any strategy provisions to underpin either a national physical strategy or the implementation of regional planning stands in stark contrast to the utterances of the Minister of the day who declared in 1965 that 'one of the most important tasks I have as Minister for Local Government is to establish a regional planning framework for our social and economic development programmes' (Blaney, 1965). So convinced of this necessity was the government that, in 1965, encouraged by United Nations Advisers, it appointed consultants to coordinate existing regional advisory reports and plans. In effect, the consultants were asked to prepare a national strategy for development, particularly with respect to the future pattern of urban development.

Having examined a range of alternative strategies, the resultant report, *Regional Studies in Ireland*, put forward a framework for regional development across the country for the twenty-year period 1966 to 1986 (Buchanan and Partners, 1969). The recommendations concentrated heavily upon a strategy for urban development in an effort to resolve a growing dispute between those favouring concentration and those arguing for dispersal (Newman, 1967). The urban strategy proposed restraining the growth of Dublin, boosting the growth of Cork and Limerick-Shannon and the promotion of regional growth centres. While these recommendations did form the basis for the government's 'planning base figures' in the 1972 statement on a review of a regional policy, no real attempt was made to implement a national physical development policy. The policy proposals lapsed and the role of the central government in respect of planning policy was to remain limited and largely passive throughout the 1970s and 1980s, with increasing emphasis upon environmental and local issues.

Looking back on these events, many reasons may be seen as contributing to the failure to implement the proposed strategy or some variant thereof. These reasons include:

- To be successful, a national development strategy must be embraced by and owned by all arms of government, not just the ministry responsible for planning;
- The dominance of sectorally-based public bodies inclined to view spatial policy as potentially restricting their freedom of action;
- An over-reliance in the strategy on the perceived benefits of one theoretical approach, i.e. the 'growth centre' concept;
- A failure to relate strategic proposals to means of implementation—resources, staffing, etc.;
- The ending of the advisory role of the United Nations;
- The problem arising from the limited mandate of consultants with no one left responsible for explaining the proposals, arguing their benefits or promoting their implementation;
- An ideological scepticism in society about strategic and long-range planning as opposed to emerging post-modern approaches to planning; and
- Lack of research into the economies to be gained through balanced regional development or the possible diseconomies arising from the 'overgrowth' of Dublin (Shannon Development, 1996).

Ultimately, however, there was a lack of political will to implement the proposals or a considered alternative strategy. With the demise of these proposals, Ireland lost the opportunity of acting as a 'demonstration model' of how a small economy might be organised and managed in terms of spatial development. The Ministry withdrew from a proactive role to concentrate largely upon matters such as the appeals process and routine, more short-term administrative issues, together with environmental considerations.

Positive signs in the 1990s

After what was largely a hiatus of over twenty years, the 1990s have witnessed a reawakening of interest in planning issues and a growing realisation that there is a need for reform and for a new commitment to the strategic development of the country. This reawakening is driven by forces for reform within Ireland and by influences from outside, principally the European Union.

Positive signs of a new approach within Ireland include serious attempts to reinvigorate the role of local government, a commission to examine the present-day appropriateness of our national constitution, the establishment of regional authorities in 1994, the drafting of a Vision 2010 strategy for enterprise development (Forfas, 1996), a policy commitment to the drawing up of a national land use policy plan, and recent work by the Department of the Environment to produce planning policy guidelines and to promote strongly both Agenda 21 and a national sustainable development strategy (Department of the Environment, 1997).

External influences come from international organisations, most notably the European Union. The EU impacts upon Irish thinking and policy in numerous ways, including:

- Discussion documents such as the reports on the future of rural society, the green paper on the urban environment, Europe 2000 and Europe 2000+, European sustainable cities, the European Spatial Development Perspective and numerous other circulars and publications (European Union, 1997);
- Networking, information exchanges and technical cooperation programmes;
- The bringing into Irish law of EU directives relating to environmental issues, e.g. the Habitats Directive;
- The direct funding by Brussels of a wide range of EU initiatives in the fields of energy, transport and urban development, e.g. LEADER, URBAN and INTERREG; and
- The implementation of a programme of structural fund investment in the EU's disadvantaged regions (Objective I areas), which include the entire territory of the Republic of Ireland for the period of 1989–1999.

In terms of this paper and the discussion of national-level planning, the role of the EU structural funds and the accompanying procedures have been profoundly important for Ireland. At the European level, the 1988 reform of the structural funds procedures introduced four major changes for participating EU regions:

- An emphasis on a five- to ten-year time-span instead of short-term actions;
- A focus upon integrated programmes rather than isolated investments;
- The creation of partnership arrangements involving the EU, the national authority, the regional body and the local private/public and voluntary sectors, as appropriate; and
- The requirement under the community support framework regulations for each EU Level II region seeking funds to prepare and submit in advance an integrated set of regional proposals—a form of regional plan.

Thus, the structural funds mechanisms have led to more focus on strategic thinking and to the preparation of partnership-based investment programmes. Even more important has been the fact that, since the Republic of Ireland has been considered a single Level II EU region, the 'regional' submissions for structural funding have been made by the Irish Government and constitute, in local terms, 'national development plans'.

The national development plans and the EU community support frameworks

Since 1988, Ireland has prepared two national development plans in response to funding opportunities offered by the EU structural funds. The National Development Plan, 1989–93 was submitted to Brussels in 1989, and after discussion and negotiation most of the proposals were taken on board and incorporated into the Community Support Framework for Ireland, 1989–93. The 'national plan' was in essence the regional submission by the Irish government for EU funding.

Further structural funding was agreed to, as part of the Maastricht Treaty negotiations. In order to avail itself of these funds, the government prepared the National Development Plan, 1994–99. The preparation of this document involved the assembly of 'sub-regional' (regional) reports and a wide range of consultation. After protracted negotiations with EU officials in Brussels, and after significant reductions and considerable changes in priorities, the Community Support Framework, 1994–99 was agreed on. It gives effect to many of the proposals in the national plan—the regional submission.

To what extent was Ireland's 'national development plan' a plan in terms of the conventional understanding of the term (Faludi, 1973)? Inasmuch as the document presented a general overview of national strengths and weaknesses and inasmuch as it formulated a series of integrated programmes for action, it deserves the title of a plan. In terms of content, the document was driven by those facets of investment that were eligible for EU support under its regulations. In conceptual terms, the plan was at times aspatial (e.g. training for investment needs), while in other instances the level of detail descended to specific mention of local projects and infrastructures for which funding was being sought, such as a road or a harbour, or, in one case, a hospital. The national development plan was essentially a short- to medium-term set of proposals. There was little evidence of vision or strategic thinking and little reference to goals or alternative strategies.

The national development plan was constrained by its purpose. For the most part it only embraced functions relating to proposals likely to come within the scope of EU regulations. Thus, major land uses such as housing, health, education and open space amenities were excluded or largely omitted. The other major problem arises from the reality that this 'plan' was a bid for EU funding. There was not, nor is there now, an ongoing process to revise the national development plan in the light of negotiations with Brussels. Thus, in 1994 there were major differences between the amounts of EU funding sought and the amount secured.

The 1994–99 Community Support Framework for Ireland

Table 3 sets out the principal proposals of the Community Support Framework (CSF) for Ireland covering the structural fund investments over the years 1994 to 1999, inclusive. These proposals have been defined in terms of four principal priority areas:

Table 2 *Main provisions of Irish planning legislation*

Act	*Main provisions*	*Actions introduced*
Local Government (Planning and Development) Act, 1963	Established the Irish planning system. Defined planning authorities and their powers. Outlined the Planning Authorities' duty to make and review development plans. Introduced the obligation to obtain planning permission. Empowers the planning authority to: • enforce planning control; • compensate in certain circumstances; • appropriate and dispose of land; • enter into agreements to regulate the development of land; • act as developers; • grant licences; and • make special amenity area orders.	
Local Government (Planning and Development) Act, 1976	Established An Bord Pleanala, the planning appeals board. Introduced a number of enforcement procedures. Amended Section 42 of the 1963 Act.	Section 25—power to secure the proper completion of housing estates. Section 26 introduced the warning notice. Section 27 provides a means for seeking an injunction in the High Court against unauthorised developments.
Local Government (Planning and Development) Act, 1982	Empowers the planning authority to: • limit the duration of planning permission; and • vary the time period and in some cases extend the life of a planning permission. Confers powers on An Bord Pleanala to declare appeals, references, etc. withdrawn. Empowers the Minister for the Environment to: • issue policy directives relating to planning; • alter penalties, including fines and imprisonment for offences committed under the planning acts; and • make regulations with regard to fees.	

Table 2 *Continued*

Act	*Main provisions*	*Actions introduced*
Local Government (Planning and Development) Act, 1983	Provided for a number of changes to An Bord Pleanala. Provided An Bord Pleanala with a number of new powers.	Defines the composition of the Board and procedure for the appointment of members. Defines procedures relating to meetings of the Board. Enables the Board to dismiss appeals as vexatious. Enables appeals to be brought against conditions of a permission.
Local Government (Planning and Development) Act, 1990	Amended and consolidated the law on compensation. Regulated rights of connections to public sewers and water supply.	Introduced wider ground for non-payment of compensation, new procedure for alternative development and new valuation rules for assessment of compensation.
Local Government (Planning and Development) Act, 1992	Amended the law in relation to planning appeals and other matters decided by An Bord Pleanala, to enable the Board to decide cases in a shorter period of time. Changed enforcement procedures. Makes provision for application for judicial review to challenge the validity of the decision of a planning authority or appeals board to be made by motion on notice within two months of the decision.	A timescale of four months from receipt of the appeal was statutorily imposed within which the appeal must be decided. Lays down new requirements relating to making an appeal which apply to all those involved in making appeals, i.e. applicants, planning authority and other parties. Introduced a five-year time limit on enforcement procedures. Permits a warning notice under Section 26 of the 1976 Act to be served where an unauthorised use is likely to be made of the land. Replaced Section 27 of the 1976 Act with a new Section 27 which allows an application for an injunction to be made to either the High Court or the Circuit Court, and also broadens the scope of this enforcement procedure. Increases fines for offences under the planning acts.

Table 2 *Continued*

Act	*Main provisions*	*Actions introduced*
Local Government (Planning and Development) Act, 1993	Regulates development by or on behalf of the state and local authorities.	Regularises the position of all development already undertaken or commenced up to June 1993. Provides for the exclusion of certain categories of development from normal planning control, i.e. those relating to national security or defence. Introduces a system of public notices and consultation for those developments not requiring planning permission and for development by local authorities which are planning authorities. Allows certain development to be excluded by ministerial order from planning controls in the event of an accident or emergency.

- The productive sector;
- Economic infrastructure;
- Human resources; and
- Local urban and rural development.

These priorities would be secured through the implementation of a range of operational programmes and sub-programmes. The CSF embraced a proposed partnership investment totalling approximately fourteen billion Ecu, with 5.6 billion Ecu financed by the EU. The reality is that those programmes covered by the CSF are likely to go ahead, while few major investment proposals not covered by the framework and involving public expenditure are likely to materialise, at least in the short term.

Operationally, the Community Support Framework is being implemented through the operational programmes and sub-programmes. As a rule, a government department or a state board was given responsibility for the implementation of a specific programme or sub-programme. Implementation generally involves a partnership with the private sector or with local authorities, community groups or voluntary bodies as appropriate. The entire CSF is subject to continuous monitoring and ongoing evaluation and there is provision for an overall monitoring committee. Furthermore, each operational programme is subject both to prior appraisal and to independent monitoring and evaluation throughout its life. At the regional level, the regional authorities have the responsibility of monitoring the operations of the CSF for their area. While the community support framework

Table 3 *The Community Support Framework for Ireland, 1994-99—Principal categories of funding by EU and total expenditure (million Ecu)*

Priorities and programs	*Total proposed expenditure*	*Total EU contribution*	*EU contribution as percentage of total*
Productive sector			
Industry	1600	817	
Agriculture, forestry, rural development	1231	944	
Fisheries	86	68	
Tourism	402	325	
Sub-total	3319	2154	64.9
Economic infrastructure			
Transport	2173	934	
Communications	1174	34	
Energy	2698	70	
Water/environment	176	81	
Health	129	74	
Sub-total	6350	1193	18.8
Human resources			
Initial education/training/youth-start	1407	719	
Continuing training—unemployed	428	237	
Continuing training—employed	81	45	
Social exclusion	1992	540	
Supporting measures of which:			
Current	72	51	
Capital	266	181	
Sectoral OPs (industry etc.)	481	361	
Sub-total	4727	2134	45.1
Local urban and rural development	170	132	78
Other			
Technical assistance	10	7	
Pre-1989 projects/miscellaneous			
Sub-total	10	7	70
Total	14 576	5620	38.6

Source: Community Support Framework, 1994–99.

requires both continuous review, monitoring and evaluation of the operational programmes and of the CSF overall, these do not add up to the notion of 'review' as might be expected in the case of a review of a regional or local physical plan.

The Community Support Framework for Ireland may not add up to a comprehensive national plan, but it establishes commitments which go at least part of the way towards a national investment strategy. Importantly, the implementation model offers an interesting way of securing the implementation of programmes in which each partner has an interest and an element of ownership, provided the partnerships between central and local authorities can be strengthened.

Conclusion

This chapter has outlined the structures for policy-making and the implementation of planning in the Republic of Ireland as of the year 1997. The chapter has outlined the nature of Ireland's relatively centralised administration, the limited nature of regional administration and the importance of the local government tier of administration in respect of the implementation of public policies. This chapter also dealt with the important issues of European Union policies, how these have been expressed in an Irish context and the beneficial impacts for Ireland of European funding.

In the late 1990s the Irish economy has experienced a long period of unprecedented growth with consequent demands for employment, housing and infrastructure provision, particularly in the Dublin region. Under the *National Development Plan, 2000–2006*, the Republic of Ireland has been subdivided into two level II regions, each having its own Regional Assembly and one having Objective One status and the other 'Objective One in Transition' status. The planning code has been consolidated and updated with the enactment of the Local Government (Planning and Development) Act, 2000. Regional Strategic Planning Guidelines have been prepared for the Greater Dublin Area and these are in the process of being implemented. The Dublin Docklands Development Authority has replaced the Custom House Docks Development Authority and is responsible for the regeneration of a significantly extended area. A set of Integrated Area Plans with objectives for social and economic, as well as physical regeneration have been completed and cover much of Dublin's inner city, while proposals for an enhanced public transport system in Dublin are at an advanced stage of preparation. Issues such as labour-shortages, house prices, increased densities and land-use intensification have replaced the traditional concerns about poverty and unemployment. Likewise, in Cork, a review and updating of the Cork Land Use and Transportation strategy is nearing completion.

Work is underway on the preparation of a National Spatial Strategy, which is due for completion in 2001. Like much that has happened in Ireland in recent years, the preparation of the NSS has been strongly influenced both by the European Union and by best practice in other European countries. European good practice has helped Ireland to get its house in order and this has enabled the country to develop its economy to an unprecedented level. In turn this has provided the opportunity for a massive resurgence of planning activity at every level and throughout the country as a whole.

REFERENCES

BANNON, M. J. (1983), 'Urbanisation in Ireland: Growth and Regulation', in J. Blackwell and F. Convery (eds), *Promise and Performance: Irish Environmental Policies Analysed*, Dublin, REPC, 261–85.

BANNON, M. J. (ed.) (1985), *The Emergence of Irish Planning, 1880–1920*, Dublin, Turoe.

BANNON, M. J. (ed.) (1989), *Planning: The Irish Experience: 1920–1988*, Dublin, Wolfhound.

BANNON, M. J. (1991), 'The Contribution of the Management System to Local and National Development', in *City and County Management, 1929–90: A Retrospective*, Dublin, IPA, 27–53.

BANNON, M. J. and M. LOMBARD (1996), 'Evolution of Regional Policy in Ireland', in Shannon Development (1996), 57–83.

BANNON, M. J. and M. GREER (1998), 'Ireland' in L. van den Berg et al., *National Urban Policies in the European Union*, Ashgate, 181–223.

BLANEY, N. T. (1965), Address at the opening of the National Conference on Regional Planning, Dublin, May.

BUCHANAN, C. and Partners (1969), *Regional Studies in Ireland*, Dublin, An Foras Forbartha.

CHUBB, B. (1992), *The Government and Politics of Ireland*, London, Longman.

COMMISSION OF THE EUROPEAN UNION (1994), *Ireland: Community Support Framework for Ireland, 1994–99*, Brussels.

DEPARTMENT OF THE ENVIRONMENT (1996), *Better Local Government: A Programme for Change*, Dublin, Stationery Office.

DEPARTMENT OF THE ENVIRONMENT (1997), *Sustainable Development: A Strategy for Ireland*, Dublin, Stationery Office.

EUROPEAN UNION (1997), *The EU Compendium of Spatial Planning Systems and Policies*, Brussels, DG XVI.

FALUDI, A. (1973), *Planning Theory*, Oxford, Pergamon.

FEEHAN, J. (ed.) (1992), *Environment and Development in Ireland*, Dublin, Environmental Institute, UCD.

FORFAS (1996), *Shaping our Future: A Strategy for Enterprise in the 21st Century*, Dublin, Forfas.

GOVERNMENT OF IRELAND (1989), *Ireland: National Development Plan, 1989–1993*, Dublin, Stationery Office.

GOVERNMENT OF IRELAND (1993), *Ireland: National Development Plan, 1994–1999*, Dublin, Stationery Office.

GRIST, B. (1983), *Twenty Years of Planning: A Review of the System since 1963*, Dublin, An Foras Forbartha.

KEANE, R. (1982), *The Law of Local Government in Ireland*, Dublin, Incorporated Law Society.

LEE, J. (1989), *Ireland: Politics & Society, 1912–1985*, Cambridge, Cambridge University Press.

NATIONAL ECONOMIC AND SOCIAL COUNCIL (1975), *Regional Policy in Ireland: A Review* (NESC Report No. 4), Dublin, Stationery Office.

NEWMAN, J. (1967), *New Dimensions in Regional Planning*, Dublin, An Foras Forbartha.

O'REILLY, L. P. (1995), 'Regional Policy in Ireland: Review and Prospects' (paper delivered to the Irish Planning Institute Annual Conference).

ROCHE, D. (1982), *Local Government in Ireland*, Dublin, IPA.

SHANNON DEVELOPMENT (1996), *Regional Policy*, Shannon, Co. Clare.

FOUR

RETHINKING SWEDISH NATIONAL PLANNING

Göran Cars and Bjorn Hårsman

The Kingdom of Sweden is the fourth largest country in Europe in terms of land area. Ten per cent of the land is cultivated and 50 per cent is covered by forest. The population is close to nine million, with a density of slightly more than 20 inhabitants per square kilometre. This low population density and the abundance of timber resources may lead one to expect sprawling settlements and spacious wooden single-family houses. This is, however, not the case. The population is largely concentrated in the metropolitan areas of Stockholm on the east coast, Gothenburg in the west and Malmö in the south, and is housed for the most part in compactly built multi-family houses. Among the factors contributing to this pattern of population distribution are transportation, strong government control over planning and building and a national goal of making social services easily accessible.

For decades Sweden has been regarded as the epitome of the welfare state—concrete proof that a sensibly governed, large public sector can make it possible to combine ambitious equity concerns with ever increasing levels of income. Some may still nurture this image of Sweden, but the facts tell a different story. As illustrated in Table 1, since 1970 Sweden has steadily regressed in the 'income league' of OECD countries. In the 1990s this trend accelerated and was accompanied by huge public deficits, rapidly increasing unemployment and widening income gaps.

The economic recession in Sweden has led to a rethinking of the role of national planning. As in other countries, the increased scarcity of resources has led to cut-backs in public undertakings and also to a questioning of the efficiency with which national planning has been carried out. In addition, the relevance of the traditional 'Swedish model', with its broad definition of social policy and planning, has been questioned and found wanting. Fundamental changes are clearly needed in the Swedish planning system.

The Emergence of National Planning

In Sweden, land-use decisions have traditionally been strongly related to land ownership. During the first decades of the twentieth century, rapid urbanisation together with industrialisation and the exploitation of natural resources demanded

Table 1 *GDP per capita, as a percentage of the OECD average*

Placing		*1970*		*1990*		*1996*
1	Switzerland	154	Luxembourg	141	Luxembourg	152
2	USA	148	USA	138	USA	137
3	Luxembourg	134	Switzerland	132	Switzerland	125
4	Sweden	115	Canada	113	Norway	117
5	Canada	108	Japan	110	Iceland	116
6	Denmark	107	France	108	Japan	113
7	France	106	Norway	108	Denmark	111
8	Netherlands	106	Sweden	105	Belgium	106
9	Australia	104	Austria	104	Canada	106
10	New Zealand	102	Denmark	103	Germany	105
11	Belgium	96	Iceland	103	Austria	105
12	UK	96	Belgium	102	France	103
13	Germany	93	Italy	101	Netherlands	102
14	Austria	92	Finland	100	Australia	101
15	Italy	90	Australia	99	Italy	100
16	Finland	87	UK	99	Sweden	95
17	Japan	85	Germany	99	Ireland	93
18	Iceland	84	Netherlands	98	UK	92
19	Norway	81	New Zealand	83	New Zealand	86
20	Spain	67	Spain	73	Finland	83
21	Ireland	55	Ireland	71	Spain	73
22	Portugal	47	Portugal	58	Portugal	65
23	Greece	44	Greece	57	Greece	63
24	Mexico	36	Mexico	34	Mexico	38
25	Turkey	28	Turkey	29	Turkey	30

Source: SAF, 1998.

more efficient methods of planning and conservation. The prevailing social conditions also called for public intervention and planning.

In the early 1930s the Social Democrats came to power with an agenda which assigned top priority to improving welfare and living conditions through planning. A Swedish planning doctrine was developed. The Swedish welfare state grew out of the concepts of 'Folkhemmet' (the People's Home), according to which all citizens should treat one another as members of the same family or home and out of the concepts of 'Det starka samhället' (the Strong Society). The strong position assigned to the central labour union and the emphasis on equity issues and social welfare goals characterised the Swedish welfare state and continue to do so. Social well-being was not a concern addressed merely by social policy. Rather, the entire focus of public policy was aimed at improving social conditions and the social well-being of the citizens.

After the Second World War, ambitious programmes were launched to develop the modern welfare state further by means of planning. The issues addressed included macro-economic planning, the labour market, housing, public services, regional policy, workplace location and, to some extent, the environment. It also became clear that 'new' methods for the coordination of the various sectors were needed. National long-term planning was launched for each of the sectors, as well as a comprehensive plan to coordinate the sectoral plans. The new Planning Act was adopted in 1947. It gave public authorities the right to decide not only where, but also when and how construction could take place. The municipalities were given the responsibility of drawing up and adopting detailed development plans. Before implementation, the plans had to be ratified by the national authorities.

This legislation was intended to ensure the local adoption of centrally determined aims for the different sectors. Thus, it established a shared public responsibility and created means for the implementation of measures, within the different sectors, which significantly improved living conditions. This is well illustrated by the housing sector. The central government identified improved housing standards and housing production as essential objectives in realizing their general welfare goals. It therefore allocated substantial economic subsidies to facilitate investments in the housing sector and invested the municipalities with the responsibility of providing adequate housing for all their inhabitants. The municipalities were also obligated to prepare five-year housing construction plans. The Planning Act was thus an important tool for implementing housing and other sectoral plans. During the post-war period, a new relationship between the different levels of public authority developed. The central government was responsible for setting overall goals and making policy, while the detailed solutions and their implementation were devolved to lower levels of government.

Swedish planning was in its heyday in the 1960s and 1970s. Housing production and improvements to the existing housing stock created housing standards that were hard to find elsewhere in Europe. Radical improvements were made in the provision of public services and cultural and recreational activities. Taxes, as a share of gross domestic product (GDP), increased rapidly, but this did not affect the high level of confidence in public planning and in the Swedish welfare model. These achievements led to new legislation. A Planning and Building Act and a parallel special Act on Conservation and Management of Natural Resources were adopted in 1987. They changed the conditions for planning by offering municipalities greater independence and stressing the importance of environmental considerations. Thus, it should be noted that the devolution of power to the municipalities was integral to the 'Swedish model'. The state provided an overall framework through legislation, guidelines and economic incentives to the municipalities. This framework allowed the municipalities considerable freedom to develop plans for the different sectors.

The following two sections describe both the current political framework for planning and the planning system.

The political and administrative framework for planning

The first paragraph of the Swedish Constitution states that democracy should be implemented by 'representative government and local self-government'. As noted above, much of the power to plan is vested in the national and local levels. The regional levels play a subordinate role. An overview of the Swedish public sector is shown in Figure 1.

THE NATIONAL LEVEL

The central government has a stronger role in Sweden than in most Western countries. Important decisions are taken by central government, which is responsible for all government decisions. The powers of the ministries are restricted, and only routine matters are decided upon by individual ministers. These decisions are later formally confirmed by central government. The principle of collective responsibility is reflected in all forms of government work. There are at present 22 ministers in the central government.

The ministries are small units of usually no more than 100 persons headed by a minister appointed by the Prime Minister. They prepare central government bills for presentation to Parliament on budget appropriations and laws. They also issue

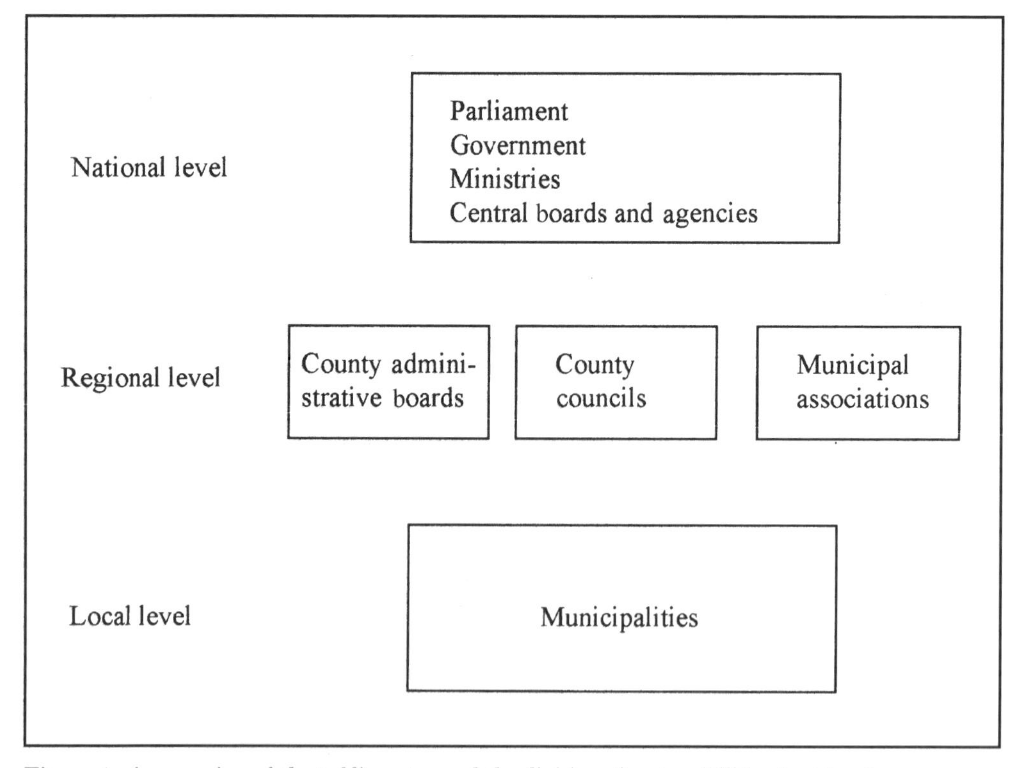

Figure 1 *An overview of the public sector and the division of responsibilities for planning*

laws, directives, regulations and general rules for the government agencies. The ministries have an outspoken role and responsibility for the implementation of central government decisions. Each minister also has political advisers, who function as the minister's political colleagues. They are political appointees, which means that they leave their posts when there is a change of government.

For each area of responsibility covered by a ministry, there are a number of government agencies. The agencies are responsible for carrying out the day-to-day business of central government administration. The government determines the agencies' objectives, focus and resource allocation, but not how each agency should apply the law or how it should decide on different issues.

Several ministries and government agencies have direct and substantial responsibilities for spatial planning. Some are responsible for establishing an overall spatial planning framework while others have responsibility for a specific planning sector. There are also other agencies which are responsible for service and other activities that indirectly impact on spatial planing. The most important ministries dealing with spatial planning are the Ministry of the Interior, the Ministry of the Environment and the Ministry of Transportation and Communication.

- *The Ministry of the Interior* is responsible for issues concerning the municipalities, the county councils and the county administrative boards; building and housing policy; and planning and land surveying. Several national agencies are accountable to the Ministry of the Interior, for example the National Board of Housing, Building and Planning, which is responsible for issues related to the management of the built environment.
- *The Ministry of the Environment* is responsible for issues concerning environmentally sustainable development, which include nature conservation, biological diversity, water and air conservation, as well as eco-cycle principles. The Ministry also has responsibility for national spatial planning and the efficient use of natural resources. The National Environmental Protection Agency and the National Licensing Board for Environment Protection are among the agencies accountable to the Ministry.
- *The Ministry of Transportation and Communication* is responsible for matters relating to road, rail, sea and air transport. It assigns a high priority to increasing the quality and performance of the infrastructure systems, developing an environmentally sound transport system and improving traffic safety. The National Road Administration and the National Rail Administration are among the agencies accountable to the Ministry.
- *Ministries with indirect roles in spatial planning* like the Ministry of Health and Social Affairs and the Ministry of Industry and Trade provide services and are responsible for economic development. The Ministry of Health and Social Affairs is responsible for welfare issues which have a significant impact on spatial planning with respect to allocation of housing

and services. The Ministry of Industry and Trade is responsible for regional development policies which have a considerable impact on spatial planning, especially in the peripheral areas such as the north. This responsibility includes efforts to facilitate and support economic development.

This structure provides national planning with strong powers for setting up the spatial planning framework and for allocating financial resources for implementation. However, these powers are not in themselves sufficient to guarantee implementation and change. The impact of national planning is to a large extent dependent on its relationship and its interaction with public planning bodies on the regional and local levels.

THE REGIONAL LEVEL

Sweden is divided into 21 counties. The county administrative boards are regional branches of the national government. Their overall objective is to ensure that national goals have an impact at the county level, consonant with the particular regional conditions and requirements. The tasks of the county administrative boards are to coordinate the different public interests within their region, to monitor closely the state of the county and to promote its development. Some national sector agencies like the National Road Administration have regional authorities which operate at this level.

There is an ongoing debate in Sweden about the role and size of county administrative boards that has been subject to government investigation. As a result, several counties were merged into larger units a few years ago. This change was intended to create counties that coincided with existing functional regions with respect to the economy, housing and the labour market.

The county councils are elected regional governments and are responsible for medical services. They also plan public transport for the county jointly with the municipalities. By a special law, the County of Stockholm is responsible for regional planning. In other counties, regional planning is optional and can be carried out either by the county council or by municipal associations.

THE LOCAL LEVEL

Sweden is divided into 288 municipalities with locally elected parliaments. The largest municipality, Stockholm, has close to 700 000 inhabitants, while small rural municipalities have populations of 3 000–6 000. The average municipality has approximately 30 000 inhabitants. The municipalities are responsible for schools, welfare services and urban planning, as well as the traditional functions of local government such as the provision of electricity, water supply, sewerage and the collection and disposal of rubbish. Both the county councils and the municipalities enjoy considerable autonomy because they have the right to levy income taxes.

The statutory planning system

As noted above, Sweden has a decentralised process for planning and decision-making on land use. Compared with other Western European countries, the municipalities have strong powers. If it is considered necessary, they can decide on the detail of new structures (i.e. layout, colours, building materials etc.) on both publicly and privately owned land. This strong public influence on land use and construction is often referred to as a 'municipal planning monopoly'. The basic law regulating spatial planning and building activities is the Planning and Building Act of 1987. The framework for Swedish planning is shown in Figure 2.

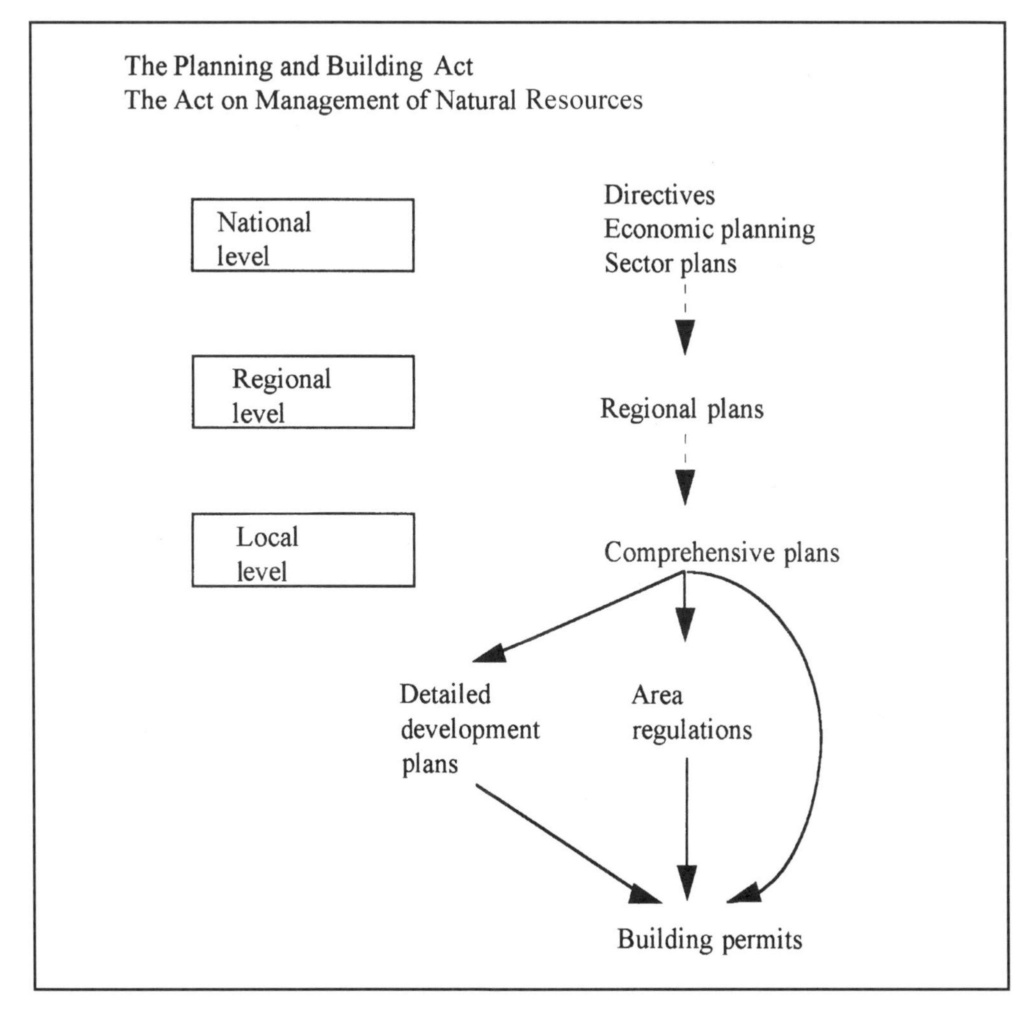

Figure 2 *The framework for Swedish planning*

THE NATIONAL LEVEL

A primary role for central government is the provision of a context and framework for planning and land-use decision-making through legislation, 'directives' and by economic incentives. Central government issues different kinds of directives. Some express aspirations and aim to underline a central government priority. This alerts ministries and government agencies to a need to rethink priorities and working methods in order to meet the directive's aims. Other directives are more precisely formulated and can be implemented through legislation.

Parliament has also decided that central government can intervene in municipal planning in some specially identified situations. In addition, the county administrative board is required to scrutinise detailed development plans adopted, amended or annulled by the municipalities. It is empowered to annul a municipality's decision in the following specific situations:

- A national interest is being ignored in contravention of the Act on the Management of Resources;
- The regulation of land and water areas which concern several municipalities is not being satisfactorily coordinated; or
- Health and accident hazards are not being adequately taken into account.

Central government and county administrative boards rarely annul detailed development plans since county administrative boards consult and negotiate with the municipalities on their plans. In addition, the very existence of such powers impacts on the way that municipal planning is carried out.

VISION 2009

In the early 1990s, the Ministry of the Environment initiated a project to identify a long-term strategy for spatial development. The National Board for Housing, Building and Planning was assigned the task of producing a proposal for a national vision. This proposal was to serve as a platform for a national plan or policy and to develop spatial planning as a tool for achieving environmental goals, regional development goals and sustainable land-use development. It was to set the framework for decisions relating to the use of land, water and natural resources, as well as urban development.

In 1994 the National Board for Housing, Building and Planning issued a research report, 'Vision 2009' (Boverket, 1994b). The first part of the report deals with the interaction between urban and rural areas and between towns and regions. It also deals with agriculture, forestry and biological diversity and it proposes a regionalised national policy. The second part of the report discusses strategies for achieving the vision with respect to spatial development. The hope was to set up procedures for the adoption and implementation of 'Vision 2009'.

The report has been subject to wide-ranging consultations with representatives of industry, business, central and local government, as well as with citizens. Some have welcomed it as a constructive and substantial contribution while others question whether its aims and aspirations are realistic and the most urgent for sustainable development. The fact that the report was produced, not by politicians, but by employees at the National Board for Housing, Building and Planning has given rise to scepticism and a questioning of the validity of the proposals (Engström, 1996; Nilsson, 1996).

The responses to the report have been compiled and are currently being considered at the Ministry of the Interior. However, the Vision has thus far had no significant influence on planning priorities. It remains to be seen whether central government will embrace the report and adopt it as a guide for spatial planning. We have reason to be sceptical.

THE REGIONAL LEVEL

There is no mandatory spatial plan for the regional level. However, if matters concerning the use of land and water impact on several municipalities and require a joint study or if the municipalities' work on comprehensive planning needs to be coordinated, central government may appoint a regional planning body to draw up a regional plan. However, a regional plan is not legally binding but is expected to serve as a guideline for comprehensive planning in the municipalities.

THE LOCAL LEVEL

It is mandatory for municipalities to draw up and adopt a 'comprehensive plan'. This plan must take note of the proposed land and water use, new development, changes to existing development or its preservation. It must also take note of how the municipality intends to take national interests into consideration in accordance with the Act on the Management of Natural Resources.

The 'detailed development plan' for continuous new development examines a site's suitability for development and controls the design of the built environment. It pays attention to individual new buildings with a significant impact on the surroundings or which are located in an area of high demand, as well as development which is going to be altered or preserved, if comprehensive control is required.

For defined areas not covered by a detailed development plan, area regulations can be adopted in order to ensure that the intentions of the comprehensive plan are achieved or that national interests are met, in accordance with the Act on the Management of Natural Resources.

NATIONAL-LEVEL STATUTORY PLANNING: A DETAILED VIEW

As noted earlier, the national government provides a context and framework for planning and decision-making on land use. In some situations, the framework is explicit. For example, Parliament designates those areas which are to be protected from development, and deviations from these designations are rare. In this respect, spatial planning has a strong position. In addition, national sector plans for infrastructure investments often have a substantial impact on spatial planning.

In other situations the aims and goals of central government are less clear and have less impact on spatial planning. This is often the case for development decisions within particular service sectors. Here the coordination between land-use planning and the political decision-making process—especially concerning services and public goods—becomes problematic. National development aims not infrequently conflict with favoured economic measures intended to stimulate a certain development. In these situations the link to spatial planning is often indirect. Within a national sector, planning is carried out by agencies aiming to meet the need for a certain type of service, and the spatial dimension often has a relatively low priority. This situation could be illustrated by the creation of an infrastructure node. It creates a potential for other kinds of development. Private and public actors take initiatives to promote their interests. When faced with strong private or public sector interests, it is often difficult to coordinate development in the way that is presupposed in the Planning and Building Act. Actual outcome and development is the result of a struggle between various private and sectoral interests rather than coordination based on holistic and long-term assessments (Cars, 1992). Increased decentralisation within the national part of the public sector, in combination with the devolution of power to lower levels of government, has sharpened the tension between local and national planning.

For authorities at the regional level, the situation is very different from that at the local level. Traditionally, the regional level has only been able to influence land-use decisions indirectly. The regional plan, which is the regional spatial instrument for coordination, has proved to be a weak instrument. The plan is prepared within the statutory system, but it does not dictate development rights and does not have to be adhered to in detailed land-use planning. Its function is a coordinative one. This can explain why it has always been difficult for county councils to coordinate decisions taken by strong national agencies such as the National Road Administration and the National Rail Administration. The county councils' aspirations to coordinate municipal planning and to foster regional development have produced few results (Broomé, 1996). These problems are in fact important reasons for the 'experiment' with new regional parliaments described above.

Thus, it can be concluded that there is a political aspiration at the national level to integrate spatial planning closely with programmes for economic development, regional development and the development of various service sectors. The basic idea is that planning should be able to provide milieux in which good conditions for living and economic activity are stimulated by effective infrastructure provision and a good natural environment. However, this aspiration is frequently hampered by conflicts of interest on the regional and the local levels. On the regional level, the political muscle for coordination is lacking. On the local level the strong planning powers allocated to the municipalities create incentives for 'NIMBY' behaviour. In the current debate it has been claimed that successful development of the planning system requires an open debate on how planning and decision-making powers should be allocated.

Structural changes and new conditions for planning

According to conventional economic wisdom, Sweden's large public sector, exceptionally high tax rates and its almost unmatched degree of egalitarianism should have slowed the rate of growth a long time ago. Tax and subsidy payments as large as those in Sweden, economic theory holds, destroy incentives and hence bring about dead-weight losses. Since Sweden's relative economic decline started in the 1970s, economists have issued warnings that the decline might accelerate if the public sector continues to grow.

Another far-reaching warning concerning the Swedish welfare model was given in 1990 by a Government Commission on 'The Distribution of Power and Democracy in Sweden' (SOU, 1990). The Commission noted that the welfare model basically presupposes a centralism and uniformity that no longer exist. Swedish society has gradually become more pluralistic and the decision-making authority has been transferred to an increasing number of power centres. One important conclusion was that the decentralisation, internationalisation and diversity currently characterising Sweden call for new administrative approaches. Social engineering and further expansion of the public sector can no longer be used to achieve sustainable economic and social development.

As is evident from Table 1, these warnings were wise. The 1990–1993 period saw an unprecedented deterioration in public finance, leaving Sweden with the largest deficit in the OECD area. In 1993, the deficit amounted to 13.3 per cent of GDP. A large decline in aggregate output and an increase in unemployment accompanied the growing deficit to levels that had been unthinkable since the 1930s.

The economic problems currently facing Sweden reflect basic structural problems rather than a temporary recession. The public sector has become too large and it is doubtful whether present institutional arrangements and incentive structures can bring about the much-needed expansion of employment opportunities in the private sector.

While Sweden's problems are exceptionally severe, most other OECD countries also have sluggish growth rates, high unemployment and government budget deficits. In fact, one of the most palpable general economic problems within Europe is the combination of a low rate of economic growth and considerable structural unemployment.

This paper claims that this development has impacted significantly on national planning. In the footsteps of the deep economic recession, the Swedish system for national planning has been questioned and challenged. In the following sections we examine the role and organisation of national planning in today's knowledge-oriented and interdependent world. First, some general observations will be reported. In the light of these observations, we will analyse and discuss the development of various forms of sectoral planning.

THE NEED TO REFORMULATE NATIONAL PLANNING

The structural changes described above have made the rethinking of old strategies necessary. The cornerstones in the debate about the future role of national planning are the need for a strengthening of the Swedish economy, improved efficiency in production of public services and increased international competitiveness.

The economic recession has eroded the foundations of the 'Swedish welfare model'. It is obvious that the previous far-reaching and ambitious welfare programmes have led Sweden into an economic impasse. Many of the welfare systems had an intrinsic characteristic—constant and sometimes uncontrolled growth. When the recession hit Sweden, the demands on the welfare system increased and, at the same time, the tax-base decreased. For financial reasons, it became obvious that the welfare system had to be reduced and restructured.

In parallel with the economic constraints, the welfare system has been questioned from an ideological perspective. A common argument is that the structure and comprehensiveness of the welfare programmes have had a negative impact on incentives and private initiatives. It is claimed that the system has promoted dependence and social exclusion, not integration and independence. Some also argue that the system promotes fraud and deception, by putting people in an unemployment trap (Cars and Hagetoft, 1988).

Another ideological starting point for challenging the traditional role of national planning is the question of 'efficiency'. Here the debate is similar to that in other Western countries. It is claimed that public services have traditionally been provided in an inefficient manner. The core of the critique is that services have been ill-targeted and costly to produce.

The question of international competitiveness has figured prominently in the Swedish debate on the role of national planning. The process shown in Table 1 has been a natural starting point for this debate. Sweden has been slipping down the 'ladder of wealth' during the last decades. Having been one of the richest countries in the world, it now has a GDP lower than the average for OECD countries. It is commonly recognised that national planning actively contributed to the growth in the economy and in welfare after the Second World War. Planning was an efficient tool for the implementation of national economic and welfare policies. At the same time, it is now claimed that government policies and intervention in housing, the economy and social policy can partly explain the financial problems and the decline in Sweden's economic competitiveness. As a consequence, there is a substantial reformulation of government policies in various sectors of society. This activity impacts on the conditions for planning. As policy changes, a parallel development in spatial planning becomes a necessity.

Thus, it is clear that Swedish national planning is in turmoil. The general picture is that financial constraints have led to a resetting of priorities, cuts and state withdrawal within many sectors of society. This might be a correct overall picture, but is does not hold true for all sectors of national planning. In some sectors, substantial growth has taken place.

SOCIAL WELFARE

Social welfare has been and still is an important feature of Swedish national-level policy. Although social welfare policy does not deal directly with spatial planning, it has indirect impacts which justify its inclusion in a description of national planning. The Swedish welfare model is characterised by a broad definition of social policy, which includes housing policy, various services and labour market policy. The main responsibility for social policy is assigned to the Ministry for Health and Social Affairs. A special agency, the National Board of Health and Social Affairs, is accountable to the Ministry. Its role is to serve the Ministry in the implementation of social policy. Besides the responsibility directly assigned to the Ministry for Health and Social Affairs, it is clearly stated that social considerations should be taken into account by other ministries as well when dealing with issues that impact on essential aspects of people's everyday life and well-being. The Ministry for Health and Social Affairs has an important task in following up efforts in which sectoral planning impacts on social conditions. The Ministry is also to initiate efforts to coordinate sectoral planning in order to meet social welfare policy goals. This means that social considerations are impacting on various types of national planning, such as construction and rehabilitation programmes for housing, regional policy, the location of public services and, to some extent, infrastructure development as well.

The Swedish social welfare model is usually defined as an institutional redistribution model. Means-tested benefits have had a subordinate role and, instead, welfare policies and public services have been designed to apply to the entire population. This has resulted in an extensive public sector providing a wide variety of social services, including health care and social insurance benefits. Redistribution becomes a characteristic feature. The general, all-inclusive character of benefits is also meant to avoid or minimise the stigmatisation of individuals.

In the aftermath of the economic recession of the early 1990s the traditional aims of welfare, and especially the role of the public sector, have been increasingly debated and questioned. Cuts in welfare programmes improved public finances and made tax cuts possible. Improved efficiency in the way traditional public services were delivered has also been observed.

Swedish welfare policy today is at a crossroads. One important discussion concerns the focus of the policy. Until now, much social welfare activity has applied to the entire population. Constraints in the public economy have led to cuts in most social programmes. Increased difficulties in financing social programmes have been used as arguments to justify shifting from the traditional general welfare policy to means-tested social benefits. In this way, scarce resources could be targeted to those individuals with the greatest need. However, in parallel it is argued that means-tested social benefits might erode the general public support for the social welfare system. Several national welfare planning programmes are facing this dilemma. Economic constraints do play a role, but perhaps even more important is the fact that these measures do not meet current social needs efficiently. It is not possible to identify a consensus on how to reconstruct or reform social policy programmes to fit the needs of today (Allen et al., 1998).

HOUSING

The general principles of the housing policy were forged by Parliament in a series of decisions taken after the Second World War. These policies were meant to provide all households with healthy and spacious homes at affordable cost. Municipalities were given responsibility for implementing programmes to develop local housing supplies in order to ensure sufficient and modern housing for the population. In order to accomplish this task, municipalities were encouraged to create their own local housing associations, set up as non-profit companies. Rents in multi-family dwellings are not free but set according to 'user value', which in practice means that they are set according to the construction cost of the local housing associations. The objective has been to ensure that the entire population, regardless of income and social affiliation, had adequate housing. This means that the public housing stock includes a great variety of dwellings in terms of size and quality. Every fifth Swede lives in a dwelling owned by a public housing association.

In the aftermath of the economic recession in the early 1990s, measures were taken to reduce housing subsidies. In combination with slow income growth, this has reduced new construction. Today the rate of housing production is at its lowest level since the beginning of this century. In parallel, costs for housing have increased. On average, households in Sweden spend over thirty per cent of their disposable income on housing. The reduction in housing construction subsidies has brought about a substantial increase in construction costs. Newly produced apartments in particular have become more expensive, and consequently low-income families can no longer afford to live in them (Åhs, 1997).

How long the administrative rent-setting system will be retained is open to question. Black market trade in attractive areas of the largest cities, in combination with substantial problems in letting existing dwellings in less popular areas, lends support to the argument for market-determined rents. At the same time, left-wing politicians and several planners fear that a transition to market rents would create increased housing segregation.

Nevertheless, it should be stressed that housing will continue to be an integral part of the Swedish planning system. Policy measures are needed in the housing sector for both efficiency and equity reasons. Public goods, aspects of the urban landscape, interdependencies between residential location and transport demand, and environmental concerns all support and encourage a strong role for housing planning at the local and regional level. Safeguarding areas of historical, cultural and environmental national interest and providing low-income families with the means to demand decent housing will continue to be important national planning interests. Another important planning role for the national government is, of course, to set the rules for an efficient and safe functioning of the housing market.

Previously, Sweden had a special Ministry for Housing. Since the early 1990s, questions concerning housing are handled within other ministries. For a period, housing was the responsibility of the Ministry for the Environment. Today, the Ministry of the Interior handles these issues. Within the Ministry a special division deals with building, housing, government support for housing provision, legislation concerning rented accommodation and tenant ownership, property taxation and other issues pertaining to housing policy. A number of government agencies are accountable to this division of the Ministry. They include the National Board of Housing, Building and Planning, the Council for Building Research, the National Organisation for Aid to Owners of Private Small Houses, and the National Housing Credit Guarantee Board. The National Board of Housing, Building and Planning is the most important and relevant to the focus of this paper. It has responsibility for issues concerning the built environment and the management of natural resources, spatial planning, building and housing. The Board is also responsible for the central administration of government support for housing in the form of housing subsidies.

The scope of national policy is restricted, compared with that of many other countries. The strong role assigned to the municipalities means that issues regarding size, density and types of housing in specific urban or regional locations are not part of national policy. Nor is the national level engaged directly in the building of housing units. Also, coordination with other spatial and non-spatial planning areas is mainly a municipal responsibility. The role of national policy is to provide a framework in which the two most important instruments are legislation and policies regarding subsidies for the construction and rehabilitation of housing.

ENVIRONMENTAL ISSUES

The Ministry of the Environment has been given overriding responsibility for environmental issues. However, a feature of Swedish planning is the strong emphasis put on shared responsibility. Thus, the Planning and Building Act and the Act on the Management of Natural Resources clearly state that all planning activities should include environmental aspects in their considerations. The achievement of a long-term sustainable society is a responsibility shared by all actors engaged in planning.

As mentioned, there is no national plan for land use in Sweden. Central government has decided that certain, specified areas are to be protected from development. Examples of such areas are coastlines, forests, rivers, mountains, parks and rural land. Decisions on protection are taken by central government, and deviations from the decisions regarding these areas are rare. In this respect, the Ministry of the Environment has a special responsibility to oversee development.

Questions regarding pollution, toxic waste, solid waste and other phenomena that have a negative impact on the environment are dealt with by special legislation. The regulations are in many cases precise and, compared with most countries in Europe, far-reaching. Within this legal framework, the municipalities are responsible for implementation arrangements that meet the standards set up by central government. There are some exceptions to this general rule. The handling of hazardous toxic waste is regarded as an issue of national concern. The licensing of enterprises with negative environmental impacts is also the responsibility of the national government.

The county administrative boards normally have a special division, which has been assigned responsibility for scrutinising local plans and planning initiatives from the perspective of environmental sustainability. A number of national agencies are also accountable to the Ministry of the Environment in this respect.

INFRASTRUCTURE

As mentioned, the Ministry for Transport and Communication is responsible for infrastructure development. The government agencies that play important roles in infrastructure development are the National Rail Administration and the National Road Administration. They are responsible for the implementation of policies and investments decided by central government. To fulfil this task, regional branches have been established to analyse conditions for implementation and to draw up tentative development plans that can be used as a base when decisions about financial allocations are taken.

In contrast to the trend of declining resources and national withdrawal which was noted in the national planning of the welfare and housing sectors, the infrastructure sector and its role in national planning has been significantly reinforced. The financial resources put into the sector have increased substantially. The motive for these investments is the recognition that infrastructure is of crucial importance for achieving other welfare and economic objectives. The aim formulated on the national level is to promote a development that is sustainable from economic, ecological and social perspectives (Snickars, 1997).

However, despite relatively generous financial support from central government, the implementation of infrastructure investments often turns out to be complicated. During recent years, the problems of reaching coordinated agreements and implementing these investments have been highlighted.

As noted, Swedish municipalities have a very strong position when it comes to planning and land-use decisions. However, the financial resources for infrastructure investments are normally placed with national authorities or regional branches of national authorities. This split of responsibilities—where planning and financing are placed on different levels of government—has more often than not created substantial problems. The core of the problem is that actors on various geographical levels have used their powers in such a way that it has not been possible to achieve coordination and effective solutions (Boverket, 1994). For example, financial resources have been used to promote solutions that satisfy the need that induced the investment. For its part, a municipality can use its planning monopoly to extort economic or other compensation for accepting an infrastructure investment that is very beneficial from a comprehensive public perspective, if it in any way has a negative impact on local conditions.

Thus the existing system for infrastructure development has not infrequently led to conflicts between national and local interest. One rather common situation is as follows. The municipality realises that a specific investment proposed by a national agency is necessary and inevitable. Despite this insight, the proposal is opposed. The development as such is desirable, but not the spatial location, which is optimal from a national or regional perspective. The municipality wants the proposed change but 'not in its own back yard'. The tension between local and regional/

national interests has fuelled a debate about the division of power and responsibility within the planning process. Voices have been raised in favour of reducing the power of local planning in projects that have considerable regional or national impacts. This debate is still at an early stage. So far, no proposals for changing policy or legislation have been made by central government.

These problems have become accentuated in recent years as planning issues have grown more complex and the interrelations between various geographical levels have increased. It is obvious that national government decisions about infrastructure often cause significant structural changes at the local level. At the same time, it is obvious that decisions taken at the local level not infrequently have a substantial impact on higher levels of government. Recent road and rail projects in Stockholm serve to illustrate these problems. The need for additional railroad tracks in Stockholm is commonly recognised. However, despite numerous investigations and proposals over the years, no sustainable solution has yet been identified. The current situation is characterised by actors representing various public-sector interests and different layers of government, taking firm positions that are impossible to coordinate. The plans for a new ring-road taking traffic from the inner city to a highway system circling it turned out to be even more complicated to implement. Here, besides public actors, private economic interests and organised interest groups also play important roles. The initial proposal, which had the support of a political majority in City Hall, has been questioned and challenged. The political support needed for implementation has eroded and different alternative solutions have been presented. The urgent need to improve the transportation system has been recognised for decades, and proposals and plans have been produced. However, the issues are still not resolved, and in the foreseeable future there will be only partial solutions to the transportation problem.

The next phase of Swedish planning

The aftermath of the economic recession of the early 1990s has in many ways changed the conditions for national planning. In some sectors, such as housing and social welfare, planning has been weakened and possibilities to intervene are more limited than they were. In other sectors, the opposite is true. National planning has strengthened its position in, for example, the infrastructure sector. While changes are taking place within various planning sectors, it is worth noting that the planning legislation remains intact. Since the 'new' planning legislation was adopted in 1987, only minor amendments have been made. No demands for radical change of the legislation have been put forward. Why has the Planning and Building Act remained intact while conditions for sectoral planning have changed radically? One explanation is that the 1987 legislation foresaw the trend of decentralisation and devolution of power. Another explanation is that the Act does not specify roles, procedures and outcomes in detail. Rather, the intention was to provide a framework within which planning and development are facilitated. Thus it might be concluded that it is not a new planning act which is needed, but a new praxis and new methods to be used within the framework of the existing legislation. Some likely and possible developments of Swedish national planning are discussed below.

NATIONAL SECTORAL PLANNING

Within some sectors, constraints on the public economy have led to decreased engagement, deregulation and privatisation during the 1990s. There is little support for a renewed national engagement in these sectors. Public budget restrictions and increased efficiency by privatisation and competition are cited in support of a permanent withdrawal of national responsibility in, for example, service production. The challenge for national planning today is to develop new planning methods for coordination that can deal efficiently with the demands of various public actors, within different sectors, and on diverse levels, as well as with the demands of private actors.

The situation is quite different when it comes to national planning for infrastructure. Here, there is increased national engagement. Hence, national planning does have a role to play. The challenge is to develop planning procedures which are efficient with respect to infrastructure investments that are desirable from a national perspective, and at the same time enable municipalities and regional authorities to influence decisions and investments that impact on the local and regional level.

NATIONAL PLANNING VERSUS REGIONAL AND LOCAL

The strong position of the municipalities in decision-making about land use and development is problematic. The possibilities for creating solutions that are anchored locally are increasing, but so is the number of conflicts between local and regional interests. The tension between local and regional/national interests has fuelled a debate on the division of responsibility within the planning process. Voices have been raised in favour of reducing the strength of local planning in projects that have considerable regional or national impacts. Previously, planning issues were mainly local and had no or minor impact on adjacent municipalities and the surrounding region. This is no longer true. Apart from detailed planning for specific and limited projects, most municipal plans have a considerable impact on conditions outside their borders. This change is due to the fact that urban development today is often about infrastructural investments which need to be regionally coordinated in order to function efficiently. The planning of services, for example the development of an external shopping mall, can also impact significantly on adjacent municipalities. In parallel to these actual changes there is also an increased awareness of the mutual dependencies of development and its environmental, social and economic impacts.

At the regional level there has been a marked increase in efforts to resolve planning conflicts within the region and to plan cooperation across county borders. Several county councils have demanded a larger say in the field of spatial planning or so-called regional development planning. Recently, a national government commission suggested that some county councils should take over responsibility for the development planning that is handled by the county administrative boards.

THE FUTURE OF NATIONAL PLANNING

A close look at Swedish planning today shows a planning process under reconstruction to meet the requirements discussed above. The most significant change is the move away from statutory planning aimed at regulating development to a planning process that enables development. Traditionally the focus has been directed towards the plan as a document, not planning as an activity. Rather than seeing planning as a bureaucratic activity aimed at the production of an adopted plan, the focus has shifted and is directed towards planning as a process, a process in which many actors with various ambitions and legitimate interests have to interact. This rather general trend has one exception—infrastructure. Here the trend is towards greater government intervention. One reason for this is that central government regards investments in infrastructure as a key element in improving Sweden's performance in the competitive international arena. Another motive for assigning a strong role to national planning is to prevent local and private interests from endangering the public good. It is commonly accepted that 'NIMBYs' would impact negatively on the efficiency of large-scale infrastructural investments if decision-making powers were entirely handed over to the municipalities.

The future role of national planning is being discussed in Sweden from the perspective of the structural and political changes in its society. The question is whether the state has lost its role in planning, when planning decisions are moved either to supra-national or regional/local levels. The emerging Swedish response to this question is 'No: national planning has not lost its role. Rather, the role is being redefined.'

On the one hand, it is clear that central government intervention in many sectors will be more restricted than previously. The role of national planning will also be reduced as a consequence of increased government activity at the regional level. To an increasing extent, Swedish regions are taking new initiatives to improve their attractiveness and competitiveness. These include coordination of local initiatives, taking responsibility for issues that were previously handled by the national government, and the establishing of contacts with other regions of Europe in order to develop mutually beneficial collaboration.

On the other hand, the national government will continue to fulfil important roles in planning. A new role which is emerging is that of intermediary between supranational planning and regional/local planning. Thus, the challenge facing national planning is to take on board supranational issues concerning, for example, the environment and global economic development and to introduce these issues to regional and local planning bodies, where the demand for autonomy and the devolution of planning powers are significant.

In conclusion, the conditions for Swedish national planning have changed significantly during the last decade. Looking into the future indicates that further change is on the way because of rapidly emerging internationalisation, and the demands for increased decentralisation and the devolution of planning powers to the local and regional levels. It is obvious that Swedish national planning is in a process of reconstruction. Perhaps the most important feature of this change is that the emerging planning system can be classified as neither hierarchical nor of the bottom-up type. Rather it is characterised by close cooperation between governments at different levels and others with interests in the planning process.

REFERENCES

ÅHS, ULLA (1997), 'Housing Policy and Social Welfare', in *Swedish Planning*, Gävle, Föreningen för samhällsplanering.

ALLEN, JUDITH, GÖRAN CARS and ALI MADANIPOUR (eds) (1998), *Social Exclusion in European Cities*, London, Jessica Kingsley.

ANDERSSON, ÅKE, BJÖRN HÅRSMAN, and JOHN QUIGLEY (1997), *Government for the Future—Unification, Fragmentation and Regionalism*, Amsterdam, North Holland.

BOVERKET (1994a), *Samverkan för bra vägar* (Boverket rapport 1994:5), Karlskrona, Boverket.

BOVERKET (1994b), *Sverige 2009 – förslag till en vision*, Karlskrona, Boverket.

BROOMÉ, HANS (1996), *Ett strategiskt förhållningssätt—om arbetet på en regional strategi för Stockholm* (Tidskriften PLAN, No 5/1996), Stockholm.

CARS, GÖRAN (1992), *Förhandlingar mellan privata och offentliga aktörer i samhällsbyggandet*, Stockholm, Tekniska högskolan, Avdelningen för regional planering.

CARS, GÖRAN and JONAS HAGETOFT (1988), *Tensta ur ett europeiskt perspektiv*, Stockholm, Svenska Bostäder.

ENGSTRÖM, CARL-JOHAN (1996), *Visioner är dialoginstrument i Tidskriften PLAN nummer 5/96*, Gävle, Tidskriften PLAN.

LARSSON, GERHARD (1994), *Spatial Planning Systems in Sweden*, Stockholm, Division of Real Estate, Royal Institute of Technology (mimeo).

NILSSON, JAN-EVERT (1996), *Visioner som analysinstrument i Tidskriften PLAN nummer 6/96*, Gävle, Tidskriften PLAN.

STATENS OFFENTLIGA UTREDNINGAR (SOU) (1990), *Demokrati och makt i Sverige*, Stockholm, Allmänna förlaget.

SNICKARS, FOLKE (1997), 'Developing Infrastructure for the 21st Century', in *Swedish Planning*, Gävle, Föreningen för samhällsplanering.

SVENSKA ARBETSGIVAREFÖRENIGEN (SAF) (1998), *Strukturrapport 98*, Stockholm, Svenska Arbetsgivareföreningen.

FIVE

NATIONAL-LEVEL INSTITUTIONS AND DECISION-MAKING PROCESSES FOR SPATIAL PLANNING IN THE UNITED KINGDOM

Malcolm Grant

The United Kingdom has no national spatial plan for its land use. In this respect at least, it resembles the USA, and its current practice is distinctly different from those countries—like Israel and Japan—which do prepare, adopt and attempt to implement national plans. But, as this paper will attempt to demonstrate, these are distinctions which are not as sharp as they might first appear. There are indeed national-level institutions, and a well-established set of decision-making processes for spatial planning, in the United Kingdom. What is different is the relationship between central government and local government in the conduct of the function of planning.

In theory, land-use planning powers in the UK today belong primarily to local government, with central government playing a supervisory and strategic role. This is a significant change from the centralised vision of town and country planning that underlay the country's innovative post-war legislation. But that is largely because the tasks themselves have changed significantly since that time, a change which has not yet occurred in many countries whose population and gross domestic product (GDP) have both been growing rapidly in recent years. The era of large-scale intervention through the building of whole new towns, slum clearance, mass housing projects, urban motorways and town centre renewal projects is over. In addition, central government has had to surrender many detailed controls over local authorities simply because it could not cope with the minutiae and proved incapable of offering efficient supervision.

Yet central government still retains so central a role in land-use planning as to allow it *de facto* to maintain the framework of a national spatial plan, not directly but through the development plans prepared by the local authorities. As always in political analysis, we need to stand back from the formal allocation of functions in legislation, because these tell us little about how the system actually works in practice. The true power relationships between central and local government, public utilities, landowners, developers and investors are constantly changing, and to understand them requires that we understand not just the legal forms but also the political and economic realities.

In this paper I explore the nature of the changes that have occurred in the character of central government's role in spatial planning in the UK since 1945, and look behind the legislative structure in an attempt to understand the political, cultural and financial forces that determine what the contemporary relationship actually is. This requires an analysis of two complementary themes: first, how the tasks allocated to the planning system have changed since those times, and how far the central–local balance has shifted with that change; second, the paradox of the 19 years in which a Conservative government, while pursuing a policy of setting back the state, succeeded in centralising power; a process which the new Labour government, elected in May 1997,[1] has not rushed to reverse.

Planning legislation in the UK makes no mention of national spatial planning. Such a reference did exist at one time. In 1943, a statutory duty was imposed upon the Minister of Town and Country Planning of 'securing consistency and continuity in the framing and execution of a national policy with respect to the use and development of land throughout England and Wales',[2] but that was quietly dropped upon the conversion in 1970 of the Ministry into a Department of State, on the curious ground that it was 'inappropriate' to subject the Secretary of State for the Environment to such a duty.[3] Yet that repealed provision continues today to provide the most appropriate statement of what the Secretary of State's[4] functions are with relation to the planning system.

Successive Conservative governments made no secret of their scepticism of attempts at regional economic and spatial planning. Indeed, one of the first acts of Mrs Thatcher's new administration in 1979 was to dismantle the regional economic planning councils and the whole structure of regional assistance. This was to be a government of markets, not of planning; of deregulation, not of regulation. Yet, while continuing to express scepticism about any form of national planning, the Thatcher government proceeded to develop an already highly centralised model of spatial planning into one which was more nationally consistent and more coherent than it was before, and one which, against the opposition of its formerly natural allies in property and development, came to encompass and espouse the principle of sustainable development with surprising enthusiasm. Though only incompletely integrated with other institutions of government and of public service (particularly in an era of widespread privatisation of public utilities), the planning system struggles to retain public support and political backing, and to maintain a working balance (albeit precarious at times) between national and local interests.

The constitutional and political context

The obvious starting point in understanding these apparent paradoxes is the historical lack of constitutional underpinning for political institutions in the UK, which means that relationships have been shaped and adjusted over time, often almost imperceptibly. The executive division of government is embedded in Parliament, under a prime ministerial rather than presidential system which ensures that all Ministers are members of either House of Parliament. The government is formed from members of the majority party in the House of Commons, and this means that it is normally able, through the processes of party whipping, to secure the passage of its legislation through Parliament. Government members rarely oppose government-initiated legislation, though they may fight for amendments on particular issues, and the government has a near monopoly on the introduction of Bills. Although individual Members of Parliament may introduce Bills, the government controls the legislative programme and hence can simply block their progress. As a matter of parliamentary convention, however, there is an opportunity every session for a handful of relatively uncontroversial private members' bills to proceed without government opposition, and sometimes with government support. Under a wholly different procedure, any citizen is entitled to promote a Bill in Parliament to achieve an end for which the general law provides no mechanism, and this process (the Private Bill procedure) has been used for centuries to secure approval for major infrastructure projects which require the expropriation of land or other interference with private rights (e.g. rights of navigation). By the latter part of this century, its purpose had been largely overtaken by public general legislation conferring the necessary powers (e.g. expropriation) on the government itself, except in the case of railways and interference with navigation, but powers in these areas were transferred to the government by the Transport and Works Act 1992, and integrated with local planning procedures, so that Private Bill procedure is no longer an instrument of securing planning approval.

But the government has one more mechanism available to it for national projects, which is to promote a Public Bill, a procedure that was employed to secure the necessary rights to construct the Channel Tunnel and which is currently being used to obtain the rights for the construction of the high-speed rail link between the Tunnel and King's Cross railway terminus in London. Since private rights are adversely affected, the usual parliamentary scrutiny processes are complemented by quasi-judicial hearings in both Houses of Parliament, where individual objections can be evaluated. But in such a case, unlike Private Bill procedure, there can be no objection to the principle of the proposal, because this will have secured parliamentary endorsement before being referred to a Select Committee, and objections therefore typically relate primarily to mitigation measures. Hence, though rarely used, this is a potentially powerful method for government to force through major infrastructure projects, which typically will cross local authority boundaries.

Parliament's legal capacity to legislate is not constrained by a national entrenched constitution, though it is constrained by constitutional conventions relating to parliamentary and judicial behaviour, and by the Treaty of Rome.[5] In particular, there is no entrenchment for any other tier of government. Local government has existed in various forms for centuries, but in its present form is a creature of parliamentary legislation. Statute law establishes the corporate basis of local authorities, and confers powers upon them. Local authorities are governed by the doctrine of *ultra vires*, which means that they may not do anything other than that which statute allows them to do. Only marginal mitigation of the rigour of this rule is provided by the limited penumbral power to do anything conducive to or incidental to the discharge of any of their functions.

These principles are important, because they explain the relative strength of central government in Britain, vis-à-vis other institutions of government. The UK is indeed a highly centralised nation, and this pattern of its government is reflected also in the organisation of its economy, its newspapers and its major industries. Given these characteristics, it would be odd if the influence of central government over local government were not to be powerful. This is best demonstrated by the structural reform of local government which has taken place over the past four years. In Scotland and Wales, it has resulted in the abolition and replacement by parliamentary legislation of all the existing local authorities, which since 1972 had been structured on a two-tier basis. There were county councils commanding an extensive territory and assuming responsibility for 'strategic' functions (i.e. those, such as highways, transportation, overall settlement pattern and economic development, for which a larger territory is appropriate); and district councils (of which there would be several in each county area) responsible for local services. The reforms, effective from April 1996, have created instead a system of single 'unitary' authorities. In England, a different approach was taken. There was to be no national blueprint of reform, and instead a special commission (the Local Government Commission for England) was appointed, both to make policy on unitary local government, and to recommend various local structures to the Secretary of State, on the basis of widespread local consultation. Notwithstanding the political and practical difficulties of employing such a consultative approach to the process, the outcome has been a relatively rational mixed structure, with two-tier local government retained in the more sparsely populated rural areas of England, but with many of the major cities (such as Bristol, Southampton, Derby and Hull) now being run by unitary authorities. This has distinct advantages for the focus and governance of these cities, though it is at the risk of fragmentation of the land-use planning framework, because the cities are no longer tied into the formal planning of their broader socio-economic areas.[6] These structures have remained in place following the devolution of power to Scotland, with the establishment there in 1999 of a Parliament with legislative competence in relation to certain devolved matters; and to Wales, in the same year, with a less powerful National Assembly.

Another powerful consideration in both national and local politics has been ideology. British politics in relation to land policy was, between 1945 and 1985, bedevilled by a simplistic dichotomy between socialism and conservatism, reflected in the party structure of the House of Commons and fanned by the adversarial character of the politics carried on there. The Labour governments of 1945 to 1950, 1964 to 1970 and 1974 to 1979 were all committed to Keynesian economic policies (as, indeed, were the Conservative regimes of 1950 to 1964 and—though with several hiccups—1970 to 1974), and to a socialist view of governance. That is not true of the Labour government elected in May 1997, which has adopted a tight fiscal policy, a belief in the ideology of the 'third way', and a 'modernising' campaign in relation to the land-use planning system and local government as a whole.

The post-war town and country planning legislation found its roots in a combination of centralised economic planning, which was thought to have served the nation well during the war years, and a Labour ambition of nationalising, if not land, then the profit that was to be made from selling it for development. Each of the previous Labour governments introduced legislation which had as its ultimate goal a tax of 100 per cent on development values in land (i.e. the difference between its market value and its value in its existing use). Each Labour government believed in the mixed economy, but was anxious to enlarge the role of the state within that mix. The Blair administration had no such ambitions, and, far from proposing a new form of land tax, committed itself to no new tax burdens at all in its early years.

The legislative framework

The relationship between Parliament and central government is important to the character of national spatial planning. There are three main components to it.

THE NATIONAL LEGISLATIVE FRAMEWORK WITHIN WHICH BOTH CENTRAL AND LOCAL GOVERNMENT MUST OPERATE

Planning legislation is, except for some minor technical differences, identical in England and Wales,[7] and the Scottish legislation[8] is a close clone. So too in Northern Ireland, though local government has no planning functions there. It is this national primary legislation that sets the parameters of land-use planning in the UK. It establishes the safeguards for citizens affected by planning controls, and confers upon them rights of objection, of administrative appeal and of legal challenge. Parliament amends this legislation regularly, and it is supplemented by extensive subordinate legislation made under powers subject to only loose parliamentary supervision.[9] Local authorities have no power to alter any of this legal framework. They are obliged to operate within it.

In many respects the framework is flexible. In particular, it requires all local authorities to make and maintain their own formal development plans. The plans are not legally binding, either on them or on citizens, leaving open a broad measure of discretion in the way local authorities are able to respond to pressures for spatial change. But this substantively broad discretion and flexibility is overlain by procedural checks and balances which significantly restrict local authorities' freedom of action.

THE CONCEPT OF CONSULTATION

National and local policy in spatial planning, as in other areas of contemporary government, can best be understood as an interlocking network, based as much upon interdependency as upon hierarchy. This network is to a large extent horizontal. For example, local planning authorities are required by law not to take decisions on development plans or on applications for planning permission unless they have first consulted other parties whose interests may be involved. These parties include, as appropriate, privatised utilities, government departments, independent agencies, amenity societies and other local authorities. The local authority is required by law to take into account any representations it receives. It need not comply with them. Indeed, it can expect to receive many which are opposed to each other. But it will be expected to consider the arguments raised, and if possible to negotiate with objectors. Successful negotiations will reduce the transaction cost of carrying through the policy or making the decision. But disappointed objectors also have the capacity to lobby government officials and Ministers to intervene, and the system allows for this vertical component of the network to provide a forum for dispute settlement. A disappointed developer (but not a disappointed applicant) can appeal to the Secretary of State against a refusal or conditional grant of planning permission. In Scotland, since devolution, this role is played by the Scottish ministers, and in Wales by the National Assembly. The Secretary of State can exercise various powers of call-in, direction, appeal, default intervention and so on. And he can also require the local authority to pay the costs of other parties if their conduct is judged unreasonable. Hence, this vertical arrangement not only imposes a premium on reasonably civilised behaviour on the part of the actors involved in the consultative process, but also provides the Secretary of State with sufficient power to exercise close supervision over its day-to-day operations. It is significant that formal powers of intervention are rarely used,[10] but their very existence helps shape the attitudes that the parties bring to the process.

THE PROMULGATION OF NATIONAL POLICY STATEMENTS

Until 1988, this was something of an ad hoc and arbitrary process. Government Ministers would announce changes in national planning policy through a variety of means: a ministerial speech, a white paper, a parliamentary answer (usually to a specially planted question), or, the most common, a circular letter addressed to all local authorities. The government circular is a strange concept. To dispatch policy statements on the ostensible basis of a mailing list that includes only local authority chief executives assumes that policy is a matter solely between central and local government. Others who might wish to know what current policy is must purchase a copy of the circular from the government stationery office. It was also a somewhat careless process: often the ministries themselves were uncertain what circulars had been issued and remained extant. But in 1988 the Department of the Environment, jointly with the Welsh Office,[11] introduced a new series of policy statements, known as Planning Policy Guidance, or PPGs. They were shortly afterwards joined by a series of Minerals Policy Guidance and Regional Policy Guidance. Most have been revised and updated, in some cases more than once. The full current list for England appears in the Appendix. The difference between a circular and a guidance note is that the latter are public statements of national policy, and they are addressed to the public at large. Circulars are still in use, but are used now to convey detailed procedural advice to local authorities on how to handle their responsibilities under planning legislation.

The PPG series has had a significant effect. It establishes a firm national policy framework, and a discipline for local authorities' planning functions. It establishes, for example, a common form for green belt policy (e.g. PPG2), a common methodology for assessing the availability of housing land in local authorities' areas (PPG3) and general guidance on transport and land-use planning (PPG12). It also has an impact in steering policy on matters for which local authorities' own development plans have proved relatively ineffective because of their inability to react quickly to national trends, for example in relation to out-of-town shopping (PPG6). PPGs do not require parliamentary approval, but their implementation and effects are investigated from time to time by Parliamentary Select Committees, to whose reports and recommendations the Secretary of State is expected to respond.[12] Local authorities might not be generally expected to comply meekly with government policy: after all, they are frequently controlled by a party of a different political persuasion from central government, and will often maintain that local circumstances are different from those contemplated in national policy. But compliance is the norm, because national policy statements have an indirect coercive element. If an authority fails to observe government policy, it runs the risk not only of losing appeals brought by disappointed developers, but also of being ordered to pay their costs.

Planning policy for housing provides a useful example of this process. There has been a policy battle between successive national governments, who have wished to see adequate land allocated for housing across the country as a whole to meet forecast demand arising from demographic change, and local authorities, under NIMBY-style pressures to protect existing amenities (such as countryside) from further development. In order to protect its policy objectives, the government has used a cascade strategy, in which the national forecasts are converted by the regional bodies[13] into targets for each county area in the region and promulgated in regional planning guidance,[14] then translated by the county structure plan[15] into targets for each local development plan,[16] which must then allocate the necessary land. The law now requires that all decisions made on applications for planning permission should be taken in accordance with the development plan, save where material considerations indicate otherwise.[17] This provides further reinforcement to the government's hierarchical control over the preparation and implementation of planning policy. Housing land requirements are imposed upon local authorities.

There have been two major shifts in housing policy over the past 10 years, and both have been led by central government through the development of policy and the issuing of policy guidance. The first is in relation to affordable housing, where some carefully crafted government circulars[18] have encouraged local planning authorities to negotiate with developers to secure the provision of affordable housing on all housing schemes above specified thresholds, where there is a need for affordable housing in the authority's area. Only on few private housing schemes can developers now expect to avoid having to provide affordable housing, yet this change has been brought about wholly by policy shifts and without any changes in the law.

Secondly, there has been an extensive policy debate[19] about the validity and implications of forecast demographic changes, implying an increase of 4.4 million households through to the year 2016. Assuming the accuracy of the forecasts, there is a major challenge to the planning system. National policy responses are still emerging through a process of public debate and attrition.[20] There has been some effective NGO (non-governmental organisation) campaigning from countryside protectionists, which sparked competition between the main political parties through the 1997 general election to see which would promise to deliver the greater proportion of new housing in existing urban areas rather than on greenfield sites.[21] The outgoing Conservative administration promised 50 per cent; the incoming Labour administration raised this to 60 per cent over the 10 years following the election, and set up an Urban Task Force under the chairmanship of the architect, Richard Rogers, to work out how best to achieve this target in the broader context of promoting a renaissance of the cities. This objective chimed well with the government's determination to launch a major programme of regeneration of urban areas.

The subjects left to national-level institutions

The absence of entrenched constitutional underpinning means that there is no *a priori* demarcation of autonomies. The role of national institutions is therefore determined in practice by three considerations: first, that some functions of government are clearly of national significance, and not for local government, quangos or privatised utilities. For example, the dividing line between central and local government in relation to linear infrastructure such as highways is reached when decisions have implications for more than one local government territory, such as the construction of a national motorway network. Local authorities may still be employed on an operational basis in maintenance and repair work as agents, but their only strategic role is to lobby central government for (or against) road proposals. Moreover, there are some functions of government which are relevant to spatial planning, but which have no basis in local government. Examples include rail transport (although light railways and tramways are now making a welcome reappearance in the larger cities, with local authorities as promoters) and health care provision and planning.

The second determinant is that of finance. Local government today raises less than 20 per cent of its revenue from local taxation, which is exclusively a property tax. Until 1989, this tax, known as the rates, was levied by local authorities on households (known as the domestic rate) and also on other occupied property, including commercial and industrial premises (the non-domestic rate). But in 1989, the domestic rate was replaced by a new tax, the community charge, which was better known as the poll tax because it was a tax on individuals rather than on property or households. Its regressive character made it, predictably, hugely unpopular, except with those—typically sole occupiers of more expensive properties—who benefited financially from the change. By 1992 it was clear that the poll tax had to be abandoned, and the Local Government Finance Act 1992 accordingly substituted for it a new simplified and hybrid household property tax, the council tax. Houses are now valued relatively crudely into eight bands of value, and the tax applies at a differential rate between these bands, so that occupiers of properties in the highest band (Band H) contribute at a rate of between three and five times that of occupiers of houses in the lowest band. In order to make the change politically palatable, a further subsidy from central funding was necessary, and this was achieved by raising the standard rate of national value added tax (VAT) from 15 per cent to 17.5 per cent. Although the council tax remains the residual funding source for local authorities, they do not have unlimited taxing power. The Secretary of State is entitled to step in and order a local authority to reduce its proposed level of council tax (called 'capping').

Another major component of the 1989 reforms, which remained unaltered by the 1992 revisions, was the nationalisation of the non-domestic rate. The right for local authorities to levy a property tax on commercial and industrial properties was removed altogether by the 1989 legislation. This was a matter of great controversy through the 1980s, when businesses complained to the government that local authorities were using them as a milch cow for local funding. A revaluation of non-domestic properties in Scotland had brought about a very significant redistribution of liability, reflecting the redistribution of actual property values that followed the massive industrial structural change in that country since the last revaluation in 1971. The fear of a business backlash in England prompted the government to opt for a replacement tax, which would be levied at a uniform national rate, and with a guarantee that any increases in rate would be held to or below the rate of inflation in the national economy. The revenues are redistributed to local authorities on a per capita basis. Their final source of income is a general government grant, which is distributed on the basis of a calculation of needs and resources of each authority in accordance with nationally negotiated formulae known as the standard spending assessment (SSA).

The outcome of these changes can be seen in Table 1. The trends have considerable significance for our understanding of national-level planning in the UK. As the table shows, government grant in 1994–95 was back at 1980–81 levels, from a drop through the 1980s, and reflecting the boost given by the addition to VAT. But the proportion actually levied and collected by local authorities themselves is now very low. There are two major implications of this for our understanding of national-level planning.

Table 1 *Funding of local government revenue expenditure in England since 1981*

Year	*Government grants*	*Non-domestic rates*	*Domestic rates**	*Proportion under local control*
1981–82	56	25	20	45
1984–85	54	28	21	49
1989–90†	44	29	26	55
1992–93‡	42	29	17	17
1994–95	57	24	17	17

Source: Adapted from *Local Government Financial Statistics England No. 5 1994* (HMSO, 1996), London, Table 9.6

* The figures for 1992–93 are for poll tax, which was levied for the three years 1990–91, 1991–92 and 1992–93; those for 1994–95 are for council tax.

† Last year pre-poll tax.

‡ Last year poll tax.

First, local authorities no longer have the financial independence they once had, and this means that they have only limited capacity to undertake positive planning on their own account. The various programmes that many local authorities had started to introduce through the 1980s for local economic development, using residual taxing and spending powers, were trimmed by the Local Government and Housing Act 1989. This has meant that local government's capacity to undertake its own investment in local economies is now severely attenuated, so that the influence that a local authority can bring to bear on securing economic and physical change in its area is no longer primarily through land acquisition and development, but through strategic partnership with other actors, including both public sector and private sector agents. This has paved the way to a new *realpolitik* of local government, which stresses the interdependency of private and public sectors in urban regeneration.

Second, the nationalisation of the business rate has severed the relationship between commercial development and local revenues. Local authorities do not benefit financially from new business development in their areas: the property tax paid by the new occupiers goes to the national treasury and is redistributed to all local authorities. Local benefits are therefore wholly indirect, and though these may nonetheless be significant (such as improving the physical fabric, stemming urban decline and contributing to job creation), the financial structure does not provide the spur to local authorities to promote growth in their areas that one sees in France and the United States. Indeed, the reverse is true, especially in the south of England, where the absence of direct financial benefit merely reinforces local anti-growth pressure. The new Labour government has so far shown little enthusiasm for change from this position. Its priority is the reform of the internal political management of local government, not least to overcome the unhealthy single-party dominance that exists in many areas, and only modest reforms to the existing financial provisions.[22] Hence, major spending decisions are inevitably reserved to central government, and within a framework of national policy.

The third determinant is that of national politics. Central government is politically and constitutionally able to make anything a matter of national policy. Indeed, the irony of Mrs Thatcher's Conservative government was that an administration that wished to roll back the state and allow private enterprise to flourish, first found it necessary to centralise state controls and to limit the functions of other political actors, particularly the only other level of elected governance, local government. There has also been a wide-reaching programme of privatisation which has given to the private sector strategic decisions over the provision of services such as telecommunications, energy, utilities and transport.

Some examples of how key issues are split between national government and other agencies may help to explain these relationships.

NATIONAL PARKS

National and devolved governments have responsibility for designating these areas, but their management is in each case in the hands of a central–local joint agency. Following the Environment Act 1995, each park now has its own national park authority, the members of which are appointed by local authorities (two-thirds) and by central government (one-third). Funding is partly from local tax, but mainly from government grant. The parks are not areas of publicly owned land, but areas whose use is primarily private agriculture.

AGRICULTURAL LAND

The power of the agricultural lobby has diminished dramatically over the past 20 years, but it remains the only industry with a ministry devoted to its affairs—the Ministry of Agriculture, Fisheries and Food. Agriculture was effectively exempted from planning controls at the outset, because of the wish to encourage agricultural self-sufficiency. The problem now is of over-production, and of strong pressure from urban interests to preserve the beauty of the British countryside during a period when farmers are anxious to change the economic use of their holdings, for example, redeploying surplus farm buildings for commercial, high-tech and craft uses.

The release of agricultural land for new urban development is tightly limited throughout the country, and a significant change of the past 20 years has been the halting of the new towns and expanded towns programmes, with a consequent reliance upon incremental growth of existing settlements in order to provide land for housing. There is a tension here between the national objective of ensuring an adequate supply of housing at affordable prices, and the local objective of resisting further growth which so dramatically changes the characters of small rural towns and villages. That tension is not overcome in national policy statements, which pull in both directions simultaneously, but with a strong preference now for the re-use of previously developed land over greenfield sites.

ROADS AND RAILWAYS

These are essentially matters of national policy, but responsibility for both is in the process of reform. For highways, this has meant the transfer of operational responsibility from central government to a notionally independent executive agency—the Highways Agency. For railways, the present policy is one of privatisation. The contract for disposal of the first railway network under the privatisation programme was signed shortly before the end of 1995, and six other networks were privatised in early 1996. Decisions as to investment in new rail facilities have been transferred to the private sector, subject to control by a statutory regulator of price and quality. However, concerns about the safety of the rail network that emerged during 2000 demonstrated the extent of the political responsibility that remained with government ministers.

WATER

A similar picture presents itself in the case of water. Until 1989, water and sewerage were functions allocated to statutory regional water authorities. Those functions, and the assets of the authorities, were privatised in 1989. Water quality is now regulated by the Environment Agency which was established in 1996. Water price is regulated by an economic regulator (OFWAT). Performance for the development industry is through the mechanisms of statutorily reinforced contracts. The water companies are under a statutory duty to supply clean water, and to take away and treat dirty water, and are statutorily liable for breaches of that duty. Yet privatisation has not wholly resolved the tension between the natural near-monopoly of water supply and treatment on the one hand, and accountability to shareholders on the other. High profits (plus directors' remuneration) has in some areas gone hand in hand with poor service to customers. The role of national government is limited to a residual function in regulating the quality of drinking water, and in maintaining the national interest in negotiations within the European Union.

HAZARDOUS USES AND LULUs (LOCALLY UNACCEPTABLE LAND USES)

Because of ministers' functions of supervision of the planning system, central government inevitably has an involvement in areas of controversial land-use change. In some cases, it has in the past taken a recommendation to Parliament and sought parliamentary endorsement. Examples include the decision to construct a reprocessing plant for spent radioactive fuel at Sellafield in Cumbria, the decision to construct the new nuclear electricity generating station at Sizewell in Suffolk, and the decision to site the third London airport at Stansted in Essex. In less hazardous cases, the decision may not go to Parliament, but it will certainly be taken by central government. In each case, however, the decision follows a public hearing of objections, undertaken by an independent inspector. Airport policy is a case in point. Successive governments have found it almost impossible to devise national policy on airport development and growth, because of the site-sensitive nature of the problem, particularly in south-east England. Hence it has effectively been public hearings which have led airport policy. A public local inquiry ran for over four years into a proposal to build a fifth terminal at London's Heathrow airport, and even two years after it closed the outcome had not been announced.

Making national spatial policy

Land-use policy in Britain is a perennially controversial issue. It highlights numerous sharp conflicts, particularly between lifestyles: between urban and rural preferences; rich and poor; the growing importance of leisure, and the decline of agricultural employment; historic conservation versus economic growth. The principal means through which these conflicts are mediated is in the design of development plans, which is where strategic decisions are taken on matters such as the designation or extension of a green belt, on policies for conserving historic centres, on focusing economic incentives, on redeveloping old industrial land and on finding a location for new housing.

Yet this is a process of local choice within a national framework and subject to national direction. At one time, national policy was elaborated wholly within a closed network of central government, and simply announced as a *fait accompli*. That style has now almost completely disappeared. It is rare today for a PPG or circular to be issued without extensive prior consultation. To some extent this is a welcome change to a more open style of government, but it is also the product of the declining power of national government as functions have been privatised.

Oversight of policy implementation

The machinery for securing that national planning policy is observed in decisions by local planning authorities is complex. The legislation secures, for local authorities, a highly discretionary system, and requires that they approach each planning application on a case-by-case basis. Yet they do so within the framework of national policy. The linkages between the two are threefold:

- There is a requirement that local planning authorities 'have regard to' national policies in formulating and adopting their development plans. Interestingly enough, this requirement has statutory force only in one instance: a requirement that metropolitan authorities should 'have regard to . . . any regional or strategic planning guidance given by the Secretary of State to assist them in the preparation of the plan'.[23] But in practice, the principle applies to other plans as well, through three means. First, the traditional British approach to judicial review, which assumes that, in order to make a decision that is legally valid, a decision-maker must have regard to all relevant considerations. Advice from central government is a relevant consideration, and plan-makers must have regard to it. Second, the right of the Secretary of State to object to a plan, and to pursue that objection at a hearing. This right has assumed new significance since 1992, when the Secretary of State announced that if he did not exercise this right, then it could be assumed that he did not object to a policy in a plan which went against national policy.[24] Third, there is the right of the Secretary of State to call-in a draft plan for his own approval,[25] a power usually reserved for extreme cases.

- There is a legal requirement on decision-makers to have regard to all material considerations also when making decisions on planning applications. Under an amendment to planning legislation enacted in 1991, they must place a premium on the adopted plan, and may only depart from it for good reason.[26] But they must also have regard to national policy. Indeed, one good reason for departing from the plan may be that national policy has changed in the meantime. Not only the local authority is subject to this duty; so is the Secretary of State. It has been known for the courts to find that the Secretary of State has misunderstood or misapplied his own policy, or failed to give adequate reasons for not applying it.
- The Secretary of State has an overall supervisory jurisdiction. This is a powerful tool for reinforcing the primacy of national policy. While a green belt may be a matter for local designation, the policies that apply in green belts are common across the country. Though there may be differences in emphasis between local authorities in their approach to green belt protection, the Secretary of State remains the ultimate arbiter.

New Labour: new planning?

Planning is essentially a political process, and the general election of 1997 which returned a Labour government with a majority of well over 200 seats in the House of Commons was expected by many to have a significant impact on the system. Change has indeed occurred, but fired more by managerialism than by the redistributive radicalism of previous Labour administrations. It is also a sign of the progress made by the former Conservative government in embracing the agenda of sustainable development that new Labour has not wished to pursue a markedly different line. It may be, however, that the institutional reforms to which the government has now committed itself may have a longer-term impact than any of its predecessors' policies. The main planks of current reforms are outlined below.

REGIONAL DEVELOPMENT AGENCIES

The government has set up agencies in each of the regions of England.[27] The momentum for this came in part as a reaction to the downplaying of regionalism under the previous Conservative governments, and in part as a carry-over into England of the government's policies for devolution in Scotland and Wales. The Regional Development Authorities (RDAs) have extensive powers to bring about economic development and urban regeneration. Although they are non-elected agencies, it is anticipated that they will provide a basis for the development of political institutions at regional level, initially through collective action by local authorities to establish voluntary chambers; ultimately, perhaps, through elected assemblies with devolved powers. It is conceivable under this scenario that, in place of the present national planning systems, there might develop as many as eight different regional systems in England, paralleling the divergence in approach which is already developing as a result of devolution in Scotland and Wales.

MODERNISING PLANNING

In a ministerial statement issued in January 1998, the then Minister for Planning pledged the government to a process of modernisation of the planning system, announcing that he wished to push to one side the 'procrastination culture' of land-use planning, to initiate clearer statements of national policy and speed up their production, as well as the production of development plans, to introduce a new European perspective to planning and to establish new economic instruments.[28] This has proved to be an ambitious programme, and stronger on diagnosis than on remedy. Three years on there was still no sign of proposals that had been promised for the reform of planning gain (the system of developers' financial contributions to public goods). Economic instruments as an alternative to regulation to promote favoured outcomes are a beguiling concept but have caused significant political difficulties: there is a predominant anti-development culture in many parts of England (particularly the relatively prosperous south-east), underpinned by sustainable development, and communities are reluctant to cede any of their planning regulation powers.

INTEGRATED TRANSPORT

One major and imaginative policy initiative was the White Paper *A New Deal for Transport: Better for Everyone* on integrated transport policy, published in July 1998. It proposed a strengthening of the links between land-use planning and transport policy, in the context of a commitment to sustainable development, and with a view to decreasing reliance upon the private motor car. The primary tools are specific transport plans, drawn up on a regional as well as a local basis. Legislation now authorises the introduction of road pricing to counter congestion, and a tax on workplace car parking, with local authorities having the right to apply the revenues to public transport and other counterbalancing measures. But these measures have proved electorally so unpopular as to force them down the political agenda.

LOCAL GOVERNMENT REFORM

There are two major initiatives for the reform of local government. The first is for the creation of a new Greater London Authority (GLA), comprising an Assembly of 25 members, and a directly elected executive mayor. The mayor is all-powerful, and the role of the Assembly is principally to hold the mayor to account.[29] The GLA is the strategic planning authority for London, and the mayor is responsible for drawing up a new spatial development strategy for the capital. This will provide a framework for the development plans drawn up by each of the 32 London boroughs. This 'top-down' approach may yet cause friction with the boroughs, which have, since the abolition in 1986 of the Greater London Council, enjoyed significant autonomy in planning matters, and have worked collectively through the London Planning Advisory Committee to develop strategic land-use policy.

There is also radical reform in sight for local government outside London. The government's decisions on this are spelt out in the White Paper *Modern Local Government: In Touch with the People* (Cm 4014) published in July 1998, which is given effect by the Local Government Act 2000. Councils will be required to move away from their historical method of conducting business through committee meetings, and to adopt instead a new model of internal political management. There is to be a choice between three models, all of which involve a separation between a powerful executive, and backbench councillors whose primary role will be in representing the interests of their area rather than participating in committee governance. There are significant implications for the way in which planning decisions will be taken for the future, and an opportunity to redefine the role of councils and councillors in the process.

How effective is national planning policy?

We start with a methodological problem. National policy statements in the PPG series are not site-specific, and they are often broadly expressed. This makes it difficult to evaluate the effectiveness of spatial and other related forms of planning in achieving the objectives held out by the policy statements such as PPGs and circulars. In some cases, the planning system is the primary vehicle for delivering the sought-for outcomes.

The green belts are an example. The success of the policies in achieving their objectives can be measured by investigating how far the designated green belts remain undeveloped, for this is the key to meeting the linked objectives of preventing urban sprawl and securing that neighbouring settlements will not coalesce (PPG2). The planning system can simply prevent all development in these areas. No compensation is payable for such a ban. But in practice, compromises are made. There may be a need to redevelop an existing building or set of buildings located in the green belt before its designation, or to meet the operational requirements of public utilities for urban fringe service locations. By these purely physical measures, we can see that the green belt has been a largely successful policy. Yet these measures tell us nothing about the development that has simply 'jumped' the green belt and been accommodated in the towns and villages beyond it, adding to commuting requirements and to traffic congestion in the town to which the green belt has added its protection. The hinterland simply becomes an inconveniently detached suburb. Nor does it tell us about the economic impact in terms of increased housing land prices, which are the product of forces other than simply the planning system but which are strongly driven by land availability.

Other spatial policies create even greater difficulties for evaluation. For example, the government changed its policies during the 1990s to forestall further out-of-town retail development, following fears that the large superstores located at the urban fringe, or adjacent to motorway junctions, were sapping the economic vitality of towns, and were reducing shopping opportunities for those who did not have the use of a private car. In the most recent version of its policy statement on retail location, the government has insisted that developers will have to establish that there is no convenient town centre or edge of town site before they are permitted to build out of town. An objective of that policy is to secure the variety and vitality of town centres. Yet, on this, the role of planning is more tangential than it is with securing the green belts. The variety and vitality of town centres can be fostered by sympathetic planning, by making town centres attractive to retail investment, by enriching the physical quality of towns, by introducing pedestrianisation schemes and by improving public transport and landscaping, and restoring historic buildings. All are legitimate public sector functions, but they are not the tools of PPG6, which is concerned largely with just one aspect—that of enhancing town centre vitality by stemming the flow of investment to commercially more attractive locations.

Conclusions

From this brief review, we can draw some tentative conclusions about the state of national-level planning in the UK today:

- National land-use planning is extensively practised by the national and devolved governments in the UK, despite the legislative allocation of first-instance planning powers to the local authorities. It is practised through the national policies promulgated by those governments, whose implementation is supervised through their reserve powers in respect of plan-making, and their role in determining appeals from local authorities' decisions on planning applications.
- It is also achieved through strategic decisions and investments in infrastructure, but as a result of the privatisation programme, the national and devolved governments no longer have the instruments they once had

to secure national objectives and the coordination of functions. This has had the effect of increasing transparency, but reducing accountability.

- National planning is necessarily hierarchical, but it is not, in the case of spatial planning, wholly prescriptive. The national and devolved governments and local authorities work alongside not only each other, but also alongside national agencies, privatised utilities, landowners and private investors. In urban regeneration, this notion of partnership has become the primary *modus operandi*. The success of regeneration schemes is measured not only by their physical, economic and social achievements, but by the extent of their leverage in securing private sector investment on the back of public sector funding.
- The ideological schism that so characterised national planning in the 40 years from 1945 has now all but vanished. All political parties are working to similar visions of the mixed economy, and the limited role of the central state, and although the Labour government has established devolution for Scotland and Wales, and even to some extent for the English regions, it is unwilling to extend additional powers to local government without first reforming their internal political management.
- Finally, the language of partnership is having a powerful effect on current public policy management. In regeneration and economic development, partnerships are emerging not only between different actors with intertwining interests on particular sites, but also between local authorities with linked interests, including some loose partnerships or groupings with French local authorities and regional bodies. Partnership is at one level a sign of weak government, and an indication of the dependency of national and local governments upon others to secure public policy objectives. But it is also a potent tool for consensus-building in generating those objectives and delivering the policies necessary to achieve them.

Appendix

Table of current national policy statements relating to spatial policy in England

Planning Policy Guidance series

PPG1, *General Policy and Principles* (3rd edn, 1997).
PPG2, *Green Belts* (2nd edn, 1995).
PPG3, *Housing* (3rd edn, 2000).
PPG4, *Industrial and Commercial Development and Small Firms* (2nd edn, 1992).
PPG5, *Simplified Planning Zones* (2nd edn, 1992).
PPG6, *Town Centres and Retail Developments* (3rd edn, 1996).
PPG7, *The Countryside: Environmental Quality and Social and Economic Development* (3rd edn, 1997).

PPG8, *Telecommunications* (1992).
PPG9, *Nature Conservation* (1994).
PPG10, *Planning and Waste Management* (1999).
PPG11, *Regional Planning Guidance* (2000).
PPG12, *Development Plans and Regional Planning Guidance* (2000).
PPG13, *Transport* (1994).
PPG14, *Development on Unstable Land* (1990).
PPG15, *Planning and the Historic Environment* (1994).
PPG16, *Archaeology and Planning* (1990).
PPG17, *Sport and Recreation* (1991).
PPG18, *Enforcing Planning Control* (1992).
PPG19, *Outdoor Advertisement Control* (1992).
PPG20, *Coastal Planning* (1992).
PPG21, *Tourism* (1992).
PPG22, *Renewable Energy* (1993).
PPG23, *Planning and Pollution Control* (1994).
PPG24, *Planning and Noise* (1994).

NOTES

1 This has, of course, taken place since this paper was presented at the Haifa Conference, and the paper has therefore been revised as far as possible to bring it up to date to 2001.

2 Minister of Town and Country Planning Act 1943, s. 1.

3 It is also true that there was at one time an explicit system of national economic planning, introduced in 1965 when the then Labour government set up a special Department of Economic Affairs, and gave it responsibility to create a national plan. But the experiment was short-lived. The Department proved to be a troublesome implant in the Whitehall machine, and it was abolished in 1969.

4 The expression 'Secretary of State' is a statutory shorthand. In practice it refers to four senior ministers exercising planning functions. In England, there are two: the Secretary of State for the Environment, Transport and the Regions, a post which was established in May 1997, arising from the government's decision to integrate the two Departments of Environment and Transport, and adding a regional dimension to them; and the Secretary of State for Culture, Sport and the Media, who has responsibility for protection of the cultural heritage, including listed buildings and archaeology. In Scotland, these functions were previously conferred on the Secretary of State for Scotland, but now fall within the area of responsibility of the new Scottish Executive; and in Wales, where the functions were previously given to the Secretary of State for Wales, they now fall to the Welsh Assembly.

5 A further constraint comes from the Human Rights Act, which came into force in Scotland and Wales in 1999, and in England in October 2000, which incorporates into domestic law the European Convention on Human Rights. Legislation is now to be read and given effect in a way which is compatible with these Convention rights (s. 3), taking into account the jurisprudence of the European Court of Human Rights and the Commission

(s. 2). If the legislation is found to be incompatible, the courts may make a declaration to that effect (s. 4). This does not affect the continuing validity of the legislation, but provides an expedited procedure for the government to introduce rectifying legislation to Parliament (s. 10). At the time of going to press, the High Court had declared the involvement of ministers in planning decisions to be incompatible with Article 6 of the Convention, and an appeal was pending against that decision.

6 This broader planning process is undertaken by the county councils through structure plans, and the cities were formerly part of these counties. The Local Government Commission insisted that, to minimise the fragmentation that was bound to follow from the establishment of the unitary cities, they should be required to continue as joint partners with the county councils in structure plan responsibilities, but these joint arrangements are necessarily fragile, especially in areas where there is strong territorial rivalry. Many of the unitary cities have tight urban boundaries, and the housing requirements of the economic growth that is taking place within their boundaries can only realistically be met outside the boundaries in areas of other local authorities.

7 The relevant legislation is the Town and Country Planning Act 1990, amended by the Property and Compensation Act 1991.

8 The Town and Country Planning Act (Scotland) 1997.

9 There are over 30 of these statutory instruments presently in force. Their principal effect is to make detailed provision for particular types of control (e.g. tree preservation, outdoor advertising and historic buildings), to prescribe procedures for planning applications and appeals, and to grant substantive development rights (e.g. the Town and Country Planning [General Permitted Development] Order 1995, and the Town and Country Planning [Use Classes] Order 1987).

10 For example, the power to call-in a planning application for determination by the Secretary of State is used annually in England in fewer than 200 cases, out of over 450 000 applications.

11 The Welsh Office, and subsequently the National Assembly for Wales, has since withdrawn from this series, in favour of its own policy guidance. This centres around two principal guidance notes, Planning Guidance (Wales), *Planning Policy* (1996) and Planning Guidance (Wales), *Unitary Development Plans* (1996), together with a series of Technical Advice Notes (TANs) paralleling the English PPGs.

12 For example, the 1996 revision of PPG6 was prompted by a critical review of it by the House of Commons Environment Committee, in *Shopping Centres and their Future* (Session 1993–94; HC 35), which identified a gap between the policies in the then version, and subsequent speeches made by the Secretary of State for the Environment in which he had suggested a further sequential test whereby there should be no out-of-town development unless there was no suitable town centre site. The government's response, *Shopping Centres and their Future* (Cm 2767; February 1995) maintained that the sequential test was already part of current policy, but undertook to revise PPG6 to encourage developers and local planning authorities to identify suitable sites through the development plan process. Over 500 responses were received in the consultation process on their draft revision.

13 These are presently the regional planning conferences, informal groupings of local authorities in each region. Only the South East Regional Planning Conference (SERPLAN) employs full-time staff.

14 Regional planning guidance (RPG) was formerly prepared and published on a non-statutory basis by the Secretary of State based on advice tendered by the regional planning

conference. It provides a framework for the preparation of statutory development plans. Its basis is significantly revised by PPG11 (2000).

15 There are two tiers of statutory development plan in the two-tier local government areas. The structure plan, prepared by the county council, is concerned with the whole of the council's territory, which may comprise several medium-sized towns and cities, and its purpose is to agree a general overall strategy, such as the major communications links which are to be developed, the balance between economic growth and historic conservation, strategic choice for housing land allocation (e.g. as between wholly new settlements and the incremental growth of existing settlements) and the location of employment-generating development. Structure plans are consulted upon in draft before being adopted by the authority. The relevant RPG must be taken into account. If the structure plan fails to give proper effect to the RPG, the Secretary of State may take it out of the county's hands by directing that they modify it, or even call it in for his own decision.

16 The local development plan is normally prepared by the district council for the area. It must be in general conformity with the structure plan. The Secretary of State also has powers to direct the modification or call-in of local plans. In the seven biggest conurbations, there are no structure plans. Instead, there is a special form of RPG, known as strategic planning guidance, and a special form of plan (a unitary development plan [UDP]) prepared by each of the councils in the area. In London, for example, 32 London borough councils plus the City of London each prepare UDPs within the framework of RPG3.

17 Town and Country Planning Act 1990, s. 54A (inserted by the Property and Compensation Act 1991).

18 The current circular is Circular 06/98, *Planning and Affordable Housing.*

19 *Household Growth: Where Shall we Live?* Cm 3471 (November 1996).

20 Most recently the 1998 White Paper, *Planning for the Communities of the Future.*

21 The theme was thoroughly explored by the Environment, Transport and the Regions Select Committee of the House of Commons in its *Tenth Report, Session 1997–98, Housing*, where at paragraph 238 it endorsed the government's view that housing and better planning were essential to urban regeneration and to 'knitting the city' back together. The committee urged the adoption of more challenging targets in regions where there was a large volume of brownfield urban land suitable for development.

22 The government's commitments to reform are spelt out in the White Paper *Modern Local Government: In Touch with the People*, Cm 4014, July 1998.

23 Town and Country Planning Act 1990, s. 12(6).

24 PPG1, *General Policy and Principles* (1992), though the promise was dropped in the revised edition that was published in 1997.

25 Under the Town and Country Planning Act 1990, ss. 18 and 44.

26 This by virtue of the Town and Country Planning Act 1990, s. 54A, inserted by the Property and Compensation Act 1991, which requires that planning applications must be determined in accordance with the development plan, except 'where material considerations indicate otherwise'. Its objective has been to increase the weight to be attached in practice to the development plan, but without making its provisions binding. Its introduction has been accompanied by measures to speed up the preparation and adoption of development plans, so as to ensure that changes in the real world were reflected in plans, rather than simply in ad hoc decision-making. However, performance in that respect has been disappointing, and there is still a significant shortfall in coverage of England by up-to-date development plans.

27 The legislation, the Regional Development Agencies Act 1998, provided the legal framework for the agencies, which then assumed their functions from April 1999.

28 *Modernizing Planning* (A statement by the Minister for Planning and the Regions), DETR, January 1998.

29 *A Mayor and Assembly for London* Cm 3897 (March 1998). The proposals were given effect by the Greater London Authority Act 1999. Elections were held in May 2000, and the mayor and assembly members took office from July 2000.

SIX

NATIONAL-LEVEL PLANNING INSTITUTIONS AND DECISIONS IN THE FEDERAL REPUBLIC OF GERMANY

Gerd Schmidt-Eichstaedt

The Federal Republic of Germany (FRG) is a federation with 16 *Länder* (states) and over 16 000 municipalities which have strong, constitutionally guaranteed powers of self-government. The German spatial planning system is typical of north European planning systems in which there is a hierarchy of authorities, each producing planning instruments, based on a system of participation and consensus. In the German spatial planning system, the laws embody the principles for planning, and some of the plans, particularly at the local level, can also be laws.

An important feature of the German system is the constitution, the *Grundgesetz*, in which responsibilities are strictly divided between the federal government and the *Länder* and in which the autonomy of local self-government is protected. The planning and administrative systems are strongly influenced by the constitutional structure and the divisions of responsibilities within this structure, according to the principle of subsidiarity by which the federal government only has authority for subjects not specifically allocated to the *Länder*. It should be noted that some of the German *Länder* are as large, or larger, in area or population than some European Union member states.

Context

Some basic statistics which are relevant to this description of national-level planning institutions in the Federal Republic of Germany are as follows:

- Germany has a land area of nearly 357 000 square kilometres, of which 12 per cent comprises settlements and roads, 54 per cent agriculture, 30 per cent forests and woodlands, two per cent water and two per cent other areas. Some 48 per cent of the land area of Germany is designated for nature protection and includes national parks, landscape protection areas, nature parks and biosphere reserves.

Figure 1 *The Federal Republic of Germany: The* Länder *(states) and the* Regierungsbezirke *(district administrations)*

▬▬▬ *Land* (state) boundary

——— *Regierungsbezirk* (district administration) boundary

Source: *Statistisches Jahrbuch 1997.*

- In December 1997, the population of Germany was 82 million.
- The average federal population density was 230 persons per square kilometre. This makes Germany one of Europe's more densely populated countries. However, it also includes a wide range of population density ranging from 527 persons per square kilometre in Land Nordrhein-

Westfalen (the Ruhr area) to 78 persons per square kilometre in Land Mecklenburg-Vorpommern.

- The population is forecast to increase to 83.1 million by the year 2000, but only as a result of net immigration. The German population is in decline and is expected to age rapidly.

The legal and political context

Three of the Federal Republic of Germany's 16 *Länder* are 'city states', that is major cities which have been granted the status of *Länder* (Berlin, Bremen and Hamburg). In keeping with the federal principle, the *Länder* are not simply regional administrative units of a central government, but have their own independent sovereignty. They have legislative competence for their own geographic areas and are charged not only with implementing their own laws, but also those of the *Bund* (federation). The division of responsibilities between the *Bund* and the *Länder* is based on the federal constitutional principle that the *Bund* is only allowed to legislate on issues specified in the constitution. In all other cases, legislation is a matter for the *Länder*. However, the administration and the implementation of federal laws are largely the responsibility of the *Länder*. The German constitution makes the following distinctions in allotting legislative competence:

- *Exclusive legislation.* This is the sole responsibility of the *Bund*.
- *Concurrent legislation.* This is the responsibility of the *Länder*, but only in so far as no legislation has been passed by the *Bund* (including land law and local spatial planning).
- *Framework legislation.* Here the *Bund* issues framework regulations or general principles, and each of the *Länder* fills in the detailed regulations through their own legislation, including supra-local spatial planning (*Raumordnung*), water resources and conservation.
- *Exclusive legislation within the competence of the Länder.* Such matters as policing, education and building regulation are the sole responsibility of the *Länder*.

Before taking effect, all federal laws must be reviewed by the *Bundesrat* (federal council), which comprises representatives of the *Länder*. Laws which affect the interests of the *Länder* or of local communities require the approval of the *Bundesrat*. Under the German federal system, the implementation and administration of all federal laws are the responsibility of the *Länder*. The *Bund* is only entitled to maintain its own administration for responsibilities specifically named in the constitution—foreign affairs, customs, defence, and to a lesser extent federal waterways. In other areas, the *Bund*'s administrative apparatus is limited to federal government ministries and their associated departments and subordinate federal agencies. These *Bund* ministries have no independent administrative representation at the regional or local level of the *Länder*.

The *Länder* form the second level of government within the German federal system. Each of the *Länder* has its own elected parliament (the *Landtag*), its own state government (*Landesregierung*) and its own state ministries and departments. Matters of local government are not within federal jurisdiction, but are the responsibility of the *Länder*. Federal legislation is nevertheless binding on all local authorities. The German constitution incorporates a guarantee for local government. The municipalities (cities, towns and villages) have their own elected councils which are responsible for all matters of a local character, including local land-use planning.

The *Bund* is severely restricted by the constitution in the provision of finance for local-level matters. The constitution provides that the *Bund* and the *Länder* have joint responsibility and shared financing for the following specific tasks:

- The construction of higher education facilities;
- The improvement of the regional economic structure;
- The improvement of the agricultural structure; and
- Coastal protection.

In addition, the constitution allows the *Bund* to provide aid for specific projects at the local level. The aid is channelled through and administered by the *Länder*. These projects include:

- Traffic projects;
- Housing programmes; and
- Urban renewal projects.

Supra-local spatial planning at the federal level

Supra-local spatial planning, also referred to as comprehensive regional planning, is essentially the responsibility of the *Länder*. The *Bund* only has the authority to pass framework legislation. At the federal level there is no binding federal-wide spatial plan. In the late 1960s and early 1970s, an attempt was made to introduce a coordinated federal spatial planning programme (*Bundesraumordnungsprogramm*) for the whole country, then West Germany. However, this programme was resisted by the *Länder*, which saw this as an attempt to dilute their autonomy in making decisions.

The *Raumordnungsgesetz* (ROG—Federal Spatial Planning Act) was originally enacted in 1965. It was basically renewed in 1997/98. It is the federal framework legislation for supra-local spatial planning. The ROG aims to create equivalent living conditions throughout the country and incorporates the 'counter-current' principle, whereby lower-level authorities must take account of higher-level plans, and participate in their preparation. The ROG also lays down the organisational and procedural framework under which spatial planning is to be implemented by the *Länder*.

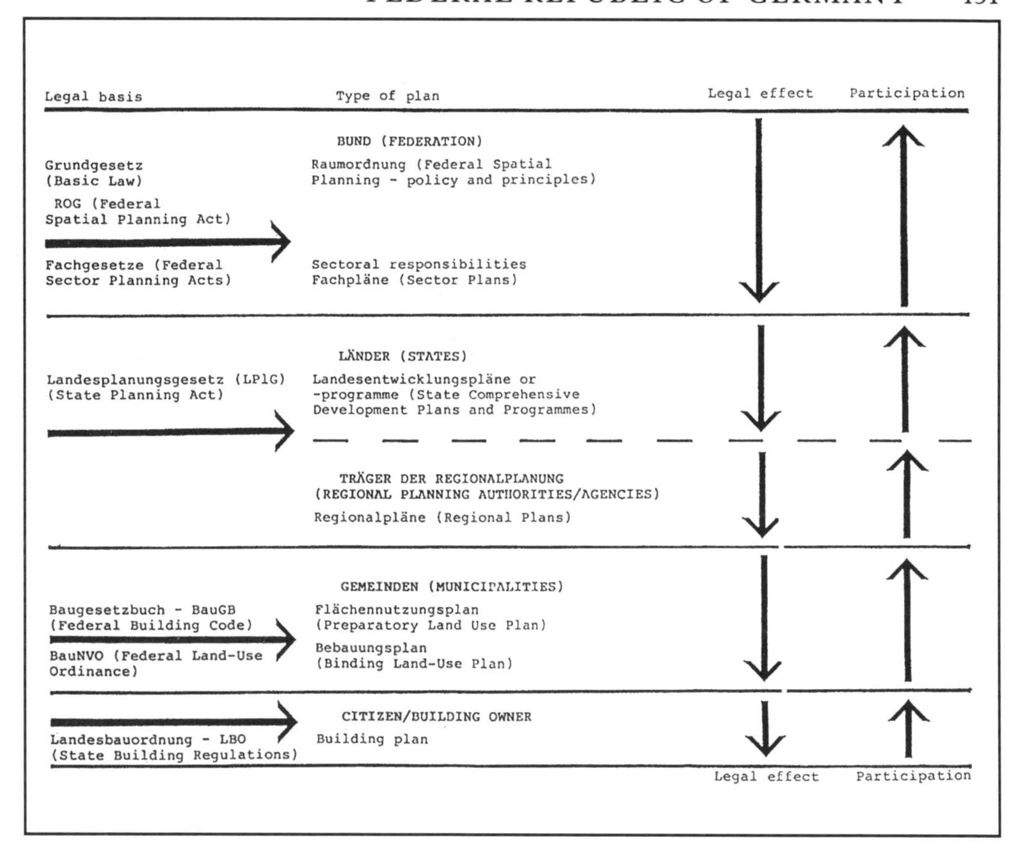

Figure 2 *Planning hierarchy and plans in Germany*
Source: adapted from *Brandenburger Umweltjournal*, No. 6/7, December 1992.

Since 1993, general outlines for future spatial development in Germany have been published by the Conference of *Bund* and *Länder* Ministers responsible for Spatial Planning (MKRO). The *Raumordnungspolitischer Orientierungsrahmen* (guidelines for regional planning) of 1993 and the *Raumordnungspolitischer Handlungsrahmen* (guidelines for the implementation of regional policies) of 1995 outline the prospects, principles and implementation strategies for spatial development in Germany in relation to the following:

- *Settlement structure*: It aims to safeguard and develop the existing system of poly-central settlement structures and to develop urban networks;
- *The environment and land use*: It aims to protect natural resources and to safeguard and improve the quality of the environment;
- *Traffic planning*: It aims to shift freight traffic from road to rail, to extend the high-speed rail network, to provide for trans-european transit routes and to implement integrated traffic concepts in the agglomerations;

- *Europe*: It aims to encourage cross-border cooperation and the implementation of a European spatial planning concept on the basis of the EU documents *Europe 2000* and *Europe 2000+*; and
- *Planning and development*: It aims to bring about equivalent living conditions and to identify areas in need of development and improved planning.

The guidelines incorporate the principles for spatial planning for the entire country, i.e. for supra-*Länder* planning. Within this framework, each *Länder* defines its general targets for development in state-wide comprehensive development plans and programmes. These targets are then made more concrete in the regional plans for the individual regions within the *Länder*. The regional plans, in turn, provide the framework for the preparation of local land-use plans by the municipalities. This multi-level process of planning is undertaken in accordance with the 'counter-current principle' established in the ROG.

Sector planning at the federal level

Fachpläne (sector plans) are prepared by the *Bundesministerien* (federal ministries). They cover the entire geographic area of the FRG and are kept continuously under review. The German constitution entrusts the preparation of sector plans in many areas of activity to the *Bund*. These include federal highways, federal railways, federal waterways, air transport and defence. The sector plans are bound by their respective 'sectoral legislation' and are coordinated between the *Bund* and the *Länder*, usually by means of ministerial committees. The federal sector plans must fit in with *Bund* and *Länder* spatial planning.

The most important sector plan in relation to spatial planning is the *Bundesverkehrswegeplan* (BVWP—federal transport infrastructure plan). The current BVWP dates from 1992. It was prepared by the Federal Ministry for Transport and sets out the *Bund*'s agenda for the planning of federal highways, railways and waterways. The sector plans are legally binding in the sense that their provisions constitute a framework to be followed by the *Länder* in the preparation of their own state-wide plans and programmes. Thus, for example, the BVWP schematically illustrates the federal railways to be built or improved, but does not pinpoint their exact location (see Fig. 3).

Federal input in the main policy areas influencing spatial planning

The following section describes the input of the *Bund* in the most essential areas of policy which influence spatial planning and in turn are influenced by it. These policy areas are:

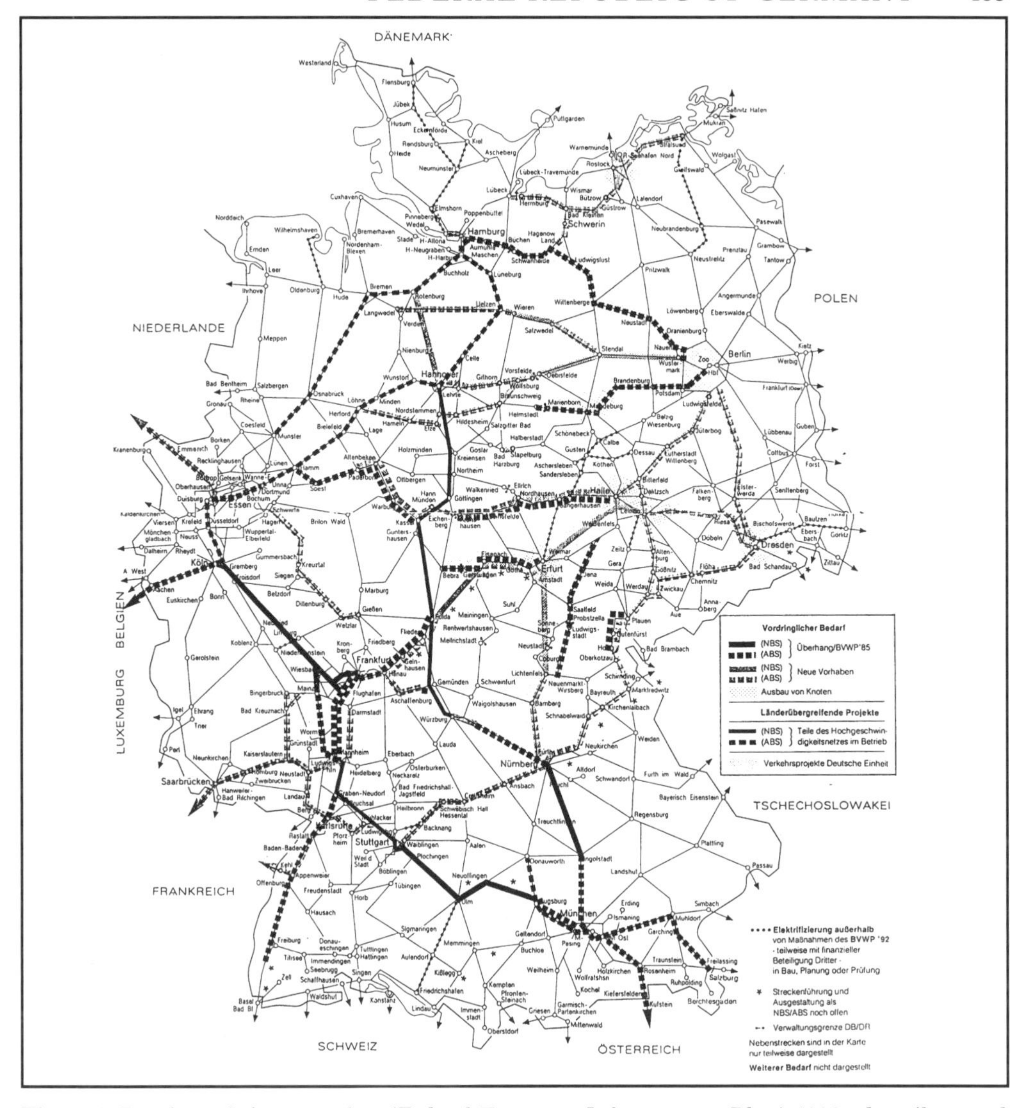

Figure 3 Bundesverkehrswegeplan *(Federal Transport Infrastructure Plan) 1992: the rail network*
Source: Bundesministerium für Raumordnung, Bauwesen und Städtebau, *Raumordnungsbericht 1993*, Bonn, 1994, p. 108.

- Regional economic development;
- Building, housing and urban development;
- Transport;
- Environmental protection, waste management and pollution control;
- Natural resources;
- Leisure and tourism; and
- Heritage conservation.

REGIONAL ECONOMIC DEVELOPMENT

Since 1969, regional economic development policy, which includes industrial and employment promotion policy, has been undertaken jointly by the *Bund* and the *Länder*. Under the Joint Task for the Improvement of Regional Economic Structure, an annual investment programme of financial subsidies is prepared by a committee of the *Bund* and *Länder* Ministers of Economics. The programme is administered by the *Länder*, each of which selects projects proposed by the municipalities, in accordance with the financial allocations made available under the programme to each *Land*. The projects selected must comply with the *Länder* spatial plans.

Subsidies under the regional economic programme are available for infrastructural improvements undertaken by the municipalities and private investors to encourage business expansion. The maximum level of subsidy is available to Assisted Areas and Promotion Areas designated under EU structural policy, i.e. the new *Länder* and areas of the older *Länder* affected by structural change.

A second major source of funding for regional economic development in the new *Länder* is provided jointly by the *Bund* and older *Länder* under the *Aufbau-Ost* Programme 1995–2005. Funding under this programme is provided to municipalities and private investors for:

- Improvements to the economic structure;
- Housing construction and urban development;
- The expansion of employment and training;
- Community and social facilities; and
- Research and development.

The *Bund* is also a significant investor in research and development by research institutions and private companies. Policy on research and development is monitored by the Federal Ministry for Research and Technology, which presents a *Bundesforschungsbericht* (federal research report) to the Federal Parliament at regular intervals, the last in 1996. This report describes the achievements in research, technology and innovation and presents the *Bund*'s policies in these areas.

Following reunification, the *Bund* established a special agency, the *Treuhandanstalt*, in 1990 to manage the privatisation of the former state-controlled industrial, commercial and agricultural enterprises in the new *Länder*. This semi-public agency, which reported directly to the Federal Ministry of Economics, completed the privatisation process at the end of 1994, and has since ceased operations.

Economic policy is monitored by the committee of *Bund* and *Länder* Ministers for Economics. Fiscal policy was the responsibility of the *Bundesbank* (federal central bank), which was granted a high level of independence by the German constitution. Together with the introduction of a new currency in the European Union, the Euro, the *Bundesbank* was replaced by the European Central Bank.

BUILDING, HOUSING AND URBAN DEVELOPMENT

In 1998 the Federal Ministry for Spatial Planning Building and Urban Development was united with the Federal Ministry for Transport. Since then the *Bundesministerium für Verkehr, Bau- und Wohnungswesen* (BMVBW) has been the federal ministry responsible for spatial planning, building and urban development. Its principal tasks include:

- The framework legislation on supra-local spatial planning—the Federal Spatial Planning Act (ROG), referred to above;
- The preparation and supervision of federal legislation on local land-use planning, planning permission, development control and measures for the implementation of urban development, which are incorporated in the Federal Building Code—the *Baugesetzbuch* (BauGB);
- Federal housing policy and the preparation of federal framework legislation for social housing construction and the promotion of housing construction in general;
- The provision of federal financial aid for housing programmes, urban development and urban renewal; and
- The coordination of spatial planning policy and other sectoral planning policies.

The Federal Agency for Construction and Spatial Planning (*Bundesamt für Bauwesen und Raumordnung*—BBR) is an independent federal research agency, subordinate to the BMVBW. It reports on spatial developments in Germany, provides information and statistics for spatial planning research and advises the Federal Minister for Spatial Planning. The BBR also provides specialist advice to the BMVBW, the *Länder* ministries and the municipalities on the implementation of model urban development and renewal projects. The BBR is also responsible for the implementation of all building matters on behalf of the *Bund*, including the provision of buildings for the *Bund* (i.e. the Reichstag in Berlin) and for the German embassies in other countries.

TRANSPORT

The Department for Transport (within the BMVBW) is responsible for:

- Framework legislation on transportation;
- Federal highways, federal (inter-city) railways, federal waterways and air transport policy;
- The preparation of the Federal Transport Infrastructure Plan (BVWP), referred to above; and
- The provision of federal aid for municipal transportation projects.

Within this context, it must be emphasised that while the Federal Ministry is responsible for the federal highway network, the actual improvement and maintenance of the highways are undertaken by the *Länder* on behalf of the Federal Ministry.

The Federal Ministry as a whole has departments covering central administration, housing, building and urban developments, spatial planning, framework transport policy, railways, road traffic, air transport, sea transport, inland waterways and road construction. There are also a large number of federal agencies subordinate to the Ministry, including those for the weather service, inland waterway construction and administration, road traffic research, the federal railway agency and the German authority for air traffic control.

ENVIRONMENTAL PROTECTION, WASTE MANAGEMENT AND POLLUTION CONTROL

In the Federal Republic of Germany, the policy areas of the environment, nature protection, waste management and pollution control are more closely integrated than in a number of other European countries. This is highlighted by grouping these policy areas together under the aegis of the Federal Ministry for the Environment, Nature Conservation and Nuclear Safety (BMU). The main tasks of this ministry include:

- The preparation and supervision of federal legislation on pollution control—the *Bundesimmissionsschutzgesetz* (BImSchG—Federal Control of Pollution Act);
- Air-quality control (Lead-in-Petrol Act);
- Control of hazardous substances and chemicals (Federal Chemicals Act);
- The preparation and supervision of federal framework legislation on nature protection—the *Bundesnaturschutzgesetz* (BNatSchG—Federal Nature Protection Act);
- Environmental impact assessment legislation;
- The protection of water resources and forests;
- Waste disposal and federal framework waste disposal legislation;
- Nuclear safety and the storage of nuclear waste; and
- The provision of federal aid for coastal protection.

The framework legislation on nature protection and waste disposal lays down the principles to be observed by the *Länder* in the preparation of their own state legislation on these matters. The Federal Control of Pollution Act provides the legal regulations covering the permissibility of industrial and commercial facilities which may cause pollution. These regulations are administered by the factory inspectorate authorities of the *Länder*.

In 1990, the BMU introduced a system of priority ecological improvement measures for the new *Länder*, to be implemented in conjunction with the joint *Bund/Länder Aufschwung Ost* economic development programme, mentioned above. The priority measures include the improvement of drinking water, the renewal and conversion of power stations, the closure of hazardous waste disposal sites and

Table 1 *Contents of the Federal Environmental Agency's Annual Report 1995*

Chapter 1:	The Federal Environmental Agency
Chapter 2:	Environmental Information and Reporting
Chapter 3:	Information Techniques and Documentation
Chapter 4:	Environmental Research and Model Projects
Chapter 5:	Global and International Environmental Protection
Chapter 6:	Instruments and Strategies
Chapter 7:	Ecological Effects and Quality Objectives
Chapter 8:	Water
Chapter 9:	Land
Chapter 10:	Air
Chapter 11:	Noise
Chapter 12:	Agriculture
Chapter 13:	Transport
Chapter 14:	Environment and Energy
Chapter 15:	Basic Issues Relating to Technical Procedures and Products
Chapter 16:	Environmentally Sustainable Technical Procedures and Products
Chapter 17:	Disposal Procedures
Chapter 18:	Assessment of Materials/Substances and Implementation
Chapter 19:	Biological and Gene Technologies

improvements to the water supply, sewage treatment and waste disposal infrastructure. A joint *Bund/Länder* programme for the identification and removal of dangerous industrial waste in the new *Länder* is also being coordinated by the BMU, as is a regional reclamation project for former military sites in the new *Länder*.

The *Umweltbundesamt* is the federal environmental agency, which although supervised by the BMU is relatively independent of it in terms of its operations. The Federal Environmental Agency advises the *Bund* and *Länder* on the scientific aspects of environmental policy and monitors policy implementation. The Agency also provides information to the public on environmental matters and publishes annual data on the environment. It publishes the *Umweltbericht* (environment report) at regular intervals, along with a comprehensive yearly update (*Jahresbericht*), which provides a review of all aspects of environmental policy (see Table 1 for contents of the yearly report).

NATURAL RESOURCES

Policy on natural resources includes not only naturally occurring raw materials, of which, apart from coal, there is relatively little in Germany, but also water resources and agriculture. The German coal-mining industry is subsidised by financial measures agreed to, between the EU and the *Bund*, and between the *Bund* and the coal-mining *Länder*, the mining companies and the mining trade unions (the *Kohlerunde*). Since coal-mining is a declining industry in Germany, the *Bund* and the *Länder* operate a joint programme for providing alternative employment and retraining former coal miners. Detailed mining policy and the identification of underground and open-cast coal-mining areas are integrated with the spatial plans of the *Länder*.

The *Bund* and the Federal Ministry for the Environment, Nature Conservation and Nuclear Safety (BMU) initiate legislation on water resources, which includes:

- The Federal Water Resources Management Act, which provides a licensing system for water usage and makes provision for the designation of Water Protection Areas; and
- The Drinking Water Ordinance, which regulates the supply and quality of drinking water.

There is no federal-wide water supply plan, since the supply of water is the responsibility of the *Länder* and their constituent municipalities. The *Länder* are also responsible for the management of water resources and the designation of Water Protection Areas in their spatial plans.

Agriculture and forestry, by far the largest land users, together account for 84 per cent of the land area of Germany. Agricultural and forestry policy is heavily influenced by the policies of the EU, which have been worked out by the *Bund* and the national governments of the other member states. The framework policy on agriculture and forestry is implemented by the *Bund* and the Federal Ministry for Food, Agriculture and Forestry (BML).

The *Bund* and *Länder* Ministers for Agriculture prepare an annual joint programme for the 'Improvement of Agricultural Structure and Coastal Protection', within the framework of the EU agricultural and structural policies. This joint programme also designates 'disadvantaged areas', including mountainous regions, and areas of poor land quality and of small landholdings, where financial aid is provided for improvements in agriculture and forestry. The main elements of this programme include:

- The transformation of agriculture in the new *Länder* to make it operate on a market-oriented footing;
- Measures and aid for the withdrawal of land from agricultural production;
- Financial aid to encourage the use of less intensive farming methods and for diversification on farms, for example tourist projects;
- The identification of new agricultural products and markets;
- The restructuring of agricultural landholdings and aid for flood protection measures; and
- Financial aid for infrastructural improvements in rural areas and for the renewal of rural settlement.

As in the other joint programmes, the *Länder* have the responsibility for the detailed planning and administration of the programme, within the agreed framework.

LEISURE AND TOURISM

The leisure and tourism policy is essentially the responsibility of the *Länder*. *Bund* policy is restricted to the promotion of leisure and tourist development under the joint *Bund/Länder* programme for the improvement of regional economic structure.

HERITAGE CONSERVATION

Heritage conservation policy is also the responsibility of the *Länder*. However, the *Bund* has established a joint scheme with the new *Länder* to save the large number of relatively intact, historic town centres in the new *Länder* from further decay, after 45 years of neglect. The main emphasis of this programme is to protect urban monuments and secure and preserve historic town centres. The first phase of the programme covered 220 towns whose centres have, at least, national importance. The programme is coordinated by the Federal Ministry for Transport, Building and Housing (BMVBW) and is administered by the State Monuments Office in each of the *Länder*.

The general urban renewal policy of the *Bund* and the BMVBW also promotes the conservation of central areas of towns and villages throughout Germany. Within this policy, the BMVBW operates a special Promotion Programme for Model Urban Renewal Projects in a number of towns and villages in the new *Länder*. The model projects are mainly funded by the BMVBW from its urban renewal budget.

The coordination of spatial planning policy with other sectoral policies

THE *RAUMORDNUNGSBERICHT*

The *Raumordnungsbericht* (Federal Spatial Planning Report), which is prepared at regular intervals by the Federal Agency for Construction and Spatial Planning (BBR), is the most comprehensive report on spatial planning and other sectoral planning policies. The preparation of the spatial planning report is a requirement of the Federal Spatial Planning Act (ROG) and is presented to the *Bund*.

German reunification and European integration have led to the publication of four reports since 1990. The spatial planning reports of 1993 and 2000 are the latest and most extensive. They cover the statistical background, problems, trends and policies of all subject areas that have an impact on spatial planning (see Table 2). They also:

- Include all recommendations of the Spatial Planning Advisory Council and the decisions of the Conference of *Bund* and *Länder* Ministers for Spatial Planning;
- Report on the most important economic and development trends and problems in Germany;
- Point out potential future problems facing spatial planning and development;

Table 2 *Contents of the* Raumordnungsbericht *(Federal Spatial Planning Report) 1993*

Part 1: Framework Conditions for Spatial Development in Germany	
Chapter 1:	Fundamental Changes in the Initial Situation
Chapter 2:	Spatially Relevant Developments
Part 2: Spatial Structure-Settlement System and Spatial Use	
Chapter 3:	Settlement System
Chapter 4:	Land Use and Land-Use Conflict
Chapter 5:	Locational Factors
Part 3: Selected Spatial Problems in the New *Länder*	
Chapter 6:	Regional Development Potential and Bottlenecks
Chapter 7:	Regional Employment Situation
Chapter 8:	Internal Migration
Chapter 9:	Regional Case Studies
Chapter 10:	Regional Planning Problems in Berlin-Brandenburg
Part 4: Policy Areas Influencing Spatial Development	
Chapter 11:	Finance Policy—Overall Situation
Chapter 12:	Economic Policy
Chapter 13:	Employment Policy
Chapter 14:	Agricultural and Forestry Policy
Chapter 15:	Urban and Village Development
Chapter 16:	Housing
Chapter 17:	Transport, Telecommunications and Post
Chapter 18:	Supply and Disposal (Energy, Water, Waste Disposal)
Chapter 19:	Research and Technology
Chapter 20:	Education
Chapter 21:	Health Facilities
Chapter 22:	Environmental Protection and Reclamation
Chapter 23:	European Spatial Planning Policy
Chapter 24:	Spatial Planning Cooperation with Neighbouring Countries and Cross-Border Regional Promotion

- Include all spatial planning policies and changes in legislation affecting spatial planning issues; and
- Provide an important and comprehensive information base for political decision-making.

THE *RAUMORDNUNGSBEIRAT*

The *Raumordnungsbeirat* (Spatial Planning Advisory Council) is a statutory advisory council established under the Federal Spatial Planning Act to advise the Federal Minister. The advisory council is made up of representatives and expert advisers from:

- The organisations representing the municipalities;
- The spatial planning ministries of the *Länder*;
- Urban development interests;
- The fields of science and technology, economics, agriculture, nature conservation and sport; and
- The employers' organisations and the trade unions.

The Advisory Council meets regularly and makes recommendations to the Federal Minister on the guiding principles for spatial planning at the federal level.

The monitoring and integration of transport and of environmental and spatial planning policies is undertaken by a conference of the *Bund* and *Land* ministers responsible for these areas. The decisions of this ministerial conference (The Krickenbeck Declaration, 1992) were included in the Guidelines for Regional Planning, 1993.

The planning of large-scale leisure and tourist facilities requires coordination with *Länder* spatial planning, in accordance with a 1992 decision of the *Bund* and *Länder* Ministers for Spatial Planning (MKRO). Because of their considerable impact on nature and the countryside, large-scale facilities, such as golf courses, marinas and adventure parks, must be coordinated with other sectors and require:

- An environmental impact assessment; and
- A *Raumordnungsverfahren* (spatial planning procedure), which is an assessment instrument to ensure a proposed project complies with the spatial plans of the *Länder*.

Financial planning

The *Stabilitätsgesetz* and the *Haushaltsgrundsätzegesetz* (framework for budget legislation) form the basis of budgetary law, which requires the *Bund*, *Länder* and the local authorities to undertake medium-term investment and financial planning. A *Finanzplan* (financial plan) is prepared for a period of five years, comprising:

- All anticipated spending on investment and other expenditures; and
- The income from various sources, which are to cover this expenditure.

The financial plans are to be extrapolated each year. Since the current year and the following year are included in the usual budget plan, the financial plan only extends the planning horizon by three years. This is a very short timescale for projects like road construction. Therefore, a number of these are planned for longer periods.

The finance plans are not binding. They are presented to the parliaments, but, unlike the budget plan, are not adopted by parliamentary resolution. Subsequent budget and finance plans can also deviate from the previous finance plans. Indeed, this happens quite frequently. The functions of multiple-year financial plans are:

- To ensure a future-oriented budgetary policy;
- To coordinate the objectives and timescale of government actions so that they fit in with the economic activities of other institutions such as the European Central Bank and with private interests; and
- To facilitate private economic planning by means of the regular publication of precise quantitative information pertaining to the government's financial plan, while at the same time placing some limits on the realisation of individual interests.

In order to support the preparation of medium-term plans, the *Finanzplanungsrat* (Finance Planning Council) was established in 1968. The Council comprises the Federal Ministers of Finance and Economics, the *Länder* Ministers of Finance and four representatives from local authority associations.

The Finance Planning Council has developed a uniform scheme for reporting the finance plans of the *Bund*, *Länder* and municipalities, in which the figures in each plan are set out so that they can easily be compared. In addition, the Council:

- Provides economic benchmark figures for the preparation of the budget and finance plans and their extrapolation;
- Coordinates plans at each level of authority; and
- May make recommendations, although these are not binding on the individual authorities.

The main problem of medium-term financial planning is the difficulty of making predictions about basic development trends in national economics and finance, and connecting these with income and expenditure groupings. This is also true for the financial subsidies and allocations which public authorities grant to each other.

Finance planning as resource planning requires comprehensive overall planning, which, as yet, does not exist in Germany. The coordination of financial planning is made more difficult by the federal structure of the country. According to the constitution, the budgets of the *Bund* and *Länder* are independent of each other. However, in practice this is only partially so because of the broad interlinkage between income, expenditure and tasks.

To date, there is a wide discrepancy between the ideas of the Finance Planning Council, the actual budget and the finance planning in progress. Nevertheless, many of the shortcomings of public financial management have been overcome by financial planning. Nowadays, nearly all investment planning, long-term financing decisions and the setting of priorities take the resulting costs into account.

Evaluation

Long-term planning not only depends on the efficiency of the institutions set up to carry out evaluation but also on objective and subjective factors, which are at least as important. *Objectively*, the following is true. Planning assumes that one can make predictions. If the conditions change unexpectedly, planning loses its basis. It is overtaken by events and relegated to the sidelines. This does not mean that planning is completely worthless in retrospect, but it does lose its effectiveness for the future. The difficulty and uncertainty involved in making predictions are illustrated not only by such unusual events as the fall of the Berlin Wall and the reunification of Germany, but also in less eventful times. The experiences in the Federal Republic of Germany demonstrate that high-level, integrated planning should not seek to be a more-or-less binding instrument for directing the relevant decision-making institutions. Rather, it should only put forward general objectives. I submit that an up-to-date report like the Federal Spatial Planning Report, with detailed information and statistics, prepared at regular, short intervals, is equally as important as the preparation of plans and programmes. A good and objective report enables its readers to draw the right conclusions.

Following on from the objective factors, there are the *subjective* factors. The control of the future by plans and programmes assumes that the public are prepared to allow themselves to be steered by such plans and programmes. This readiness of the public is not just a matter of public participation in the preparation of plans—that is, participation in the formulation of objectives with the aim of increasing the acceptance of the plans—but is also dependent on the spontaneity and temperament of the people. Do they tend to abide by norms and laws or do they tend towards individuality and a disregard of the law?

The German people are still relatively law-abiding and 'obedient'. Social conflict is dealt with by working towards consensus and not by perpetuating confrontation. This is clearly seen in the relationship between employees and employers and their representatives in the trade unions and employer associations. On this basis, there is a willingness, in principle, to accept plans; for example the *Bebauungspläne* (binding land-use plans) or, in terms of the municipal–state relationship, the *Regionalpläne* (regional plans).

However, spatial plans in general have only a limited role. This role becomes more limited, the further the planning level is from the persons affected. Federal spatial planning is of marginal interest to most people, including politicians. It has little observable impact at the level of the individual. This is also true for the development plans and programmes of the *Länder*.

By contrast, the 'hard' sector plans such as those for motorways, airports, railways, inland waterways, power stations and waste incineration sites are of interest. The construction of universities (a *Länder* responsibility), nature protection areas (*Länder*), the conservation of monuments, coastal protection and the protection and improvement of rivers and lakes are of a more local concern. The physical plans for these subjects are far more interesting to the people and the politicians as 'political plans' than general spatial planning, regional planning and the medium-term financial plans which everybody knows may be changed later.

The problems in implementing urgent projects

The problem of implementing plans is most clearly shown by the 'hard' sector plans mentioned above. Spatial plans, regional plans and also the preparatory land-use plans (*Flächennutzungspläne*) are only 'internally' binding on the administrative authorities. They are implemented in accordance with the hierarchical principle in which the lower authorities must follow the instructions of the higher authority.

However, this is somewhat different in the case of 'hard' sector plans, which require the provision of areas of land for their implementation, and these plots of land are mostly in private ownership. These projects also have neighbours, who may be adversely affected by pollution such as aircraft noise, potentially harmful emissions and radiation. These neighbours will use every available appeal process and delaying tactic to fight the plans for such projects. The reaction of the legislature is to restrict the possibilities for appeal and to limit the length of the appeal process. Appeals against building permission have had no stalling function since 1998, and since 1997 applications for a judicial review of the regulations included in *Bebauungspläne* (binding land-use plans) must be made within two years of the official publication of the plan.

However, planning decisions in Germany are often delayed, and in some cases overturned, in the courts. Plans and planning decisions are carefully examined by the courts. They scrutinise the material legitimacy of the plan and the planning decision. This means that planning decisions must be carefully thought through, but it also results in a slowing down of the planning process. Because of this, there may indeed be good reason to reduce the number of stages in the appeal process. At present, normal planning decisions are reviewed in the following stages:

- 1st stage: Appeal against the decision to the authority which made the decision.
- 2nd stage: If the original authority does not grant the appeal, the appeal must be submitted to the next higher authority. This authority then issues a decision on the appeal.
- 3rd stage: Appeal against the administrative act, in the form of an administrative appeal on the decision to the local Administrative Court.
- 4th stage: Appeal against the decision of the Administrative Court to the Higher Administrative Court.
- 5th stage: Application for a review of the decision of the Higher Administrative Court to the Federal Administrative Court, if allowed; if not, an action for non-allowance is brought.
- 6th stage: Constitutional action against the decision of the courts to the Federal Constitutional Court.
- 7th stage: Legal action before the European Court.

Concurrently with the various stages of this process, one may also take the following steps:

- The procedure of interim injunction, in order to halt construction, is undertaken together with the main appeal procedure in all cases. This procedure is begun in the administration and can be pursued in up to two stages—the Administrative Court and the Higher Administrative Court;
- The procedure of direct judicial review can be instituted against plans which comprise regulations and local statutes (i.e. all regional plans and binding land-use plans) in up to two stages—the Higher Administrative Court and the Federal Administrative Court).

The many attempts at reforming this process have met with limited success. The public are very sensitive to any changes, because they consider the possibilities of appeal and the independent control function of the courts to be the cornerstone of the state under the rule of law.

Conclusion

The planning and decision-making systems in the Federal Republic of Germany are influenced by the federal structure and the relatively strong position of the municipalities within the context of local self-government. The top level of this system comprises the Federal Republic, that is the Federal State. This is followed by the 16 *Länder* (states). The *Länder* are divided into *Regierungsbezirke* (district administrations). Below these are the local authorities, which are of two kinds: the 115 *kreisfreien Städte* (independent cities) and the 322 *Landkreise* (counties), with about 16 000 *kreisangehörigen Gemeinden* (municipalities belonging to a county). The independent cities have such a strong administration that they require no other local administrative support. The *kreisangehörigen Gemeinden* are small municipalities which require institutional support. This is provided by the counties.

The counties serve as a link between the state administrative system and the system of local government. The *Landrat* (county council) serves both the state (the *Land*) and the municipalities which belong to the county. The counties are a typically German institution and do not exist in this form in any other administrative system. They are not similar to the county councils in England, but are a combination of local government and state-internal administrative levels.

The system of land-use planning and spatial planning is organised by two federal acts: the *Baugesetzbuch* (Federal Building Code), and the *Bundes-Raumordnungsgesetz* (Federal Spatial Planning Act). The Federal Building Code unifies the procedures, not the contents, of local land-use planning. In Germany, the cities and smaller municipalities are responsible for local land-use planning, not the *Länder*. Thus, the Federal Building Code is not aimed at the *Länder*, but directly at local government, i.e. the independent cities and the other smaller municipalities. Each city, town and village must prepare its own local land-use plans, the *Flächennutzungsplan* (preparatory land-use plan) and, where necessary, the *Bebauungspläne* (binding land-use plans). The Federal Building Code does not prescribe any specific contents for these plans, but it provides the limitations and the procedural rules for their preparation. The preparatory land-use plan is a local, administrative-internal plan, without external binding powers. Binding land-use plans are issued as legally binding local statutes and make up a uniform building law.

The Federal Spatial Planning Act is aimed only at the *Länder*. It is a framework act and includes only broad guidelines for supra-local, spatial planning. The *Länder* are authorised and required to fill out the federal framework in their own spatial planning legislation and to incorporate the details. The Act requires the *Länder* to undertake *Landesplanung* (state-wide spatial planning). With the help of state-wide spatial planning, the *Länder* provide the organisational framework for local land-use planning at the city and municipal level. They must ensure that local land-use planning is coordinated at the supra-local level. Thus, each of the 13 *Länder* that are not city-state *Länder* (Berlin, Bremen and Hamburg) have their own *Landesplanungsgesetz* (state spatial planning act). The acts generally provide for two types of planning instrument: the *Landesentwicklungsprogramm* (state development programme), which presents a very broad spatial organisation for the entire state, and *Regionalpläne* (regional plans), which cover the spatial proposals at the level above municipal planning and below state-wide planning.

The regional plans comprise a written presentation of the system of central places (higher-order centres, middle centres and lower centres). They include the subdivision of the region for particular purposes (recreation, industrial locations, housing, etc.) and they broadly identify the areas for which new development is envisaged. Thus, they provide a systematic framework within which local land-use plans are to be prepared by the cities and municipalities.

Thus, the Federal Republic of Germany has two systems of spatial planning. One system, which focuses on the local level, is independent in its implementation at this level and is governed by the Federal Building Code. This is a 'bottom-up' system, because the local land-use plans are prepared at the lower level. This system is known as *Bauleitplanung* (local land-use or building guidance planning). The second system is a 'top-down' one and is known as *Raumordnung und Landesplanung* (spatial organisation and state spatial planning). Federal spatial planning by the *Bund* and state spatial planning by the *Länder* set the requirements from above for the cities and smaller municipalities. These systems are linked with each other by the 'countercurrent principle'. The 'top-down' planning from above must allow for participation by the cities and municipalities. The cities and smaller municipalities must ensure that their local plans fit in with the objectives of federal and state spatial planning.

This system of integrated spatial planning is extended by sector planning at the different levels. Responsibility for sector planning generally follows the same pattern as that for other plans, at the various levels, in particular the legislative levels. The *Bund* is responsible for (among other things):

- The federal transport routes on land (roads, railways) and on water (inland waterways);
- Telecommunications (post, telecom);

- Energy (including nuclear energy);
- Waste disposal; and
- Air pollution control.

Federal spatial planning, as integrated planning, is a relatively weak instrument, especially when compared with sector planning. It functions best in the form of an informal, non-binding publication—the *Federal Spatial Planning Report, Guidelines for Regional Planning*. The *Länder* are responsible for (among other things):

- The law on building regulations;
- The protection of monuments;
- The detailed law on nature and countryside protection; and
- Higher education and schooling.

State spatial planning, particularly in the form of regional plans, is binding on the cities and other municipalities. The cities and smaller municipalities are responsible for matters affecting the local community, but only within the legislative framework. In order to organise local matters, the cities and other municipalities may issue local statutes (e.g. binding land-use plans) and deliver administrative rulings and specific regulations.

This system of coordinated division of regulatory authority and the power to implement (including spatial planning) functions well in principle. All three levels—*Bund*, *Länder* and the municipalities—interact fairly well with one another. No system-threatening, constitutional conflict has occurred as yet between the three levels. There is another power broker, the judiciary, which functions as the third authority. The courts in the Federal Republic of Germany occupy a very powerful position. This can, in individual cases, lead to planning processes being drawn out over long periods or being halted completely. To date, no effective reform has dealt with this problem.

SEVEN

NATIONAL-LEVEL PLANNING IN THE DANISH SYSTEM

Stig Enemark and Ib Jorgensen

A little Danish planning history

Denmark covers a land area of 43 000 square kilometres, not including the self-governing regions of Greenland and the Faroe Islands. The country's population is 5.2 million, of which one-third lives in the capital region—Greater Copenhagen. Sixty-seven per cent of the country is devoted to agriculture, 12 per cent to forests, 11 per cent to semi-natural areas and 10 per cent to urban zones and transport installations.

The first Danish town planning act was passed in 1925. It was, however, not widely applied because the planning regulations entailed an economic risk since the municipal councils were held responsible for paying compensation. In 1938 a new planning act was passed which obligated the municipal councils to adopt a town plan for any built-up area with more than 1000 inhabitants. The town plan was subject to the approval of the Minister of Housing. A considerable number of town plans were then adopted, as the planning regulations did not require the municipal councils to pay compensation to landowners. The plan, however, only related to towns and there was no legal basis for ensuring the separation of urban and rural areas or for limiting urban growth. The resulting urban sprawl created the need for a bill, passed in 1949, which set up urban development committees for all expanding urban districts. The committees provided urban development plans, which divided the expanding areas into zones and preserved areas of open country. This forms the basis for the present zoning division of the whole country.

As in many other countries, urban development in Denmark accelerated between 1945 and the mid-1980s. The resulting new urban districts, which include residential and commercial areas and centre and service functions, now encompass 75 per cent of the developed urban land and half of the population. This huge urban development was regulated mainly by the zoning provisions, and by the urban development committees. The municipalities developed master plans for the cities and towns, and district and regional plans were voluntarily prepared in several areas.

These master plans indicated the need for overall comprehensive planning and led to the planning law reform of the mid-1960s. This was based on the local government reform which came into force in 1970 and reduced the number of counties from 25 to 14 and the number of local authorities from 1388 municipalities to 275. The reform created the basis for transferring a number of responsibilities and powers to the counties, and especially to the municipal councils, thus bringing about decentralisation.

The planning law reform was implemented between 1970 and 1977 and included the Urban and Rural Zones Act (1970), the National and Regional Planning Act (1973), and the Municipal Planning Act (1977). A number of acts were then repealed and many administrative agencies abolished, including the urban development committees. The planning law reform was reviewed and revised and was finally integrated into a single planning act which came into force in 1992.

A very important innovation in this new law was that one of the main objectives of planning was changed from uniform and equal development of the country to appropriate development (see below). Fifteen years of experience and new environmental challenges helped to simplify and modernise the legislation. The result is a planning system with not only a highly political and effective planning process, but also a planning system where the quality of the plans depends on the quality of the political and democratic process at the local level. The planning process, then, in a way expresses the strengths and weaknesses of the democracy. Recently, the Danish planning system has been challenged by new developments resulting from Denmark's joining the European Union and particularly from the Maastricht Treaty. Some of these will be dealt with below.

The Danish planning system

Danish planning law delegates responsibility for spatial planning to the Minister for the Environment, the 14 county councils and the 275 municipal councils. The key feature of the planning system is the requirement that both the regional (county) and local (municipal) authorities draw up, adopt and revise comprehensive structure plans and a set of land-use regulations, covering all of their respective areas. This means that since 1980 we have two sets of new, comprehensive plans covering the entire country. The plans are revised every four years.

The declared objectives of Danish planning law are to ensure that the overall planning, future spatial structure and land use synthesise the interests of society and contribute to the protection of the country's natural resources and environment, so that sustainable development particularly with respect to people's living conditions is achieved. Obviously, the planning process is highly political since it is concerned with shaping the future human environment through public debate and the balancing of different interests. Table 1 summarises the key elements of the Danish planning system, with particular emphasis on the national level, showing how that is linked with the two lower levels—the regional and the local. Figure 1 presents diagrammatic examples of the four levels of plans.

Table 1 *The Danish planning policy framework*

Policy institutions			*Policy instruments*		
Level	*Planning authority*	*Number of inhabitants*	*Type of plans*	*Description*	*Legal power*
1. National	Ministry of the Environment, Department of Spatial Planning	5 million	*Landsplan-perspektiv* (National Planning Perspective)	Policies, maps and general guidelines	Advisory guidelines
			Landsplan-redegörelser (National Planning Reports)	Written statements	Advisory guidelines
			Landsplan-direktiver (National Planning Directives)	Maps and legal provisions/circulars	Binding for the regional and local authorities
2. Regional	14 county councils	about 350 000 on average	*Regionplaner* (Regional structure plans)—revision every 4th year	Policies, maps and land-use guidelines	Binding for the regional and local authorities
3. Local	275 municipal councils	about 20 000 on average (wide deviations)	*Kommuneplaner* (Municipal structure plans)—revision every 4th year	Policies, maps and land-use regulations	Binding for the local authorities
			Lokalplaner (Binding local/neighbourhood plans)	Maps and detailed legal land-use regulations	Binding for the landowners

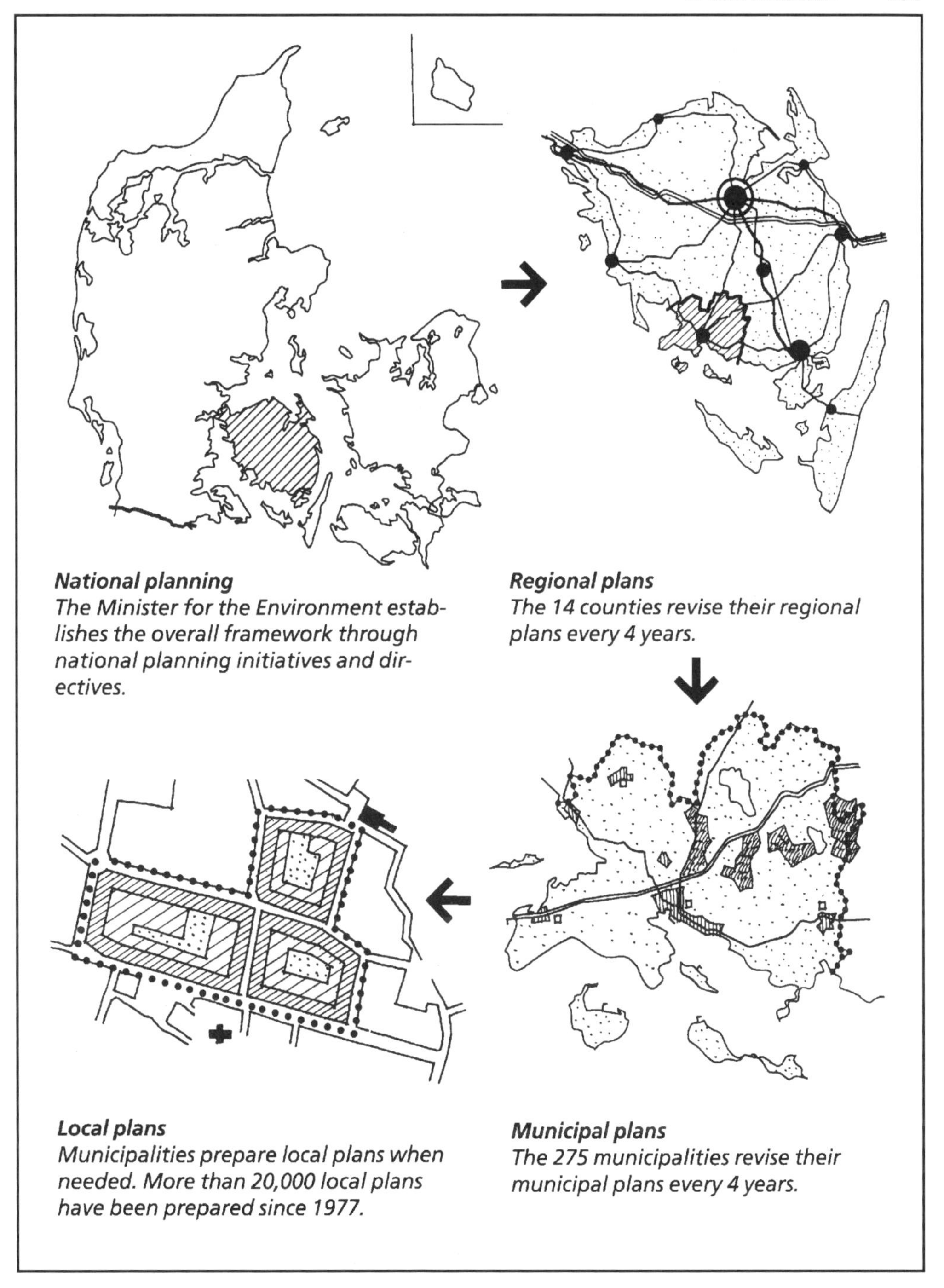

Figure 1 *The Danish four-level planning system (The Planning Act is based on the principle of framework control in which plans must not contradict decisions at higher levels)*
Source: Ostergaard (1994)

THE ZONING SYSTEM

The basic element of the planning system is the division of the entire country into three zones—urban, recreational and rural. In the urban and recreational zones, development is controlled by the current planning regulations.

In the rural zones, which cover about 90 per cent of the country, development or any changes of land use for purposes other than agriculture and forestry are prohibited, unless they receive special permission in accordance with the planning and zoning regulations. The conversion of a rural area into an urban zone requires a binding local plan. A land-use tax is levied on the landowner.

The planning system is based on the principle of framework control. The plans at lower levels must not contradict planning decisions at higher levels. But the objectives and the contents of the plans are different at the three administrative levels.

NATIONAL PLANNING

The National Planning Department, which is part of the Ministry of the Environment, is the national administrative authority for spatial planning functions. In addition, the Ministry of Housing and Building, the Ministry of Agriculture, the Ministry of Transport and the Ministry of Energy also deal with spatial planning. The National Planning Department represents Denmark in international cooperation forums dealing with spatial planning and environmental impact assessment of projects, policies, plans and programmes. These fora include the European Union, the United Nations, the Council of Europe, the Organisation for Economic Cooperation and Development, and planning authorities in the Nordic and Baltic countries. The Spatial Planning Department is also consulted on planning projects outside Denmark.

The National Planning Department is responsible for promoting and facilitating the planning system and for monitoring the planning carried out by the counties and municipalities. At the national level there is no actual national plan or blueprint prepared by the government, as it is not the purpose of national planning to manage in detail or approve the planning decisions of the counties and municipalities. But national interests may require intervention in the decisions of the regional and local authorities. The Planning Act therefore gives the Minister for the Environment the power to intervene through national planning instruments, which comprise a framework and a supplement to the spatial planning activities carried out by the counties and municipalities. The national planning instruments include national planning reports, national planning directives, specific powers, orders, vetoes, call-ins, pilot projects, guidelines and information.

National planning reports

The government's spatial planning policies are mainly expressed in the mandatory national planning report which the Minister for the Environment submits to Parliament after each new national election. The objectives of the national planning reports are to provide guidance in a persuasive way to the counties and municipalities and to present the national planning policies on specific topical issues. The reports are prepared by the *National* Planning Department in cooperation with other ministries and national agencies.

The 1992 national planning report for Denmark was issued as a national planning perspective entitled 'Denmark towards the Year 2018'. It deals with the spatial structuring of Denmark in the future Europe. The perspective expresses the current national planning policies and contains the goals for the development of Denmark's cities and towns, the transport system, rural districts and tourism. Its approach is new in that it views Danish planning within the context of the spatial development of the European Union and focuses strategically on common themes.

Its successor, the 1997 national planning report, signified an intensified preoccupation with the European aspects and influences. This concern is mirrored in the report's title, 'Denmark and European Spatial Planning Policy', and in the fact that this report—the first ever—has been published in its entirety in English also. The national planning report is a reference framework for the decisions that have spatial effects. It is not binding for county and municipal authorities, but it is intended to encourage counties, municipalities and the private sector to promote high-quality development.

National planning directives

The Minister for the Environment may issue national planning directives to implement specific major projects or to promote specific trends. These directives are binding on county and municipal authorities. Only a small number are issued each year. The main use of the directives is to locate specific activities and thereby determine the specific content of planning at county and municipal level. About twenty had been published by 1993. Three-quarters of these deal with the location of specific activities.

Specific powers, orders, vetoes and call-ins

In special circumstances, the Minister of the Environment may order the county or municipal council to produce a plan with a specific content to ensure overall national interests. This provision is not used often, because negotiated solutions are normally reached, but it can be used, for example, for locating a state facility. The Minister for the Environment may also veto a proposed regional plan on behalf of the state. If the proposed plan is vetoed, the plan cannot be adopted by the county council before agreement is reached with the Minister for the Environment, who also acts on behalf of those state authorities involved. The power of veto then replaces the need for approval of the regional plan by the national authorities. Finally, the Minister for the Environment may decide to assume the authority granted to county or municipal councils by calling-in a planning proposal prepared by the councils. The Minister may then make the necessary alterations or may negotiate changes. This power is seldom used.

Pilot projects, guidelines and information

The Minister for the Environment may provide financial support for pilot projects that promote the intentions of the Planning Act, and may exempt county and municipal councils from complying with regulations governing procedure and competence. Such projects can test new forms of regulations and planning methods. Through the Spatial Planning Department, the Minister disseminates information and advice, participates in conferences and is directly consulted. Spatial planning methods are developed, in conjunction with research institutes and consultants, and applied to specific development projects.

NATIONAL PLANNING POLICIES

From a long-term perspective, spatial planning seeks to contribute to economic development, to the environment and to the quality of life in urban and rural areas. Today, regional differences are of far less importance than in former decades. Regional policies and development have, over the years, eliminated many inequalities between the regions. Nevertheless, some regions and islands are still structurally weak. At the national level there is no direct funding for regional and local economic development. The EU Commission, however, does fund such activities. The national regional policy is based on the idea of a free competitive market. Regional and local economic development are the responsibility of the counties and municipalities.

The aim with regard to urban areas is three-fold—to create functional, aesthetic and environmentally sound cities. The urban planning at the municipal level carried out in the 1960s and 1970s mostly took account of the environment by setting standards for new construction and by separating the functions causing environmental conflicts, such as dwellings versus industry and motor vehicle traffic versus pedestrians and cyclists. However, in recent years planning has been oriented towards integrating urban functions by more actively incorporating environmental considerations into the planning of urban development, urban restructuring and urban renewal.

The urban system plays a key role in spatial planning policy, and is considered the 'backbone' of the country. It consists of the capital, four national centres, regional centres, municipal centres and local centres. New urban functions and developments must be located in the respective centres. As a result of the increasing influence from European urban planning and development, this 'traditional' attitude towards Danish urban development is now under scrutiny.

Accessibility and infrastructure linking towns in the urban system in a sustainable manner, and a comprehensive planning of rural areas and landscapes, are other key elements in national planning policy. At present, the location of parks and wind turbines, massive afforestation, the safeguarding of groundwater resources and the counteracting of retail developments outside urban areas near motorways are all hot planning topics in the debate about the uses of non-urbanised space.

Over the last decade, planning in Denmark has increasingly taken account of environmental considerations. Planning, in combination with other instruments such as sectoral policies and sectoral land-use legislation, plays an important role in improving general environmental conditions. The main environmental problems are related to urban activity, transport and agriculture. The Government of Denmark has initiated several plans of action to solve these problems. The strict tradition whereby sectors work independently is slowly being replaced by a more integrated and environmentally sound approach. Ministers responsible for sectoral policies are obligated to integrate environmental considerations into policy-making.

In 1994, the Minister for the Environment and Energy presented a national policy on urban environment and planning. Its central aims were:

- To prepare a national environmental action plan;
- To prepare local Agenda 21 plans;
- To make the planning and management of cities more central to environmental policy;
- To recognise the need for intensive decentralisation;
- To initiate demonstration projects;
- To support community initiatives;
- To develop mixed-use urban districts and promote resource awareness;
- To green the cities;
- To exploit the development opportunities of new urban districts; and
- To promote urban architecture and to preserve the historical heritage.

The instruments and policies of national planning form a framework for comprehensive spatial planning and land-use decision-making to be carried out at the regional and local levels.

REGIONAL (COUNTY) PLANNING

The counties are responsible for assessing the future development of the region and preparing the regional plans with policies, maps and land-use guidelines for all of the county. The plans must contain guidelines for the designation of urban areas, rural land use and recreation and environmental protection. They should also deal with the location of large public institutions, large shopping areas in the cities and out-of-town retail centres, major traffic and infrastructure facilities, as well as the location of major projects and enterprises having special environmental requirements. The county authorities are responsible for hospital services, which account for about 50 per cent of their total expenditure. They are also responsible for high schools (gymnasiums) and adult education, health insurance and social welfare, public transport and major roads, environmental protection, regional planning and rural land-use administration.

There are 14 counties in Denmark and the county councils are elected for a four-year period by direct election using the system of proportional representation. The number of councillors varies between 17 and 31, which is the maximum number allowed. The county council is headed by the mayor, who is elected by the council from among its members. The political process normally takes a local approach and works by striving for consensus.

The Danish environmental impact assessment (EIA) procedure is integrated into the planning process at this level. The regional-level plans thus reflect the administrative responsibilities of the county authorities, and the guidelines constitute the legal basis for permits in conformity with the sectoral land-use provisions for agriculture, and nature and environmental protection. Each individual county council decides on its own regional development policy, and decides too on the legal, economic and political means by which it should be implemented. The policies are presented in the regional plans and can be supported by consultative centres which offer public consultancy and disseminate expertise on, for example, technological development. This form of support aims at creating jobs in the private sector and it should maintain and expand the infrastructure of the region, for transport and energy as well as education and training programmes. The county council also may promote regional development by expanding local public services and by locating public institutions in the areas to be developed. Regional development is financed by taxation, administered through the general budget of the individual county council.

MUNICIPAL PLANNING

The municipal plans, with policies, maps and land-use regulations for the total municipal area, are prepared on the basis of an overall assessment of the present and future land use, the economic resources and the local sectoral plans. The municipal-level plans determine the future development of the urban communities with respect to housing, employment, environmental conditions, the infrastructural facilities and the supply of public and private services. The guidelines and regulations determine the administration of land use and are implemented by building permits, etc. The plans therefore serve as both a political and a legal tool. As a political tool they provide a strategy for controlling development and environmental adaptability. As a legal tool they provide a pattern for land-use administration.

The municipalities have the right and duty to provide binding local plans, with maps and detailed land-use regulations for smaller, designated neighbourhood areas, and to ensure implementation. These binding local plans are mandatory before the inception of any larger developments and investment projects. The local authorities may also prepare such plans whenever they wish to issue detailed planning regulations. The plans are binding on the landowners, but they regulate only future transactions. About 20 000 local plans have been drawn up since 1975.

The Danish system of local government comprises 275 municipalities which influence a good part of the daily life of citizens. The internal organisation of the municipalities and the autonomy in governing and managing their own affairs are similar to those of the counties, as described above. The aim of the local government reform, adopted in 1970, was to establish viable administrative units of a size, and with a tax base, which was adequate for the responsible administration of local communities. The size of the population of the municipalities varies from about half a million in the city of Copenhagen to a few thousand. They average about 20 000 inhabitants.

The municipal councils are responsible for facilitating local economic development using the municipal and local plans as their basic means of control. Each one decides on its own local development policy. They must strive to implement the municipal plan by stimulating local economic development. The methods and the level of activity are decided by the individual council. Local economic development is financed by taxation through the general budget of each individual council.

REGIONAL AND MUNICIPAL PLANNING TASKS

Many responsibilities have been devolved from the state to the county councils and municipalities. The municipal authorities are responsible for social security benefits, health services, primary and secondary schools, libraries and cultural activities, local and private roads, public utilities, as well as local spatial planning and development, which includes urban renewal and building permits.

The regional- and municipal-level plans as well as the binding local plans have to be submitted for public debate and for public inspection and objection before final adoption. This means that public participation is ensured in the planning process at these levels. However, once a plan has been adopted, there is no opportunity for appeal or inquiry. The adoption of a plan is conclusively determined by the county and the municipal council, and there is no compensation to landowners for any development limitations thereby incurred.

The procedures for public participation, noted above, are regarded as an adequate democratic basis for the political decision. Thus, if a project or a development proposal is consistent with adopted planning regulations, there will be no bureaucratic delay in its implementation. The building permit granted by the municipality ensures that the development is consistent with the approved planning regulations.

Since the beginning of the 1970s there has been a significant movement towards comprehensive planning and environmental control. The municipalities have played an active role in designing and controlling urban development while the counties have been in charge of controlling rural land use and countryside protection. The county and municipal councils are also responsible for implementing and monitoring planning policies. Development may thus be characterised as plan-regulated. The planning regulations, however, are mainly restrictive, providing the spatial framework for future development, while the implementation of the planning policies is dependent, in large measure, on the private sector.

In order to fulfil the planning policies, the county and especially the municipal councils must allocate public activities and investments so that they will attract private development. The councils may also try to attract specific private investments by negotiating with the developers about the conditions for implementation within the overall planning framework. The county and municipal councils implement their plans through legal, economic and political means.

- The legal means of planning control seem sufficient to ensure that undesirable development does not occur, since more than 90 per cent of the building activity is consistent with the planning regulations, and is implemented without any changes or amendments to the plans. The planning system therefore is in control. However, this does not guarantee the implementation of all planning policies.

- Economic means are used for implementing public works, public institutions, public roads etc., and providing local urban infrastructure. The municipal council may purchase any land deemed necessary for urban development by the municipal plan. The council may also expropriate land if it is required in order to implement a local plan.
- Political means have a more strategic approach since they attempt to coordinate spatial planning, sectoral planning and budget planning, in order to achieve comprehensive control. This may lead to a strategy for local spatial and environmental development. The private investors and/or developers will then be sure of present and future conditions and their investments will be secured. Furthermore, the councils may look to the EU development funds and other higher authorities for subsidies and support.

By using these means of implementation, councils are responsible for their own activities of public implementation in accordance with the aims and objectives of the regional and municipal plans. For the private sector, the county and municipal councils act mainly as facilitators and approvers of private development proposals.

The municipal councils are active in urban regeneration. The previous focus on demolishing buildings for rebuilding multi-storey housing and new streets has changed. During the 1980s and 1990s there has been more focus on the protection and preservation of buildings and urban features. Urban renewal and comprehensive planning aim to create an attractive and sustainable urban environment.

The county councils have responsibility for administering development in rural areas and for nature and environmental protection. Environmental issues are becoming increasingly important in the political agenda and a number of new measures have been instituted. The protection of nature and the countryside is based partly on the rural zone provisions and the planning regulations of the regional plans, and partly on the Nature Protection Act. This act provides protection for a number of landscape features and makes provision for nature conservation and nature management. During the last decade there has been a significant movement towards multi-functional nature management.

The regulations established by this planning system are mainly restrictive. The system may ensure that undesirable development does not occur. However, it cannot ensure that desirable development actually occurs at the right place and the right time, as the planning intentions are mainly realised through private development. When there is a development proposal that is not in line with the plan, a minor departure may be allowed, or the plan itself would need to be changed prior to implementation. This process includes public participation. The development opportunities are ultimately determined by the municipal councils. The system therefore ultimately determines the control of appropriate development through building permits, permits for subdivision and for sectoral land use. These permits are issued only where the development is consistent with the planning regulations.

The Danish planning system has been developed over a number of years. While it has changed as conditions have changed, it has also remained rooted in its understanding of the need for functional cities and the regulation of land use.

Sectoral planning

Besides the National Planning Department, a number of agencies under the Ministry of the Environment and Energy are responsible for overseeing and implementing planning in several sectors.

THE MINISTRY OF THE ENVIRONMENT

The National Forest and Nature Agency

The aim of the National Forest and Nature Agency is to combine sustainable development with the protection of natural resources and the cultural heritage. The Agency is responsible for implementing national policies on forestry, nature protection and conservation, maintenance of ancient monuments, outdoor recreation, building preservation, raw materials exploitation, and hunting and wildlife management. It also manages the state forests and other properties owned by the Ministry of the Environment.

The Agency administers a number of acts—the Nature Protection Act, the Forestry Act, the Raw Materials Act, and the Preservation of Buildings Act. The Agency also monitors forests, fauna and flora.

The Danish Environmental Protection Agency

This Agency is responsible for implementing national policies on pollution and environmental control of air, water and soil, waste management and environmental technology. It administers a number of acts—the Environmental Protection Act, the Watercourse Act, the Water Supply Act, and the Recycling Act. The Agency also administers a number of studies and research projects on environmental technology and know-how.

The Nature Protection Board of Appeal

The Nature Protection Board of Appeal is a quasi-judicial organ and consists of a chairman, two supreme court judges and a member appointed by each of the parties in the Parliamentary Finance Committee. It is the final court of appeal for decisions pursuant to the Planning Act and the Nature Protection Act.

The Environmental Board of Appeal

The Environmental Board of Appeal is a quasi-judicial organ and consists of a chairman, one or more deputies and a number of experts who are appointed by the

Minister for the Environment on the recommendations of ministries, boards, agencies and other interest groups. The Board is the final court of appeal for administrative decisions pursuant to the Environmental Protection Act.

THE MINISTRY OF ENERGY

The Ministry of Energy was created in 1979 when energy issues were separated from the Ministry of Trade as a result of growing political interest. The Ministry is responsible for the country's overall energy policy and energy management. After the national election held in the autumn of 1994, the Ministry was integrated into the Ministry of the Environment and named the Ministry of the Environment and Energy.

The Danish Energy Agency, as part of the Ministry, performs the basic administrative and professional work for the Ministry. The Agency collects and processes data, publishes statistics and carries out inspections. It also monitors and administers research and development allocations and support schemes. The Agency administers its particular areas of responsibility, such as the inspection and approval of power stations and offshore installations, and provides a range of consulting and information services.

The Ministry prepares the national energy plan to be adopted by the government. In 1990 the 'Energy 2000' plan was adopted. It espoused the policy of fulfilling international recommendations on environmental improvement by reducing emissions and increasing the use of renewable energy.

Apart from the above-mentioned Ministry of the Environment and Energy (actually two ministries in one), other ministries that have an important influence on planning include the Ministry of Housing and Urban Affairs, the Ministry of Food, Agriculture and Fisheries, the Ministry of Transport and the Ministry of Business and Industry.

THE MINISTRY OF HOUSING AND URBAN AFFAIRS

The Ministry of housing and Urban Affairs is responsible for Denmark's housing and building policy, including the detailed rulings on construction activities, housing subsidies, housing credits and urban renewal. It is divided into three agencies: the National Building and Housing Agency, the Palaces and Properties Agency and the National Survey and Cadastre. It administers a number of acts related to the process of spatial planning and implementation—the Building Act, the Urban Renewal Act and the Act of Subdivision and Land Registration.

THE MINISTRY OF FOOD, AGRICULTURE AND FISHERIES

The Ministry of Food, Agriculture and Fisheries formulates agricultural policy in cooperation with Parliament and the agricultural organisations. Since Denmark joined the European Union in 1973, one of the major tasks has been that of securing the country's interests within the formulation and implementation of Europe's agricultural policies.

The Ministry is divided into a number of directorates and research institutions.

The Directorate for Agricultural Development administers several schemes for promoting agricultural development and for facilitating structural adjustment. It also prepares the regulations for these schemes, and subsidises a number of activities related to agriculture. It administers the Agricultural Holdings Act and the Land Consolidation Act, which aim to promote structural development in agriculture. In addition, it handles agricultural interests related to planning in rural areas under the aegis of the county authorities.

THE MINISTRY OF TRANSPORT

The Ministry of Transport is responsible for transport policy and management. Its responsibilities include roads, passengers and goods, transport by motor vehicle, ports and harbours, civil aviation and airports. The Ministry is divided into a number of directorates and authorities. The Road Directorate is responsible for overall road planning, and for construction and maintenance of motorways and trunk roads. The Danish state railways (DSB) form part of the Ministry and are responsible for the transport of passengers and goods by rail. The state railways also operate a number of car-ferry and coach services.

The Ministry prepares the national transport plan which has to be adopted by the government. In 1993 the plan 'Transport 2005' was adopted. This outlines the objectives and strategies which should be pursued within the area of transport in order to achieve a balance between development and the environment on the basis of sustainable growth. The plan of action also contains a summary of the necessary infrastructural investments until the year 2005.

THE MINISTRY OF BUSINESS AND INDUSTRY

In order to stimulate economic activity in the country, the Ministry of Business and Industry has developed a policy of 'industrial nodes' or regional industrial development. Regions, municipalities, private companies and local organisations are encouraged to form associations in order to promote economic growth locally and regionally. Since this is a very recent trend, it is too early to judge the repercussions for regional and municipal planning. But some preliminary research seems to indicate that such activities may very well counteract other planning objectives.

In connection with this, it should be remembered that when the new planning law was passed in 1992, the former objective of *uniform and equal* development for the country was changed to *appropriate* development in the country as such and in the individual counties and municipalities. Local development, then, is the responsibility of the regional and local councils, which use comprehensive planning for control and tax collection. But the policy of the Ministry of Business and Industry and other similar measures could undermine the regional and local councils.

The elimination of national subsidies for regional development has also contributed to the creation of lopsided planning conditions. These subsidies have been 'replaced' by the European Union programmes that provide funds for promoting regional development. Even though Denmark is an affluent country with only minor regional disparities compared with other European countries, a number of regional development projects have been funded by the European Commission's structural funds.

COORDINATION OF SECTORAL PLANNING

The various policy areas described above should ideally be linked together to form a global planning approach. The means to achieve this are not precisely identified in the legislative framework. Some areas, such as pollution control, agriculture, and nature and heritage protection, are mainly organised vertically. Standards and regulations are set at the national level and are administered at the regional or local level. Other areas, such as transport and energy, provide a firm framework at the national level to be further detailed through sectoral and comprehensive planning at the regional and local level. Areas such as housing, the environment, nature and resource management, tourism, and economic development fulfil national policies through comprehensive planning which is based on regional and local considerations and on their needs.

In general, the system of public administration in Denmark can be described as a mix of vertical and horizontal connections. The vertical connections implement each sectoral policy by a top-down approach and the horizontal connections link the different sectoral policies on the same level through comprehensive spatial planning. A global planning approach then is mainly achieved through the principle of framework control within the spatial planning system, in which plans must not contradict decisions at higher levels. The principle of framework control ensures that planning decisions at the regional and local level—in principle—will be in conformity with the overall national policies.

The national planning policies, however, are not formally linked together to form a general national plan or 'blueprint'. The National Spatial Development Perspective 'Denmark towards the Year 2018' was *not* a plan, but a vision, serving as a reference framework. There were *no* requirements or stipulations binding regional and local decisions. Instead, the system of framework control operates through two means—dialogue and veto.

The 1997 national planning report *Denmark and European Spatial Planning Policy* reveals a strong attempt to coordinate Danish planning with the type of European spatial development envisaged in the European Spatial Development Perspective. As an innovation in the national planning approach, this 1997 national planning report was presented to the Danish Parliament, counties, municipalities and the public as a *proposal*, and only after several months of hearings and two revisions was it issued in its final form. In EU fora it is now being hailed as an example of how national planning can be integrated in the larger European context.

The process for revising regional-level plans every four years is based on a comprehensive national report which presents the current preconditions for managing the national aims and objectives within specific and topical policy areas. The report is prepared by the Ministry of the Environment and Energy and is based on negotiations with relevant ministries and national agencies. It should thus prevent the use of veto against the proposed regional-level plans, because the national interest is considered, discussed and dealt with in advance. The regional plans adopted have a binding effect on local-level planning. The preparation of regional-level plans is also based on dialogue and negotiations with local authorities. Within the system of framework control, the Minister for the Environment and Energy may veto a proposed regional plan and a binding local plan when national interests are at stake. The power of veto based on national interest thus leads to negotiations in order to balance the interests of the three levels of administration.

Planning in Denmark today—some impressions

The seemingly tightly knit planning system described above was mainly conceived and constructed against the background of the prosperous 1960s. As we all know, this situation didn't last. The oil crisis of the 1970s was followed by economic recession and stagflation, and much of the dynamism of that period's physical development has petered out. As far as planning was concerned, this meant that the very ambitious planning apparatus which had been constructed seemed somewhat overblown. In particular, the emphasis on public participation has been moderated. Nonetheless, the system has allowed for much good and pragmatic planning so that today it would be unfair to say that Denmark has experienced large and damaging planning failures.

This, at least, could be maintained until very recently. However, during the last decade we have seen the planning and implementation of some very large projects, particularly in transport infrastructure, the repercussions of which cannot yet be assessed. A new car/rail link between the eastern and western regions of the country, hitherto divided by the Great Belt and requiring a ferry trip of about an hour for all travellers, was opened in June 1998. And in July 2000, a new car/rail link between Denmark and Sweden, which joins the Danish coast near to Kastrup, Copenhagen's airport, was opened. On the drawing board and still not politically confirmed is a third link between Germany and Denmark across the Fehmar Belt in the Baltic Sea.

These infrastructure innovations will mean that not only national but also European regional and economic development will have considerable effects and consequences for future development planning in Denmark. Indeed, we have already witnessed a few not insignificant examples of this trend in the physical and urban planning of Denmark. Among these are the Triangular City and the industrial nodes.

THE ØRESTAD

In the Copenhagen region, the Danish state and the municipality of Copenhagen have joined forces in a very ambitious and groundbreaking attempt to create an entire new city extension on the island of Amager, close to the airport and the new highway to Sweden. This plan is linked to a decision to build a new mini-metro in Copenhagen, connecting part of densely populated Amager and the new city extension, and jointly owned by the state and the municipality. The catch is that construction of the metro and the city will be financed through the sale of the land in question and by future property taxes. In order to pull this scheme off, it has been necessary to establish a semi-public/private organisation, which has been entrusted with the planning, construction and running of the new city and the mini-metro. It finances its activities through loans guaranteed by the state and the municipality, on the assumption that the income from the sale of land will cover these huge loans. Any losses will have to be borne by the Danish taxpayer. It should be noted that the entire scheme represents a very flexible (to put it mildly) application of the normal planning system.

THE TRIANGULAR CITY

Another example of a 'flexible' and centrally inspired experiment in planning is taking place in Jutland, where two east/west and north/south transport corridors intersect. Around this central location, three middle-sized cities, each being the main city of three different municipalities in three different counties, have united to establish a new city network intended to become a powerful centre, the Triangular City, which is now vying to become the fifth national centre in Denmark. Clearly, such a planning scheme is bound to create disturbances and tensions vis-à-vis the 'normal' municipal and regional planning system set down in Danish planning law.

INDUSTRIAL NODES

The policy of setting up 'industrial nodes' espoused by the Ministry of Business and Industry is subject to the same criticism. The networks of municipal and/or regional authorities, local private corporations and organisations which it sees as essential for regional development, economic growth and employment are likely to threaten the existing system in the same way. If regional and municipal authorities become increasingly involved in such activities, it might prove difficult for them to live up to the responsibilities entrusted to them by the planning law, for their areas and their populations.

Conclusion

The above examples illustrate a more general and gradually emerging situation which in political science has been interpreted as a transition from government to governance. It has been linked to theories of socio-economic development described as a movement from Fordism to post-Fordism, from standardised mass production to specialised, flexible production. In this development some theorists see a change in the role of the national state. Allegedly, the growing complexity of society and the globalisation of the economy have rendered the hierarchical state obsolete. Decisions are being moved from the national level to the supranational and local/regional levels. But, according to the German political sociologist Helmuth Willke, this has created an even greater need for the national state to play an intermediary role between global needs for regulation of, for instance, the environment, and local and regional urges for autonomy and self-determination. He even speaks of the state as 'local hero' and he emphasises the territoriality of the political systems (Willke, 1992, 362). Willke ends his book on this moderately optimistic note:

> Und es liegt mehr als ein Anflug von Paradoxie darin, dass gegenüber der Modernität supranationaler und globaler Relevanzen es die klassische Form des Verfassungsstaates ist, welche gegenüber der voranschreitenden Organisiertheit der Welt die Idee kollektiver Solidarität in die Zukunft rettet. Vermutlich braucht die Politik zur Erfüllung dieser zutiefst widersprüchlichen Aufgabe eine innere Leitidee, eine Form des Staates, die Engagement und Distanz gleichermassen provoziert und über eine reflexive Brechung ihrer Identität andere Optionen aufhebt. Wenn in dieser Weise Ironie als öffentliche Tugend gelänge, dann wäre 'Ironie des Staates' weder Abgesang noch blosse Hoffnung, sondern Grundlage einer gesellschaftsadäquaten Modernität des Staates und insofern Ausgangspunkt der notwendigen Revision der Staatstheorie. (372)[1]

This new 'ironic state' would be engaged and involved at many more levels of society and would fulfil the function of a reflective and distancing element in political life.

Some years ago a group of influential Danish politicians and civil servants, including the then EU commissioner for the environment, Ritt Bjerregaard, published a book analysing the ongoing changes in the Danish political and administrative system (Pedersen et al., 1994). In this book, *Demokratiets Lette Tilstand* (*The Lightness of Democracy*), the authors try to demonstrate what they call the 'explosion of politics'. Instead of being anchored in the good old representative institutions like parliaments and elected councils, politics has moved out and away from these and is now increasingly being conducted in new institutional settings like committees, semi-private organisations (quangos) and corporates bodies. The authors are manifestly inspired by theories like that of Helmuth Willke.

In one sense, this development has meant an opening up of the political system, a drawing in of circles outside the political establishment and also, to a certain extent, making these new partners co-responsible for the societal situation. At the same time, some of these developments have led to a diffusion of political responsibility, making it more difficult for the politically elected to penetrate and influence the more complex webs and networks of political decision-making processes. We would suggest that the few examples from Danish planning we have noted above can be interpreted as indications of such changes in politics and their embodiment within the planning field.[2]

A criticism, or rather an extension and completion, of the socio-economic interpretation noted above has been put forward by Amin and Thrift (1995). Following among others Hirst (1994), they label this tendency *associationism* and point to the dangers inherent in the development of such institutional innovations:

> The question is whether practical associationist measures oriented to economic success are sufficient for democracy and social equity. Associationism could be seen as a means by which the powerful, however defined, maintain or extend their sphere of influence. (Amin and Thrift, 1995, 58)

They add that such associationist institutions do not (and cannot, we would add) 'see it as their mission to put social goals before profit-based goals' (Amin and Thrift, 1995, 58). Since planning can be said to be placed squarely in the interface between economics and politics, it should be obvious how perilous the involvement of public planning authorities with private interests could be.

REFERENCES

AMIN, A. and N. THRIFT. (1995), 'Institutional Issues for the European Regions: From Markets and Plans to Socioeconomics and Powers of Association', *Economy and Society*, **24**, 41–66.

ASSOCIATION OF COUNTY COUNCILS IN DENMARK (1993), *The Danish Counties*, Copenhagen.

AUKEN, SVEN (1994), *A National Policy on the Urban Environment and Planning*, Copenhagen.

CHRISTIANSEN, O. (1986), 'Comprehensive Physical Planning in Denmark' in J. F Garner and N. P. Gravells (eds), *Planning Law in Western Europe*, Oxford, North Holland.

EDWARD, D. (1989), 'Denmark' in *Planning Control in Western Europe*, London, HMSO, 82–148.

ENEMARK, STIG (2000), *The EU Compendium of Spatial Planning Systems and Policies–Denmark*, Regional Development Studies 28C, Brussels, European Commission.

GAARDMAND, A. (1993), *Dansk Byplanlaegning 1938–1992*, Copenhagen, Arkitektens Forlag.

HIRST, P. (1994), *Associative Democracy: New Forms of Social and Economic Governance*, London, Polity Press.

NATIONAL ASSOCIATION OF LOCAL AUTHORITIES (1990), *Local Government in Denmark: Open to the World*, Copenhagen.

ÖSTERGAARD, NIELS (1994), *Spatial Planning in Denmark*, Copenhagen.

PEDERSEN, O. K. et al. (1994), *Demokratiets Lette Tilstand*, Copenhagen, Spektrum.

SPATIAL PLANNING DEPARTMENT (1990), *The Present and Future Situation in the Regions of Denmark* (the 1989 National Planning Report for Denmark), Copenhagen, Ministry of Environment and Energy.

SPATIAL PLANNING DEPARTMENT (1991), *The Contribution of Planning to a Better Environment*, Copenhagen, Ministry of Environment and Energy.

SPATIAL PLANNING DEPARTMENT (1992a), *The 1992 Planning Act in Denmark*, Copenhagen, Ministry of Environment and Energy.

SPATIAL PLANNING DEPARTMENT (1992b), *Denmark towards the Year 2018* (the 1992 National Planning Report for Denmark), Copenhagen, Ministry of Environment and Energy.

SPATIAL PLANNING DEPARTMENT (1993), *The Öresund Region—A Europole*, Copenhagen, Ministry of Environment and Energy.

SPATIAL PLANNING DEPARTMENT (1995), *The Urban Environment and Planning—Examples from Denmark*, Copenhagen, Ministry of Environment and Energy.

SPATIAL PLANNING DEPARTMENT (1997), *Denmark and European Spatial Planning Policy* (the 1997 National Planning Report for Denmark), Copenhagen, Ministry of Environment and Energy.

WILLKE, H. (1992), *Ironie des Staates*, Frankfurt, Suhrkamp.

NOTE

1 There is more than a touch of paradox in the fact that it is the classical constitutional state, opposed to the modernity of supranational and global urges, and confronted with the increasing organisation of the world which brings the idea of collective solidarity safely into the future. In order to fulfil this deeply contradictory task, it must be presumed that politics will need an internal lodestar, a form of state which at one and the same time provokes involvement and distance, bringing to a new level [*aufheben*] other options through a reflexive fractionisation of its own identity. If in this way irony were successfully installed as a public virtue, then 'the irony of the state' would neither be a swan song nor mere hope, but the foundation for a socially adequate modernity of the state and, as such, the starting point for the necessary revision of the theory of the state.

2 Interplan, the Danish Association for International Urban and Regional Planning, Peder Skramsgade 8, DK-1054 Copenhagen K, publishes a series in English on Danish planning issues (tel +45 33919360).

EIGHT

NATIONAL-LEVEL PLANNING INSTITUTIONS AND DECISION-MAKING IN FRANCE

Gérard Marcou

After the Second World War, France developed an economic planning system based on a wide industrial public sector and financial and regulatory instruments targeted at private firms. This reflected the leading role of the state in the economy and in society. The policy of *aménagement du territoire* (regional development planning) emerged in the early 1950s to support regional economic development, to balance the dominance of Paris and to develop backward regions. The economic planning system and the policy of *aménagement du territoire* each gave rise to specific government institutions which were responsible to the head of government—the Commissariat Général au Plan (CGP) (general planning office), established in 1946, and the Délégation à l'Aménagement du Territoire et à l'Action régionale (DATAR—interministerial regional and territorial development agency) established in 1963. *Aménagement du territoire* was devised to balance the impact of the First Plan, which was expected to create sharp regional inequalities, since it focused on the geographically concentrated key industry sectors. It soon became closely integrated with economic planning. Major development plans responded to the needs of the big industries of that time, like the steel and chemical industries. The investment budget of the state was broken down on a regional basis. Regions had been mapped in the late 1950s to match the requirements of the economic plan and to enable the planning and implementation of regional policies (de Lanversin et al., 1989; de Montricher, 1995).

The French concepts of 'planning' and 'development' are based on this experience. They do not denote the identical meanings attributed to them in the UK and the USA. This may cause confusion. 'Planning' encompasses a series of actions and means combined to achieve given objectives over a fixed period of time. These objectives are defined on the basis of specific forecasts. It covers primarily the planning of economic, social and cultural development. The *aménagement du territoire*, an expression that can hardly be translated into English, has developed as a central government policy which aims to correct economic imbalances in the country through policies relating to the location of business activities and the provision of infrastructures. It is primarily a strategy applied to the spatial dimension of economic development. Consequently, 'planning' is usually not understood as physical planning, i.e. as an expression of the organisation of a spatial area achieved either by its own binding force upon decision-makers or by policies implemented by public authorities. Until recently, physical planning existed only at the local level as urban planning, and in a small number of regions which are subject to specific planning provisions, such as Ile-de-France (the region around Paris) and Corsica (Marcou et al., 1994).

The French planning experience has often been considered abroad as a model, despite the fact that all plans have been upset by political changes (Mescheriakoff, 1995). However, since the late 1970s, this model has been undermined by the process of economic internationalisation and globalisation, and even more directly by the process of European integration. In this new context, state economic planning and the usual instruments of *aménagement du territoire* have become irrelevant, or simply less effective. This has necessitated a revision of the current conceptions in both fields. Local development based on the mobilisation of local resources has been considered as a possible way out of the crisis. The excesses of centralisation have been criticised in government commissions (*Vivre ensemble*, 1976). This prepared the way for the decentralisation of 1982 and the rise of the regional paradigm in French planning.

In retrospect, this new paradigm represents an important change, which has been taking place since the early 1980s. It has been energised by the decentralisation reform as well as by European policies. An important step in this regard has been the Act of 4 February 1995, which laid down the framework of a comprehensive regional planning policy. The decentralisation reform had caused a decrease in regional development planning as a central government policy. At that time it was disregarded because of its inability to bring any benefit to local communities. However, it was revived in the late 1980s because of the increased regional inequalities which resulted from decentralisation, and the pressure of the European single market on exchanges and on the location of businesses (ARL/DATAR, 1994; Marcou, 1995). This turning point was marked by a government report entitled *Pour une politique d'aménagement du territoire* (1986). The report led to the laws which have developed the policy-making instruments of the state for regional development planning. These laws were enacted by both the socialist and the right-wing governments.

This paper focuses on the 1995 reform. I will preface my discussion with a short analysis of the rise of the regional paradigm in French planning. I will then review the main national institutions involved in planning and their relationship to regional development planning. I will show that the recent reform did attempt to introduce comprehensive land-use planning, but planning institutions and policy-making are more fragmented than in the past and require more coordination. This is also the reason for the partial failure of the 1995 reform and of the need for new legislation.

The rise of the regional paradigm in French planning

The regional paradigm grew out of the institutional and planning reforms of 1981–82 introduced by the socialist-communist government and was influenced by the increasing importance of the European dimension for planning decisions.

THE INSTITUTIONAL AND PLANNING REFORMS OF THE EARLY 1980s

The three main reforms were intended to be interdependent. The nationalisation programme was designed to support an industrial policy led by the government to foster economic growth. The decentralisation reform was expected to give local governments new opportunities to develop economic initiatives aimed at keeping or creating jobs, and to open new avenues for citizen participation in the management of local affairs. The planning reform was devised both to organise and implement a comprehensive national economic strategy and to decentralise the planning processes of the regions which were elevated to a new level of self-government.

This reform programme had a pervasive impact on the administrative and the planning system, albeit not quite as expected. The nationalisations were effective in recapitalising firms which had had too low a level of investment and too high a debt. They improved the nationalised companies so that they were stronger when they were returned to the private sector under the right-wing governments of 1986 and 1993.

The decentralisation reform has had a more permanent impact on policy-making. It has, in fact, been an ongoing reform process which has been modified several times. The following are the main institutional changes introduced in 1982–83:

- The region was elevated to a new local government level;
- Executive functions at the *département* and regional levels were transferred from the prefect to the chairman of the general or regional council;
- Numerous state responsibilities were devolved upon the various local government levels with concomitant administrative and financial resources; and
- The main legal instruments of the prefect's tutelage on local government were replaced by new procedures of supervision involving administrative courts.

In the sphere of planning and *aménagement du territoire*, the following new responsibilities were given to local government:

- Municipalities were made fully responsible for town and urban planning. Structure plans (*schémas directeurs*) and, since 1991, housing plans (*programmes locaux de l'habitat*) have to be enacted by joint local government authorities, or by municipalities with specific deprived urban areas. The mayor is empowered to grant building permits in accordance with the local municipal land-use plan if one is in force in the municipality;
- *Départements* are responsible for planning social care institutions which provide services ensured by law and for planning local public transport services which cover areas outside urban public transport areas; and
- Regions are responsible for job training and provision as well as all employment planning, since the Five Year Law for Employment Development of 20 December 1993. They are also subject to general or

> experimental training schemes which the state may still undertake. In addition, they are responsible for secondary education planning and the management of high schools at the upper level. Regions may also adopt regional higher education development plans, in accordance with a national plan, and research development plans, and they usually do so. They have to adopt regional public transport plans, in conjunction with the plans of the SNCF (the national railway public corporation). Devolving responsibility for the organisation and provision of regional rail passenger services is being seriously considered. In fact several regions have volunteered to run an experimental programme which is still going on (the Act of 4 February 1995, as modified by the Act of 13 February 1997).

In addition to having responsibilities devolved upon them, local governments have been granted the right to intervene in new areas concurrently with the state. They may support economic development through direct or indirect aid to businesses, according to the provisions of the Act approving the national plan. Regions and departments are allowed to support businesses in difficulties. In practice this is rare. The amount of local economic interventions has increased from two to 14 billion francs (current value) between 1982 and 1997 (Ministère de l'Intérieur, 1999). These local policies are subject to EC law. Regions are granted wide responsibility for *aménagement du territoire* and regional economic development planning, but their plans are not binding for lower local government levels.

The decentralisation reform, however, retained major characteristics of the French local government system. The prefects have been maintained as representatives of the government and of all ministers at the regional level and at the *département* level. They oversee the local implementation of state responsibilities and policies. Since 1982, *déconcentration* has been a major counterpart to decentralisation. The prefects had to be buttressed by full decision-making powers, which were delegated to them in all matters of local interest within the competence of the state, to enable them to be reliable partners of local government-elected officials who would otherwise continue to turn to Paris to solve some of their problems. This policy received a major impetus from the Act of 6 February 1992, which laid down that all state responsibilities had to be performed locally by the state field services (*services déconcentrés*) under the prefect's authority, except for those which, by their nature or by their scope, still had to be carried out by the central government services. Since 1 January 1998, all administrative decisions within state responsibility are taken by the prefect of *département*, unless they are specifically devolved by law to another public authority.

Special legislation for various policy fields in which local governments are now involved has been revised to adapt them to the new institutional contexts. The new provisions allow for a shared responsibility of state and local government on a number of key issues, thus freeing central government to concentrate on national policies. This applies in particular to physical planning, namely to the master plan for Ile-de-France, which has to be issued by the prefect of the region. The *département* prefect will permit projects of general interest on the authority of local planning documents. Similar regulations can be found in the fields of education, social benefits, social housing[1] and urban policy. The laws of 30 May 1990 and 13 July 1991 have reinforced this approach for social housing and urban policy.

The French local government system remains very fragmented, both horizontally and vertically. Horizontally, there are more than 36 500 municipalities, of which only 841 have more than 10 000 inhabitants, and these comprise 50 per cent of the whole population. Vertically, there are three local government levels. Each has the same legal status and equal powers derived from the constitution. The only way to overcome fragmentation in order to match planning needs has been to develop cooperation among municipalities (there are almost 19 000 joint authorities of all kinds, of which only a minority exercise planning responsibilities), or, more recently, to have contracts signed (with the state in most cases) on development projects.

The planning reform was passed by the Act of 29 July 1982. The purpose of this reform was to restore economic planning after a number of years when the governments had ceased to see in it a relevant economic policy instrument. In accordance with the decentralisation reform, the new planning law had to be decentralised. The then planning minister, Michel Rocard, felt that public planning had to be consistent despite the shift towards decentralisation. The reform was innovative in two respects. It sought to achieve better implementation of the national plan by establishing closer links to budget planning, and by relying on decentralisation and contracts to supplement and to implement the national plan.

The planning procedure involved the regions at an early stage. It called for a first parliamentary act approving the main planning orientations, followed by a second parliamentary act to approve the plan itself, as well as a number of priority programmes with fixed financial amounts allocated to them. These were binding for the yearly state budgets. However, only the IXth Plan (1984–1988) was adopted and structured according to the Planning Reform Act. It was not completely implemented because the government changed after the general elections of March 1986, and the new government rejected economic planning. Nevertheless, state-region plan conventions were respected (see below), and there were negotiations on contracts for the next planning period. After the presidential election of May 1988, the socialists returned to power, and the preparation of the next national plan was undertaken. However, the time was too short to comply with the procedural requirements of the Planning Reform Act, and the new government itself had reservations about economic development planning. It preferred to turn the national plan into a comprehensive economic strategy to prepare the French economy for the Single Market. The Xth plan (1989–1992) was approved by a parliamentary act of 10 July 1989. It did not include any programme for implementation. The preparation of the XIth Plan (1993–1997) was initiated within the same framework by the government in 1991. The Prime Minister's circular of 31 March 1992 did not refer to the procedure established by the Planning Reform Act, and required all the prefects of the regions to prepare plans for the development of the region in conformity with the state's ideas for the next state-region convention. The idea of a national economic development plan based on executory programmes seemed then to have been abandoned. The new right-wing government elected in March 1993 gave up the preparatory work on the XIth Plan and the idea of a national plan. It nevertheless undertook the negotiation of the state-region plan conventions for the period 1994–1998. All plan conventions were signed in 1994. Their completion year was postponed to the end of 1999 for budgetary reasons and to comply with the commitment to the EC to keep the budgetary deficit under three per cent of GDP. The next planning period will cover seven years (2000–2006) to be consistent with the schedule of EC structural policies, and again the state-region plan conventions will be prepared without any national plan (in accordance with the Prime Minister's instruction to the region prefects of 31 July 1998).

The plan implementation was based, according to the law, on contracts—mainly state-region plan conventions, and plan conventions between the state and public enterprises. In fact, within a few years, the state-region plan conventions have become the core of the French planning system, and the main instrument of the policy of *aménagement du territoire*. Initially, the state-region convention was instituted by the Planning Reform Act as a way to reconcile national priorities laid down in the national plan, and regional priorities laid down in the regional plans. The convention had to be signed for the period of the plan and the content was ensured by the law of contracts, i.e. binding for both parts (*contra* Mescheriakoff, 1995). As a result, state-region plan conventions were intended to be based on the national plan and the respective regional plans. Surprisingly, whereas the state withdrew from economic development planning on a national scale and only a few regions adopted a document which resembled a plan, the state-region plan conventions have been very successful and have resisted the decline of economic development planning (Commissariat général au Plan, 1989; Conseil Economique et Social, 1997).

THE EUROPEAN DIMENSION

European integration has affected the planning system through its policy of open economic borders, its regional planning policies and its structure funds. The abolition of economic borders between national member states has resulted in the formation of a single market. Here the European Single Market Act of 1987 has been the major influence. In addition, the implementation of a number of Community-wide policies has had pervasive regional impacts.

Less visible in terms of policy-making, the abolition of economic borders is probably the most important. Most of Western Europe is now a wide area open to full economic competition, and public intervention in the economy has been considerably entrenched as a result of the application of the EC Treaty rules and of EC legislation, enforced by the Commission and the European Court of Justice. This has had major consequences. With the increased mobility of businesses and capital, locational advantages may be more important in firms' investment decisions.

There is also a shift in public economic intervention towards the improvement of land values by development schemes or infrastructure provision, making some locations or areas more economically attractive. This is becoming the major area in which public authorities may be active in order to support economic development. The effectiveness and the quality of physical planning, to prevent congestion costs and to achieve a good balance between various functional areas, can help to increase the land value of an area (ARL/DATAR, 1994; Marcou, 1995). A quasi-market of locational areas is thus emerging at the European scale, or, more exactly, a number of quasi-markets, since not all places are competing in the same market, to develop or attract the same kinds of activities. The ranking of these markets is not determined solely by geographical location, or by the existing agglomeration effects, or by the level of infrastructure provision, which are all important factors, but also, and to an extent which is difficult to evaluate, by the legal and administrative system operating in each area (Gerken, 1995). System competition is thus part of location competition, and may lead to the improvement of the legal and administrative system. Thus regionalisation can enhance the scope of planning. The Xth and XIth Plans included, for the first time, commissions on state and local government structure, on their responsibilities, and on the reforms that would be needed (Commissariat général au Plan, 1993).

The EC regional policy and the structure funds have an increasing impact on planning, because of the amounts of money involved and because of their focus on deprived areas or on areas facing economic restructuring in member states. The EC regional policy was introduced by way of compensation for the UK, which was reluctant to join the common agricultural policy. The European Fund for Regional Development (EFRD) was created in 1975. Major reforms and increases in the funds made available for implementing regional policy followed the addition of new members. Examples of this are the reforms instituted in 1988 to accommodate Spain and Portugal, and the financial instrument created to support fishing after the Nordic countries joined the EC. But all member states have some access to the funds. The areas eligible for Community funds may be a whole country (Ireland, Portugal, Greece). Some of its objectives are not regional (see Objective 3: to fight unemployment, facilitate young people joining the labour market).[2] It can be argued that regional policy, and more specifically the structure funds, stimulate European political integration through multi-level policy-making, more than they contribute to overcoming regional disparities. For all local governments, the EU means access to additional resources. For central governments, the definition of the objectives and of the conditions for eligibility, as well as the choice of beneficiary projects, are crucial issues of national policy. For the EU Commission, the structure funds and regional policy provide it with the opportunity for addressing local governments, even if, by law, still indirectly (Balme, 1995).

For the French policy of *aménagement du territoire*, the EU funds are integrated in the formulation of the objectives and in the evaluation of the means available. In eligible areas, almost all important projects are based on joint financing by EU, state and various local governments. France has obtained a 50 per cent increase in the areas eligible for EU grants for Objective 2 for a period of three years, and for Objective 5b for a period of six years. France received 83.1 billion francs for the period 1994–1999, and a further 43.6 billion francs were committed for the financing of programmes listed in the state-region plan conventions of 1994–1998, whereas financial commitments by the state and local governments in the plan conventions are treated as the public counterpart required by the EC to benefit from its grants (but for Objectives 3 and 4).

Another matter for negotiation was the exact delineation of areas entitled to aid for businesses and for economic development. Because of restrictions to regional aid in the EC Treaty (article 92), the Commission has to approve the areas eligible for these grants as well as the amounts. The government resisted the Commission's request to reduce the proportion of the whole population covered by these areas from 42 per cent to 36 per cent and succeeded in keeping the eligible areas at 40.9 per cent of the total population (Pépin, 1994). This issue was again at stake in the negotiation of the EC regulations on structure funds for the next period.

Regions do not take part in these negotiations, the results of which are major preconditions for the whole planning system. The French position is represented by the Secrétariat général pour la Coopération internationale (SGCI). The DATAR takes the lead as far as structure funds and objectives are concerned, except for the European Social Fund (ESF) and the European Fund for Agricultural Orientation and Guaranty (EFAOG) for which the Ministry of Labour and the Ministry of Agriculture are responsible. The regions lobby at Brussels, jointly with other interested associations or separately. However, they cannot bypass central government in their relations with the Commission (Balme, 1995; Marcou, 1988). This means that, today, much French planning activity is dependent on the participation and performance of the country's representatives in the EU. They decide on the orientation of the EC regional policy.

Furthermore, a European Spatial Development Perspective was prepared by the Spatial Development Committee under the direct responsibility of the central governments of the EU (not of the Commission, but with its participation), on the basis of proposals submitted by the various governments *and* of policy guidelines issued by the Commission (Marcou, 1995). The final project was adopted by the 'informal meeting' of ministers of regional planning in Potsdam on 10/11 May 1999. However, it is only a political document, without binding or even prescriptive effect.

The Guidance Act on Spatial Planning and Development 1995: a comprehensive spatial planning policy

The Guidance Act on Spatial Planning and Development of 4 February 1995 reflects a new emphasis on the legal framework of the *aménagement du territoire*, and on physical planning. This is a result of the decentralisation reform, which gave more autonomy to local governments, and of European integration, which raises fears of greater regional inequalities. The main purpose of this Act regarding the *aménagement du territoire* was to affirm the overall responsibility of the state in this matter. It did not relinquish its prerogatives to the European Community and local government, especially not to the regions.

The new Bill which is being prepared to amend the Guidance Act will be more moderate (see Addendum). It will propose that this policy should promote development in a consistent European ensemble and recognise a wider role in implementation for local governments. The linkage between planning (*aménagement*) and development is underlined by the law. The new Bill will add the qualification 'sustainability' and will reflect a stronger concern for environmental problems in planning policies. In the present law, the spatial planning and development policy (*politique d'aménagement et de développement du territoire*) is defined as a central government policy, carried out 'in association' with local governments. It is required to contribute to 'national unity and solidarity'. Its goal is to secure equal opportunities for all citizens in the whole country and to create conditions for equal access to education (article 1). The scope of this policy is extended to financial equalisation between local governments. It includes a commitment to equal access to public services without regard to location. The management of public services is thus required to deliver the service to the people in order to overcome locational inequalities. A number of sectoral policies and related sectoral planning instruments help to facilitate the realisation of spatial planning and development policy goals. The Act endeavours to set up a comprehensive national planning policy, which covers strategic planning schemes, sectoral planning schemes, new instruments of physical planning, instruments of policy implementation, and local finance equalisation. I will deal with each of these aspects in detail, and point out the main changes that can be expected from the present reform process.

STRATEGIC PLANNING SCHEMES

Strategic planning schemes are spatial planning documents which set out major long-term orientations for spatial planning and development. The law provides for the adoption by Parliament of a national spatial planning scheme (*schéma national d'aménagement et de développement du territoire*) and the adoption by regional councils of regional spatial planning schemes (*schéma régional d'aménagement et de développement du territoire*) (Manesse, 1998). Although not laid down by the law, the period of time covered by the schemes should extend to 2015, and, according to the law, the schemes are subject to reappraisal and revision every five years. Strategic planning schemes should be the basis for investment planning and programming in the future. This law, therefore, attempted to replace traditional economic planning by spatial planning as a way for the government to adjust to an open economy and to retain its influence on the only factors of the economic potential which are still dependent on public policy. However, this design could not be implemented and was replaced by another one in the Bill which was submitted to Parliament early in 1999.

The national scheme (articles 2 and 3) had to be based on a three-level spatial-economic structure, which *does not* correspond to administrative units: propinquity areas (*bassins de vie*), districts (*pays*), and city networks (*réseaux de villes*). The districts are relatively small areas corresponding to existing interactions at the local level, and to traditional or historical links among the population. They are about the size of German *Kreise* and are expected to provide new opportunities for the state to support local economic development, and to encourage cooperation between municipalities. The present *arrondissement*, a subdivision of the *département* headed by an appointed sub-prefect, could be adjusted to conform to the district divisions. The justification for city networks is based on the idea that economic development will depend even more on cities in the future, and that the only chance for the majority of French cities to compete at the European level and to develop locational advantages is for them to cooperate and seek complementarities in agglomeration effects and infrastructure provision.

The national scheme had to set out basic long-term orientations regarding spatial planning, environmental and sustainable development, principles for the location of the main infrastructures and of the main national public services, as well as guidelines for the role of sectoral policies in the implementation of these orientations and principles. The government had to consult with regions, *départements* and municipal associations before the adoption of the national scheme. Indeed, after more than two years' work, it was only possible to submit a 'draft project' to the interministerial committee on *aménagement du territoire*, on 10 April 1997. This draft had then to be submitted to regions, *départements*, joint authorities and various representations of local governments, and then to the national council of *aménagement du territoire* established by the law, before being approved by Parliament (Manesse, 1998).

Despite the fact that progress had been made, the defeat of the government at the general election of spring 1997 brought about a critical appraisal of this process by the left-wing government which was then elected. The new Bill will give up the idea of a national scheme, but not of a long-term comprehensive national policy (see Addendum). Article 2 will set out basic long-term strategic choices, which are thought to be appropriate for the next twenty years (they can be compared to the *Leitsätze* of the German federal law on *Raumordnung*), and these choices will be implemented through a number of coordinated 'collective service schemes' (*schémas de services collectifs*), which will replace the 'sectoral schemes' of the present Guidance Act, though with no significant, substantial change. This means that the strategy of the state will rely on long-term sectoral policies (for 20 years). The purpose of this is to steer regional or local development and planning policies promoted by local governments. In fact, this is what has really been happening for many decades. These policies will be discussed with interest groups, local governments and others, and submitted to a number of consultations beginning with the consultation of regions, presumably before the end of 1999. Most of these schemes already exist. They will have to be reconsidered or replaced, to reflect more general planning and environmental concerns, and coordinated.

Regional spatial planning schemes (articles 6 to 8) were obliged to set out basic long-term orientations for spatial and environmental planning, sustainable development and the main infrastructures and services for the region. These regional planning schemes had to take into account the national scheme, and the investment programmes of the state and of local governments as far as they have an impact on regional spatial planning. The *départements*, main cities and joint planning authorities were to be involved in the preparation of the regional scheme. Coastal regions were permitted to join together to adopt an interregional coastal development scheme. These strategic spatial planning schemes were not binding by law.

In fact, no region adopted such a scheme, since there was no national scheme. Only some regions undertook some preparatory work. The new Bill (see Addendum) will maintain regional schemes adopted by regional councils and is aimed at making their content more substantial and more comprehensive. Regional schemes will have to comply with 'collective service schemes' and 'take into account' development projects at the level of the European area. They would include a prospective analysis and a planning perspective illustrated by maps (this is a new requirement), and providing for the location of main infrastructures, for city developments, for environmental planning and protection, and for the regeneration of derelict areas. They could recommend issuing specific instruments at the town planning level, or for environmental planning. They would replace existing regional transport plans, and the regional plans for economic and social development. However, they should not be binding either, and it remains to be seen whether development programmes of any city, for example, would have to comply with orientations formulated in the regional scheme. There is also no way to compel a regional council to prepare a regional scheme.

Indeed, the most successful experience in this matter was rather unexpected. The Guidance Act provided for a new policy scale, called *pays* (districts) (articles 22 to 24). This was not devised to become a new constituency or a new local government level, but rather a project territory to be defined between municipalities without regard to existing administrative borders (Portier, 1997). The *pays* was supposed to be based on existing economic, historical or cultural links. This was devised as an experiment, but by 1998 about 250 *pays* had been recognised. The innovation was that the territory was defined with the development project. This also gave an impetus to cooperation between municipalities, which was in fact one of the hidden objectives of the *pays*. Therefore, the new Bill will attempt to stabilise and generalise this experience. For this purpose, giving the *pays* some institutional identity is being considered. A development council should be established with representatives of local governments which would be involved in the elaboration of a development perspective for the *pays*. The *pays* should then be linked to the state-region plan convention once it brings about the creation of joint authorities of involved municipalities. The same strategy should be used in an attempt to restructure urban agglomerations. According to the Bill, a development project of the agglomeration should be adopted in all agglomerations with over 50 000 inhabitants and with a city centre of over 15 000 inhabitants. This could give rise to a contract with the state or the region if municipalities decide to create a joint authority with a unique business tax rate.

SECTORAL PLANNING SCHEMES

Sectoral planning schemes are not new in French planning practice. But the 1995 law gives them a clear legal basis and makes their preparation mandatory in areas where they did not exist before. It prescribes clear and quantified objectives to be achieved by the implementation of these schemes (see articles 10 to 21). Some of these objectives will be listed here. More remarkably, they correspond to major powers retained by the state and not relinquished to local government and the EU. The following sectoral planning schemes had to be adopted:

- The higher education and research scheme: The goal is to relocate 65 per cent of research and academic personnel outside Ile-de-France (the region of Paris);
- The culture provision scheme;
- The transport infrastructure schemes: by 2015 no place in the country should be further than 50 kilometres or 45 minutes by car from a motorway or a TGV (high-speed railway) station. Schemes have to be adopted for roads, railways, seaports and airports;
- The telecommunication schemes: by 2015 the whole population should have access to high-capacity interactive telecommunication networks. This scheme calls for the development of equipment and software as well as the setting out of industrial and research policies for that purpose. It has to evaluate the amount of private and public investment which is needed, and to set the tariff policy to be applied by operators; and
- The health organisation planning scheme.

The new Bill would basically keep this structure but with some changes in the definition and content of the schemes. Most provisions on transport would be shifted to the existing guidance act on domestic transport, and only two multi-modal transport infrastructure schemes should continue to exist. A power scheme, and a scheme on natural and rural areas, are among the new schemes that would be added. All sectoral planning schemes are enforced by government decree and are binding for current government decision-making.

NEW INSTRUMENTS OF PHYSICAL PLANNING

Until recently the only physical planning documents, mandated by town planning law, were the *schéma directeur* or *schéma de secteur* (structure plan) and the *plan d'occupation des sols* (POS—local land-use plan) and, under certain conditions, the *plan d'aménagement de ZAC* (PAZ—the zoning land-use plan). Regional plans existed only for Ile-de-France, Corsica and the overseas regions. These plans still exist, and their regulations are binding. The proposal to extend this solution to all regional schemes was again discussed for the new Bill, but has been finally ruled out by the government. However, only POSs and PAZs are directly binding for landowners. A number of other planning documents like water management plans also exist. These cover specific areas and are authorised by the state authorities.

Some, like local housing plans—*programmes locaux de l'habitat*—are within the province of local government. Others, such as waste management plans (see the Environment Protection Act 1995, 2 February) may be called for by local governments. These will not be dealt with here.

When planning powers were devolved to municipalities in 1983, the state authority (the *département* prefect) retained a number of prerogatives to protect wider general interests from purely local considerations. But these supervisory powers proved to be insufficient to enforce consistency in planning, or simply to bring about the adoption of any planning regulation in the face of fragmentation and conflicting interests. The law now provides for a new instrument, the planning spatial guide line (*directive territoriale d'aménagement*—DTA), which will be

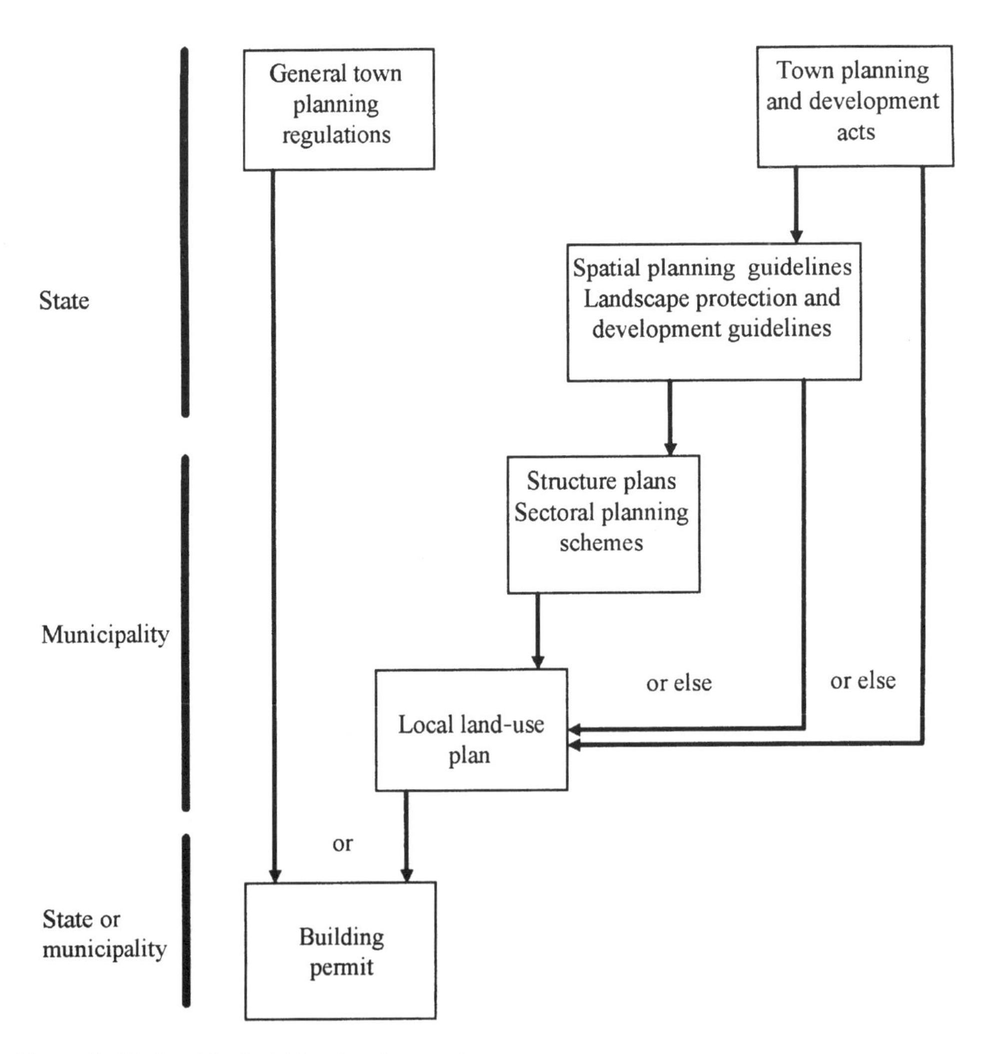

Figure 1 *Binding physical planning documents*

endorsed by a government decree for any area where such a planning document is deemed necessary. The guideline will be initiated and prepared by the state authority. The local government will be involved in its preparation, but the prefect will take the leading role. These guidelines are binding for the lower local planning document. They may furnish precise prescriptions for town and development planning laws, which, like the guidelines, are binding on owners (Perinet-Marquet, 1995). Six DTAs are now being prepared, for example for the metropolitan area of Marseilles and for the estuary of the River Seine, and for the Alps. This is a long process due to numerous consultations, but several DTAs are nearing completion, and thus far this reform seems to have been successful. Figure 1 shows the relationship between the different binding physical planning documents required by the Town Planning Code (before basic changes introduced by the new Act on Urban Solidarity and Regeneration of 13 December 2000).

LOCAL FINANCE EQUALISATION

The Guidance Act on Spatial Planning and Development sets local finance equalisation as a major goal of the spatial planning and development policy (Muzellec, 1995). The goal is that by 2010 all local government resources within a region (debt not included) will be neither below 80 per cent nor in excess of 120 per cent of the national average per inhabitant. This goal must be achieved by an overall reform of all state budgetary transfers to local government budgets, and by a reform of the business tax which would introduce a unified equalisation mechanism reallocating a much larger proportion of the tax yield than is the case at present. Some steps towards implementing these two reforms were introduced in the 1995 budget law, but the overall implementation plan of both reforms is still pending. The general goal of this reform is controversial because of the disparities among the regions, and because of the lack of differentiation among municipalities.

This part of the Guidance Act is now being reconsidered, and other reforms are being prepared on local taxes. This issue will not be touched by the new Bill, which will be much more focused than the Guidance Act. Several local government tax revenues were removed in 1999 and 2000 (for instance, the salary basis of the business tax for all local government levels, the regional housing tax and the annual car tax of *départements*), and replaced by grants; this will soon put equalisation on the public agenda once more.

IMPLEMENTATION INSTRUMENTS

The two instruments for implementation are planning instruments and financial instruments (Némery, 1995).

Planning instruments

There are basically three planning implementation instruments. These are five-year programming laws, regional plans, and state-region plan conventions. Only the first of these three instruments is new. The two others are not new, but the way they are integrated into the whole framework is.

Five-year programming laws will be adopted by Parliament for the implementation of the national spatial planning scheme (article 32-I). These laws have to fix priorities and set the amount of public funding which will be made available for them. The new Bill would keep this provision, which means that they would implement the 'collective service schemes' which will replace the national spatial planning scheme.

Regional plans, adopted by regional councils, have existed since the Decentralisation Reform Acts of 1982 and 1983. In reality, however, only a few regions have designed a proper plan. The reform involved regional plans as a major instrument for the implementation of regional spatial planning schemes. The regional plan had to set out the priorities for the implementation of the regional scheme for each five-year period. It should also have provided for the resources to be allocated to these priorities, despite the fact that the law is silent on this subject. This has been necessary because the regional plan should have been a preliminary to the state-region plan convention. However, with the new Bill, learning lessons from the practice, the regional plan as such would disappear and would be replaced by the regional scheme as described above.

The core role of state-region plan conventions in the present French planning system is confirmed by the reform. It will be maintained by the new Bill, and the government has already addressed instructions to the prefects to undertake the preparation of the next plan conventions (instructions of 31 July 1998). According to the law, state-region plan conventions have to take into account the orientations set out in the national spatial planning scheme, and those of the regional spatial planning scheme (see article 2, al. 6 and article 6, al. 9). Formally, the national and

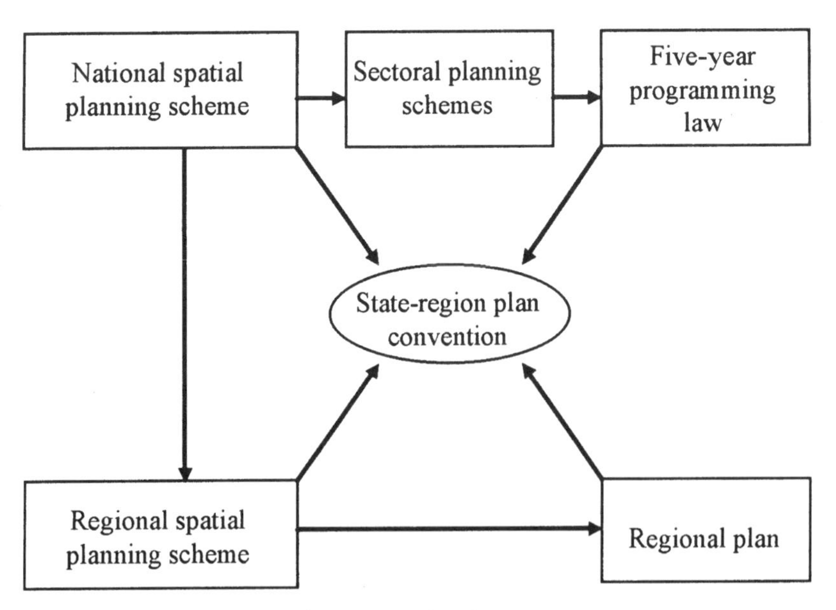

Figure 2 *Medium-term financial and investment planning*

regional spatial planning schemes seemed to be a substitute for the former national and regional plans of the Planning Reform Act, 1982. The reality is that the plan convention is in essence a joint programming instrument, usually for five-year periods, but seven years for the next period (2000–2006). This makes it possible to overcome the obstacle of the 'annuality' principle in budgetary law, even if payments may be committed only on the basis of the yearly budget, as referred to in the state-region plan conventions. The state five-year programming law and the regional plan have to lay down the five-year financial commitments of both parties. If the Bill is passed as expected, detailed 'collective service schemes' could be more binding than a vague national scheme, but the region should have a financial planning instrument in which to ground its commitments. The relationship between all these planning instruments is shown in Figure 2. If the new Bill is passed, this figure should be modified as follows: i) in the upper left box, 'collective service schemes' would replace the national spatial planning scheme; ii) the present sectoral planning schemes would disappear as such; and iii) the regional plan would disappear, and be replaced by the regional spatial planning scheme.

Financial instruments

Since its beginnings in the 1950s, French *aménagement du territoire* has used financial instruments as incentives. The targets have been private businesses. The aims have been to influence their locational decisions or to favour investment and job creation in certain areas, and local governments. Furthermore, these incentives were to support the development of public infrastructure in deprived areas or to respond to the needs of expanding activities in more dynamic areas.

The 1995 law merged a number of funds into a single national fund for spatial planning (*Fonds national d'aménagement du territoire*) (article 33). Through subsidies, this fund supports the implementation of public investment which accords with spatial planning goals. More than half of these resources are delegated to the prefects of regions, particularly in areas where agricultural land has been converted to other uses.

A separate fund has been established for the development of rural areas (*Fonds de gestion de l'espace rural*), which contributes to the upkeep or the regeneration of rural areas through subsidies to collective projects. However, appropriations have diminished from year to year, and in 1997 the law restricted its purpose to projects in which farmers are involved. The Bill would add a new fund for the management of natural milieux, and some purposes initially devised for the former could be shifted to it.

A fund to support small businesses in priority areas by subsidised or guaranteed loans has also been set up. A number of new tax or social security exemptions can be granted in these areas. This fund is controversial and its future is uncertain. It received its first appropriations only in December 1997, and these were quite low. Until that time, it was able to work with some resources from the Caisse des Dépôts et Consignations and other institutions. However, the creation, in 1996, of the Small Business Development Bank (BDPME) with a loan capacity of 30 billion francs opens a better perspective for small businesses and makes less attractive the provision of the Guidance Act of 1995.

The law delineated new categories of priority areas for the policy goals of the *aménagement du territoire* (Labia and Bernard-Gélabert, 1997). Of these, three categories are characterised by indicators which can be quantified:

- Spatial Planning Zones (*zones d'aménagement du territoire*): These are areas with low economic levels and insufficient industrial and tertiary structure. They are entitled to the spatial planning grant (*prime d'aménagement du territoire*—PAT), which is the main incentive for the location of plants and offices. As noted earlier, after difficult negotiations, the boundaries of areas eligible for this grant have been revised to comply with EU requirements;
- Priority Rural Development Areas (*territoires ruraux de développement prioritaires*): These are deprived rural areas in which 12.7 million people live. Within these areas there are revitalisation zones (*zones de revitalisation rurale*) inhabited by about four million people who have suffered a more severe decline and who are entitled to additional support measures; and
- Weak Urban Areas (*zones urbaines sensibles*): These areas have large social housing settlements, derelict buildings and sharp imbalances between housing and employment. Urban Regeneration Zones (*zones de redynamisation urbaine)* comprise about 2.2 million inhabitants. These are the most deprived urban areas and they are therefore entitled to increased support, especially for the creation of low-skilled jobs. A new law of 14 December 1996 has provided for the creation of free zones in even more deprived areas, with additional tax and social charge exemptions. Today there are 750 *zones urbaines sensibles*, of which 416 are *zones de redynamisation urbaine*, and of these 44 are free zones.

Urban areas that are more deprived are subject to a specific central government policy, called *politique de la ville*. This policy is carried out through contracts authorised by the state authority (prefect of the *département*) with municipalities for multi-faceted programmes combining social care, housing renewal, training and education, crime prevention and economic development. In 1994, 214 three-year contracts were authorised (Donzelot and Estebe, 1994; Raoult, 1994). A new programme has been introduced in a number of extremely deprived areas, which have locations that present opportunities for economic development. This is the 'Major Urban Project' (*grand projet urbain*—GPU), which is a long-term regeneration programme based on huge public investments in housing and infrastructure provided by the state with the financial participation of the municipality. It is formalised in a contract between the state and the municipality. Thirteen major urban projects have been decided upon, and are now established. This policy uses existing legal instruments, and is not bound by the Guidance Act on Spatial and Development Planning. However, the areas covered by the priority zones and by these 'contracts' do not coincide. In addition, 26 areas were selected for 'concerted urban development programmes' (*programmes concertés de territoires urbains*), and contracts have been set up between the state and local government to deal jointly with economic industrial restructuring and urban deprivation. The *politique de la ville* is again under review, and a recent report commissioned by the government recommends a stronger commitment on the part of the state and more cooperation between municipalities to overcome social crises in urban suburbs (Sueur, 1998). A new procedure is being experimented with to set up joint city development programmes which include the whole agglomeration in a contract signed with the state. Fifteen cities are involved (*Monde*, 12 September 1998; Marcou, 1997). In 1997, Corsica was declared a free zone for five years to foster the economic development of the depressed economy of the island.

Another financial instrument is the domestic air transport equalisation fund (*fonds de péréquation des transports aériens*) established by the law of 4 February 1995, to subsidise domestic airlines' operating routes which support spatial planning and development needs, but are not profitable. This fund is a response to the full deregulation of air transport which made cross-subsidising between profitable and non-profitable services unfeasible. From 1 January 1999 this equalisation scheme has been included in the newly created intervention fund for airports and air transport (*fonds d'intervention pour les aeroports et le transport aérien*). It is funded by part of the revenue yielded by the civil aviation tax, paid by airline companies according to their passenger and freight traffic in France (Dupéron, 2000).

A similar mechanism ensures that the managements of public corporations or services take spatial planning and development goals into account. Here, too, contracts are signed with the state. The contract has to lay down the goals to which the management of the corporation or service is committed, and the financial commitment of the state (eventually local government) to cover the additional costs resulting from these obligations. The contract is binding. A dispute settlement procedure has been established by law. It gives the responsible minister the right of final decision as to whether to cancel or maintain a service, at the request of the management and in accordance with the contract. These procedures have not yet been applied. The new Bill would make a specific application to contracts between municipalities and the Post Office to retain agencies in sparsely populated areas. The purpose of these provisions is more to authorise affirmative steps in order to redress or soften excessive disparities, than to enforce the equality principle (Madiot, 1995).

Central planning institutions and policy-making

In France, economic planning and *aménagement du territoire* developed within a centralised state structure. As a result, these functions are still vested in central government institutions directly subordinated to the Prime Minister. However, European integration and the decentralisation reform have made matters much more complex. The DATAR, together with a small number of sectoral ministries and the Ministry of Finance, and their field agencies, remain key institutions. However, difficulties and delays in the implementation of the Guidance Act on Spatial Planning and Development reflect the fact that central planning institutions can no longer control policy-making in this field, and that they are dependent on their partners.

CENTRAL PLANNING INSTITUTIONS

The decline of macro-economic planning and the rise of spatial planning have brought about a shift of influence from the CGP in favour of the DATAR. Both are, as they have been since their inception, directly subordinated to the head of the government, in order to benefit from his or her authority in their relations with government departments. Usually, however, the Prime Minister delegates his or her powers regarding these agencies to another minister.

The main weakness of the CGP, in this context, is that it is linked to no executive organisation or procedure, and since there is no longer macro-economic planning it is no longer able to take the lead in long-term policy formulation. Nevertheless, the CGP remains an influential reference because of the reports published by its working groups on a number of key issues. Their distance from public action makes them objective authorities, and their reports are consulted in implementation-oriented policy-making. Public policy evaluation is also a new activity for the CGP. The Commissariat must carry out decisions taken by the interministerial committee on public policy evaluation. It manages the national fund for the development of evaluation (Decree no. 98–1048, 18 November 1998).

The reform of the CGP has been under discussion since the late 1980s, and has given rise to a number of reports. A principal subject of controversy is whether or not the responsibilities of the agency should include financial and budgetary forecasts, or a multi-annual financial framework (Dollé, 1993; de Gaulle, 1995; Moreigne, 1995; Raignoux, 1995; Velitchkovitch, 1987).

The DATAR created in 1963 is much more influential for four reasons:

- It is directly involved in government decision-making, through the interministerial spatial planning committee (Comité interministériel de l'Aménagement du Territoire—CIAT). The DATAR prepares and implements the decisions of the CIAT;
- The DATAR has an intervention budget, made up from the funds noted above, and the development grant (PAT);
- The DATAR has its own sphere of influence throughout the country. The prefects of regions, who are in charge of the implementation of the economic and spatial planning policy of the state and of the European Union, receive instructions from the DATAR. In addition, the Secretary General of Regional Affairs (SGAR) in each regional prefecture is in charge of *aménagement du territoire*. Furthermore, the DATAR indirectly controls a number of development associations involving regional economic circles; and

- All applications for EC funding are channelled through the DATAR by the prefectures of regions, so that EC policies have made the DATAR a central government institution for resources.

The DATAR is a small agency in terms of personnel and budget, but it acts through others because of its interministerial function. The CIAT is the major decision-making body with which the DATAR works. The main decisions for the state-region plan conventions are taken by the Prime Minister in the CIAT, after they have been prepared by the DATAR. The total amount of the state commitment in the conventions, and its division among regions, the main orientations for negotiating plan conventions like determining the kind of discretionary scope left to the prefects, the approval of state-region plan conventions, the sectoral planning schemes—all of these are usually approved in the CIAT. The influence of the DATAR depends very much on the political support given to it by the Prime Minister and on the political profile of the minister in charge of the *aménagement du territoire*. It can be backed by recommendations issued by the National Spatial Planning and Development Council (Conseil national de l'Aménagement et du Développement du Territoire—CNADT), composed of Members of Parliament and of local government, of representatives of economic, social and cultural agencies, as well as of representatives of the ministries involved in *aménagement du territoire*. The 1995 law reflects the intention of elevating CNADT to an influential body, consulted on all important decisions like sectoral planning schemes (later probably called 'collective service schemes'), planning spatial guidelines, and the reorganisation of the state administration. The new Bill will again upgrade the Council, which will appoint a permanent commission working between its full sessions. This will give more continuity to its activity, and it will be especially charged with evaluating planning and development policies. Such a council did exist before, albeit with a lower profile. The Council has met irregularly in recent years.

The main challenge for the DATAR is to enforce its interministerial role in pursuing spatial planning and development policy goals in the face of strong sectoral ministries and conflicting interministerial policies organised on the same model. In recent years, the major conflict has been with the urban policy agency (*politique de la ville*), which has been allotted a budget, a *Délégation interministérielle à la Ville et au Développement social urbain* (DIV), and like the DATAR directly subordinated to the Prime Minister. It is also connected with the interministerial committee on urban policy (*comité interministériel de la ville*), and supported by the National Council of Cities (*conseil national des villes*). It has intervention funds and relies for implementation on the prefects, using contracts with municipalities to ensure policy implementation. Basically the DATAR believes that there is no solution to the social crisis in deprived suburbs without economic development, while the DIV puts the emphasis on programmes directly addressing the population living in these suburbs. These approaches are not contradictory when formulated in general terms, but they can be when the question of priorities is raised.

Another major challenge is the relationship with the ministries in charge of public works and transport, since decisions taken and investments carried out by these ministries will structure the organisation of the relevant regions, with or without the DATAR's approval. These difficulties are reflected in the way the DATAR has been shunted around within the government structure. In the Balladur Government (1993–1995), the DATAR was the responsibility of the Minister of Home Affairs (in charge of local government). In the first Juppé Government (May–November 1995), there was a special ministry for spatial planning and development, public works and transport, with the minister having direct control of the DATAR. The DIV was separate from the DATAR in another ministry. In the second Juppé Government (from November 1995), there was a unified ministry for spatial planning and development, urban policy and social integration, which included both the DATAR and the DIV. By including both in one ministry, an attempt was made to reconcile their policies. Thus, in the second case, the *politique de la ville* was a separate ministerial jurisdiction, whereas in the third case there is an attempt to reconcile the interministerial policies of both. Finally, in the Jospin Government elected in spring 1997, the DATAR has come under the Ministry of Environment, Planning and Development, which has been assigned to Dominique Voynet, the leader of the Green Party, whereas the DIV comes under the Ministry of Employment and Solidarity (social affairs).

In addition to such organisational measures, conflicts of this nature are solved mainly through political arbitration at the top and regulation at the bottom. The Prime Minister is empowered by the constitution to take the final decision in cases of conflict between members of the government, and in conflicts relating to the *aménagement du territoire*. This does not happen often. For example, in 1994 the minister in charge of urban policy decided to sign the *contrats de ville* without regard to the state-region plan conventions under negotiation. This also ignored an earlier decision to link the *contrats de ville* to the state-region plan conventions and to make them contracts implemented by the latter (DIV, 1993; Donzelot and Estebe, 1994). This decision could not have been taken without the consent of the then Prime Minister. There have been cases in which there has been no arbitration at all, and plans were imposed by the responsible body. This happened in the case of the controversial TGV station built by the national rail company between Amiens and Saint-Quentin, which was distant from both. But more often, conflicts can be solved through negotiations at the local level, where it is easier to design a package with gains for all parties. Over time, the practice of state-region plan conventions has nurtured the cooperative experience of the state field services and the regional council offices. They have learned to work together. In some regions this has resulted in very stable policy communities which are able to overcome disagreements between sectoral bureaucracies and settle political conflicts if they do not become too heated.[3] For example, in the region of Nord-Pas de Calais, the prefecture of the region and the regional council office were able to reach an agreement on the relation between the spatial planning and development policy and the *politique de la ville* once the *contrats de ville* were signed as ordered by the minister.

Such policy communities are even more important as far as European funds are concerned. Once the policy framework has been agreed—and this is negotiated at the government level—all projects have to be prepared locally. Proposals by local governments are necessary, since their financial contribution will be required, but the prefecture of the region is the interface with the DATAR which will be responsible for the projects and will support them before the relevant division of the EC Commission (the *Direction générale*—DG—XVI—responsible for regional policy). Then, the EC money is paid to the state budget and delegated to the prefect of the region, who orders the payment to the owner (*maître d'ouvrage*). The payment is then made by the state regional accountant. In fact, the expected EC contribution is anticipated in the state-region plan convention for the five-year period. This means that the prefecture of the region and the regional office have to agree on the projects which will be presented under each EC objective within the limits of the amount that can be expected. This could be a purely regional project that would be supported at the EC level by the state. The state-region plan convention procedure has absorbed the 'regional development plans' and the 'community support frameworks' that the member states are required to prepare to benefit from the structure funds, and which must involve a designated local government partner (EC Council Regulation no. 2081/93, 20 July 1993: articles 8 to 11bis). The EC regulation did not necessitate any change in the French procedure. No change in this respect is expected from the current revision of the EC regulations on structure funds.

IMPLEMENTATION PROBLEMS OF THE GUIDANCE ACT ON SPATIAL PLANNING AND DEVELOPMENT

The content of the Guidance Act is heterogeneous and requires a differentiated assessment. Some provisions, like the ones on spatial planning guidelines (DTA), were immediately applicable. Others become applicable after the publication of the statutory instruments required by law. This is the case with provisions on domestic air transport equalisation and provisions on new tax exemptions in deprived areas. A number of decrees have been published. But the main provisions, which form the core of a new comprehensive planning system, were to be enacted by political bodies. They were thus subject to political uncertainties, and several were dropped or are still being prepared for implementation.

As noted above, the national spatial planning and development plan has been abandoned in spite of all the preparatory work. Initially, a project was attached to the Bill, as prepared after a long national debate, on the *aménagement du territoire* in each region, but later it was separated from it because of political opposition which could have undermined the adoption of the Bill before the end of the parliamentary session. However, the Act declared that the national scheme had to be adopted within one year. Because of the presidential election, it was only in August 1995 that the Prime Minister announced the procedure and the timescale for the preparation of the national scheme. Jointly with the presidents of the regional councils, prefects of the regions had to prepare a regional discussion and present the results before 30 November. The ministries had to formulate their policies and programmes looking towards 2015, before 30 October. Despite the short notice this deadline was met because important work had already been done in 1993 and 1994 in preparation for the great national debate on '*aménagement du territoire*'. A number of working groups and special committees were established to develop policy guidelines. They had to report before 30 October. An interministerial pilot committee was set up to assist the DATAR in directing the whole process. After all the legally required consultations, the DATAR was expected to present the national scheme for parliamentary approval in the spring of 1996. This timeframe was too short and could not be met.

Indeed, the main problems are political—the place of France within Europe, and the opportunities available to the various regions. The approach elaborated by the DATAR is based on a three-tier country-wide system, comprising seven large areas (called '*grands chantiers*') structured around big cities or city networks which are able to develop some functions of international scope, and on a number of policies for specific areas. The seven *grands chantiers* are the Bassin Parisien (Ile-de-France and seven surrounding regions), Northern France (four regions), the Grand Est (eastern region, from the Moselle and the Rhine to the Rhéne), the Centre-Est (part of the Massif Central, Dijon-Besançon, Lyon, Grenoble and the Avignon-Nîmes-Arles area), the Midi-Méditerranée area, the South-West (Bordeaux and Toulouse) and the Loire-Bretagne area. They are designated as strategic planning areas; they are not deemed to become new tiers of government. The Bassin Parisien was the only 'grand chantier' for which a specific planning instrument could be elaborated—an interregional plan convention with the state, for projects of common interest in the associated regions. The Midi-Méditerranée area ought to become the pivotal point of a Mediterranean axis. The South-West area is due to develop itself on the basis of cooperation between Bordeaux and Toulouse. The Loire-Bretagne area should benefit from improved relations between its cities and improved east-west communications with the establishment of parallel links between estuaries. Specific policies are necessary for mountain areas and coastal areas. But the policy implications of such a system, which I have simplified here, have to be developed with the regions, and within each region as well as with the main cities and *départements*. Diverging views and interests make agreement difficult to reach. A number of mayors and presidents of *départements* or regions find it unacceptable that the central state assigns to them the role they have to play, and decides which opportunities are best for their territory. This is why it would be unrealistic to expect a national spatial development scheme that would be both relevant and acceptable. It would simply engender too much contention.

The future of regional spatial planning schemes was also subject to uncertainty. There was nothing in the law to compel a region to adopt a regional spatial planning scheme. It was clear that regions would not initiate such a plan before the national scheme was adopted, but there was no reason to believe that they would do so after its adoption. The reasons why only five regions among them have adopted regional plans in the past are probably not only ideological, since these include a socialist region, Nord-Pas de Calais and a conservative one. Planning requires the clarification of goals, on the basis of analysis, and the formulation of strategic choices. It is likely that only a few regions are able to reach agreement on these matters. In other regions, conflicts of interest between *départements* or between cities make it more difficult. In addition, regions are reluctant to commit themselves further at present because they suspect that central government plans to transfer new financial burdens on to them.

The success of the DTA as a new physical planning instrument at a mid-level is now very likely. The new institution is certainly needed to impose more consistency in the development of certain regions. However, since the central government in France is dependent on local government and, similarly, national politics is dependent on local politics, central government is usually reluctant to exercise this kind of power, for it ensures the loss of support from one side or the other. Neutrality, on the other hand, makes it possible to keep the support of both. Furthermore, when structure plans (*schémas directeurs*) exist, the DTA is not directly binding on the local land-use plan, unless it is related to the precise provisions of a town and development planning act. Once again, the result of the reform depends on political factors and on local conditions. Nevertheless, the current experience in the preparatory work is encouraging. It is a rather long process, requiring numerous studies, and much discussion with all interested parties. The task of the prefect and of the state field services has been to provide expertise and to bring all parties together. In practice, the prefects have opened the process much more than strictly required by law, in order to generate more understanding and make agreement possible. However, only when the first DTA is issued by government decree will it be possible to evaluate the results of this process.

The future of national-level planning is again uncertain, since a new Bill will be discussed in Parliament, probably from January 1999, in order to enable the new Act to be in force when the next state-region plan conventions have to be signed, i.e. before the summer holidays of 1999 (see Addendum). However, this schedule cannot be taken for granted. Since Members of Parliament are very sensitive to this issue (most of them are also elected local officials), it is likely that the parliamentary discussion will have a substantial influence on the final text.

Addendum

The new Bill which is referred to in the text became a new Act on 25 June 1999, heavily amending the Act of 4 February 1995 (Marcou, 2000a). This new law has to be considered jointly with two other pieces of legislation, which as a whole are expected to change the French planning system in depth: these are the Act of 12 July 1999, reforming cooperation bodies of municipalities (Marcou, 2000b), and the Act on Urban Solidarity and Regeneration of 13 December 2000, which has amended essential parts the town planning code.

It is only possible to give here a very brief outline of the laws.

Anticipations based on the Bill in the text have been confirmed. However, some basic orientations and innovations of the Act of 25 June 1999 should be emphasised. The European context is given much more attention in the new law, although no binding relationship is established. For example, the new regional schemes should be the reference for the implementation of programmes co-financed by the EU through structure funds; EC regulations on transport and environment might have an impact on national planning documents. Nevertheless, the new law still provides for a national planning policy covering a wide scope of matters.

Priority areas have largely been maintained with some changes. According to the new law, there are at present two kinds of areas subject to development projects to be elaborated at the local level and articulated with national policies. For rural or less urbanised areas, the *pays* is confirmed as a project-based area, e.g. an area designed according to a development strategy which has to be elaborated jointly by local self-governments, state agencies, and economic and social actors; it is not expected to become an administrative constituency or a new government tier. For urban settlements of higher importance, the agglomeration (defined as an urban area in excess of 50 000, with a core municipality in excess of 15 000), as resulting from economic and demographic processes and recognised by statistical surveys, is also expected to become the basis for a development strategy, including housing and infrastructures as well as job creation, land provision, orientations for planning and so on. The Act of 12 July 1999 provides for new cooperation forms between municipalities which are focused on economic development, planning, housing, and environment protection. The main institutions proposed (not imposed) to municipalities are the *communauté d'agglomération* (with demographic thresholds as quoted above), the *communauté urbaine* (these existed before but new ones can now be created for agglomerations in excess of 500 000) and the *communauté de communes* (these also existed before, but may now be created with larger responsibilities and new tax powers). A major change is that they are granted exclusive tax powers on businesses, in order to secure unity and solidarity within the agglomeration, instead of competition between municipalities; this is an option for *communautés de communes* subject to certain conditions. The next step of the reform will probably be to introduce the direct election of the councils of these joint authorities. The new *communautés d'agglomération* (80 at the end of 2000) and *communautés urbaines* (19) should initiate and support the development strategy for the agglomeration.

The development strategies for *pays* and for agglomerations (being elaborated during 2000) will be formulated in a development project, which will form the basis for a plan convention agreed upon with the state (represented by the regional prefect) and related to the state-region plan convention. Financial commitments from the state budget and from the budget of the region (as local government) are written down in the state-region plan convention, and form its so-called 'territorial section'. This will give an opportunity to reconcile local development strategies and national priorities.

National planning instruments are also newly devised. The idea of a comprehensive spatial national planning scheme has been abandoned, as already reported in the text. Nevertheless, national spatial planning has not been given up. It will be achieved through eight so-called *schémas de services collectifs* (collective service schemes), focused on the service provided to the public rather than on infrastructure planning. They absorb the former sectoral planning schemes, but they focus long-term (20 years) national planning on matters for which the national government still has the necessary authority and resources to secure implementation. These are regarded as sufficiently essential to justify the national government's maintaining a significant influence on the spatial organisation of the territory.

At the regional level, regional plans are suppressed, and the regional planning scheme to be adopted by each regional council is now the unique planning document at that level. Regional planning schemes have to formulate a long-term strategy for the region in terms of the organisation and the development of the regional area, as well as protection measures. They have to be compatible with national collective service schemes, but they are not binding. Nevertheless, they may recommend that binding planning documents be adopted: territorial planning directives (DTAs) for some areas, or local planning documents which are the responsibility of municipalities. Normally, state-region plan conventions have to be based on regional planning schemes, but in fact the former have been signed whereas the latter are far from being elaborated; nevertheless, the law provides for an evaluation of the impact of planning documents after some years, and in particular in 2004, two years before the term of the current state-region plan conventions expires. This will give an opportunity for adjustment if necessary.

The new Act on urban solidarity and regeneration is also coordinated with the new planning legislation. Existing local planning documents will be replaced by two new planning documents: the territorial consistency scheme (*schéma de cohérence territoriale*), which is a comprehensive planning document at the intermunicipal level embracing all local policies that have an impact on spatial development, not just land use and infrastructure; and the local plan (*plan local d'urbanisme*), which will have a similarly extended scope, but which will have to be compatible with the territorial consistency scheme, and will be binding for building permits. National collective schemes will have to be considered during the elaboration or the revision of these local planning documents. The prefect will still have to ensure the respect of projects resulting from national schemes, or from higher local governments, although the Act limits the prefect's powers more strictly with respect to municipalities (Jacquot and Lebreton, 2001).

Paris, June 2000

REFERENCES

AKADEMIE FÜR RAUMFORSCHUNG UND LANDESPLANUNG (ARL)/ DATAR (eds) (under the direction of G. Marcou and H. Siedentopf) (1994), *Conditions institutionnelles d'une politique européenne de développement spatial*, Hanover, ARL.

BALME, R. (1995), 'La politique régionale communautaire comme construction institutionnelle' in Y. Meny, P. Muller and J. L. Quermonne (eds), *Politiques publiques en Europe*, Paris, L'Harmattan.

BALME, R., Ph. GARRAUD, V. HOFFMANN-MARTINOT and E. RITAINE (1994), *Le territoire pour politiques: variations européennes*, Paris, L'Harmattan.

COMMISSARIAT GÉNÉRAL AU PLAN (CGP) (1989), *Xème Plan et planification régionale*, Rapport pour la préparation du Xième plan, Paris, La Documentation française.

COMMISSARIAT GÉNÉRAL AU PLAN (CGP) (1993), *Décentralisation: l'âge de raison*, Rapport pour la préparation du Xième plan, Paris, La Documentation française.

COMMISSARIAT GÉNÉRAL AU PLAN (CGP) (1993), *Pour un Etat stratège, garant de l'intérêt général*, Rapport pour la préparation du Xième plan, Paris, La Documentation française.

COMMISSION DES COMMUNAUTÉS EUROPÉENNES (1993), *Fonds structurels communautaires 1994–1999. Textes réglementaires et commentaires*, Luxembourg, Office des Publications officielles des Communautés européennes.

CONSEIL ECONOMIQUE ET SOCIAL (1997), *Le suivi et la réalisation des contrats de plan*, rapport presenté par Jean Billet, Paris, Journaux Officiels, 4335.

DATAR (1994), *Débat national pour l'aménagement du territoire. Document d'étape*, Paris, La Documentation française.

DIV (1993), *Les contrats de ville du Xième Plan* (2nd edn), Paris, DIV.

DOLLÉ, M. (1993), *Un nouvel horizon pour la planification* (report presented to the Commissaire général au Plan) (not published).

DONZELOT, J. and Ph. ESTEBE (1994), *L'Etat animateur. Essai sur la politique de la ville*, Paris, Editions Esprit.

DUPÉRON, O. (2000), 'Desserte aérienne du territoire et service publique', *Actualité juridique. Droit administratif*, November, 873–82.

L'Etat en France. Servir une nation ouverte sur le monde (1995) (report to the Prime Minister by the Mission on state responsibilities and organisation, presided over by Jean Picq), Paris, La Documentation française.

FRANÇOIS-PONCET, A. (1995), 'Acquis et carences de la loi sur l'aménagement et le développement du territoire', *Revue française de Droit administratif*, **5**, 871.

GAULLE, J. de (1995), *L'avenir du Plan et la place de la planification dans la société française* (report to the Prime Minister), Paris, La Documentation française.

GERKEN, L. (1995), 'Nicht alles über einen Leisten', *Frankfurter Allgemeine Zeitung*, 2 December, 17.

JACQUOT, H. and J.-P. LEBRETON (2001), 'La refonte de la planification urbaine', *Actualité juridique. Droit administratif*, January.

LABIA, P. and M.-C. BERNARD-GÉLABERT (1997), *Zones d'aménagement du territoire, mode d'emploi*, Paris, LGDJ, coll. 'Systèmes'.

LANVERSIN, J. de, A. LANZA and F. ZITOUNI (1989), *La région et l'aménagement du territoire dans la decentralisation* (4th edn), Paris, Economica.

LE GALLES, P. and M. THATCHER (eds) (1995), *Les réseaux de politique publique. Débat autour des policy networks*, Paris, L'Harmattan.

MADIOT, Y. (1995), 'Vers une territorialisation du droit', *Revue française de Droit administratif*, **5**, 946.

MADIOT, Y. (1996), *Droit de l'aménagement du territoire* (3rd edn), Paris, Masson.

MANESSE, J. (1998), *L'aménagement du territoire*, Paris, LGDJ, coll. 'Systèmes'.

MARCOU, G. (ed.) (1988), *Regional Planning and Local Government Confronted with Economic Change*, Brussels, International Institute of Administrative Science.

MARCOU, G. (1995), 'L'analyse comparative des cadres institutionnels de l'aménagement du territoire dans six pays de l'Union européenne et ses enseignements pour un aménagement du territoire communautaire', *ARL Beiträge*, Hanover.

MARCOU, G. (1997), 'Politiques de la ville et aménagement du territoire' in G. Loinger and J.-Cl. Némery (eds), *Construire la dynamique des territoires. Acteurs, institutions, citoyenneté active*, Paris, L'Harmattan.

MARCOU, G. (2000a), 'La loi d'orientation du 25 juin 1999 pour l'aménagement et le développement durable du territoire', in *Annuaire Français de Droit de l'Urbanisme et de l'Habitat 2000*, Paris, Dalloz, 9–87.

MARCOU, G. (ed.) (2000b), 'La réforme de l'intercommunalité', *Annuaire des Collectivités locales 2000*, Paris, editions du CNRS.

MARCOU, G., H. KISTENMACHER and H.-G. CLEV (1994), *L'aménagement du territoire en France et en Allemagne*, Paris, La Documentation française.

MESCHERIAKOFF, A. S. (1995), 'La planification française entre centralisation et décentralisation', *Revue française de Droit administratif*, **5**, 999.

MINISTÈRE DE L'INTÉRIEUR. DIRECTION GÉNÉRALE DES COLLECTIVITÉS LOCALES (1999), *Les collectivités locales en chiffres 1999*, Paris, La Documentation française.

MONTRICHER, N. de (1995), *L'aménagement du territoire*, Paris, La Découverte.

MOREIGNE, M. (1995), 'Rapport spécial sur les services du Premier ministre: IV Plan', *J. O. Doc. Sénat* no. 77, session ord. 1995-96, annexe no. 36.

MUZELLEC, R. (1995), 'La péréquation financière entre les collectivités locales: perspectives et réalités', *Revue française de Droit administratif*, **5**, 923.

NÉMERY, J.-C. (ed.) (1994), *Le renouveau de l'aménagement du territoire en France et en Europe*, Paris, Economica.

NÉMERY, J.-C. (1995), 'Les nouveaux instruments d'intervention en matière de développement et d'aménagement du territoire', *Revue française de Droit administratif*, **5**, 914.

PÉPIN, J. (1994), 'Avis au nom de la Commission des affaires économiques et du plan du Sénat, sur le projet de loi de finances pour 1995', *J. O. Doc. Sénat*, 1ère session 1994-95, no. 81, tome X 'Aménagement du territoire'.

PERINET-MARQUET, H. (1995), 'Le nouvel article L. 111-1-1 du code l'urbanisme (l'intégration des directives territoriales d'aménau sein des normes d'urbanisme)', *Revue française de Droit administratif*, **5**, 905.

POLA, G., G. MARCOU and N. BOSCH (eds) (1994), *Investissements publics et régions. Le rôle des différents niveaux de collectivités publiques dans six régions fortes d'Europe*, Paris, L'Harmattan.

PONTIER, J.-M. (1993), 'Contractualisation et planification', *Revue du Droit public et de la Science politique*, **3**, 641.

PORTIER, N. (1997), 'Le pays: un territoire pour le développement local', in G. Loinger and J.-Cl. Némery (eds), *Construire la dynamique des territoires. Acteurs, institutions, citoyenneté active*, Paris, L'Harmattan, 109–21.

RAIGNOUX, R. (1995), *Quel avenir pour la planification française*, Conseil Economique et Social, Rapport et avis adopté.

RAOULT, E. (1994), Rapport au nom de la Commission des finances de l'Assemblée nationale sur le projet de loi de finances pour 1995, 1ère session 1994-95, no. 1560 'Affaires sociales, Santé et Ville: Ville'.

SUEUR, J.-P. (1998), *Demain la ville, rapport présenté au ministre de l'Emploi et de la Solidarité*, Paris, La Documentation Française.

VELITCHKOVITCH, J. (1987), *L'avenir de la planification*, Conseil Economique et Social, Rapport.

Vivre ensemble. Rapport de la commission de développement des responsabilités locales (1976), Paris, La Documentation française.

NOTES

1 Social housing in the European sense means any housing stimulated or subsidised by the public sector, not necessarily built directly by government. It is a broader term than public housing in the US sense.

2 According to the framework regulation on structure funds of 20 July 1993, the objectives pursued by this policy are the following: 1) development and structural adjustment of backward regions; 2) restructuring regions affected by industrial decline; 3) to fight against chronic unemployment and include young and socially excluded people in the labour force; 4) retraining workers so that they can adapt to industrial changes; and 5) promotion of rural areas by a) adapting agricultural structures, and b) facilitating the development and structural adjustment of rural areas.

The various structure funds contribute in a coordinated way to one or several objectives. The structure funds are: the European Fund for Regional Development (EFRD), the European Social Fund (ESF), the European Fund for Agricultural Orientation and Guaranty (EFAOG), which is divided into two sections for Objectives 5a and 5b respectively, and the Financial Instrument for Fishing Orientation (FIFO). Other financial instruments are involved in pursuing these objectives: the newly created Financial Instrument for Cohesiveness (at present for only four countries: Greece, Ireland, Portugal and Spain), the European Investment Bank, the interventions of the European Community for Coal and Steel.

The amount of grants totals 141.5 billion Ecus for the six-year period 1994–1999 (of which 70 per cent has been earmarked for Objective 1). This compares with 43.8 billion Ecus allotted during the four-year period 1989–1992 (Commission des Communautés Européennes, 1993).

3 On the concept of 'policy community', see the recent discussion in Le Galles and Thatcher (1995).

NINE

NATIONAL-LEVEL ECONOMIC AND SPATIAL PLANNING IN JAPAN

Paul H. Tanimura and David W. Edgington

Context and institutional framework

Japan comprises four main islands, Hokkaido, Honshu, Shikoku and Kyushu which, together with more than 6800 small islands, have a land area of 377 800 square kilometres (Fig. 1). Japan's population, at some 125 million, is ranked seventh in the world and second in the OECD (Organisation for Economic Cooperation and Development), and it now has the highest standard of living as measured by its per capita GNP (Foreign Press Centre Japan, 1996). The density levels of population per habitable hectare and GDP per habitable hectare are the highest in the world. These indicators point to the scarcity of habitable land in Japan and the need to manage this important resource effectively (National Land Agency of Japan, 1993).

While Japan is a mixed economy, many scholars have characterised it as a 'developmental state'; one in which government has assumed greater responsibility for fostering national economic growth than is normally the case in either Europe or North America (see Huber, 1996; Johnson, 1982; 1995; McMillan, 1996; Sakakibara, 1993).[1] Consequently, a commitment to national planning has been a material part of the nation's successful economic reconstruction after defeat in the Second World War. Moreover, the need to link spatial policy to the overall 'development-oriented' goals of the state has been an important consideration. As will be shown more clearly later, national development plans in the initial post-war period were driven by the need to 'catch up with the West' and to maximise economic growth. In the last thirty years or so, they have been tempered by social and environmental goals, and the need to stabilise income levels among Japan's various regions.

This commitment to national-level economic and spatial planning has been assisted by continuity of the ruling LDP (Liberal Democratic Party) for more than 40 years[2] and the tradition of a strong bureaucracy (Abe et al., 1994). In addition, Japan has fostered a widespread consensus between industry and the bureaucracy over the need for planning, and while in many countries any debate over planning *per se* often takes on a deep ideological flavour, in Japan very little ideological debate

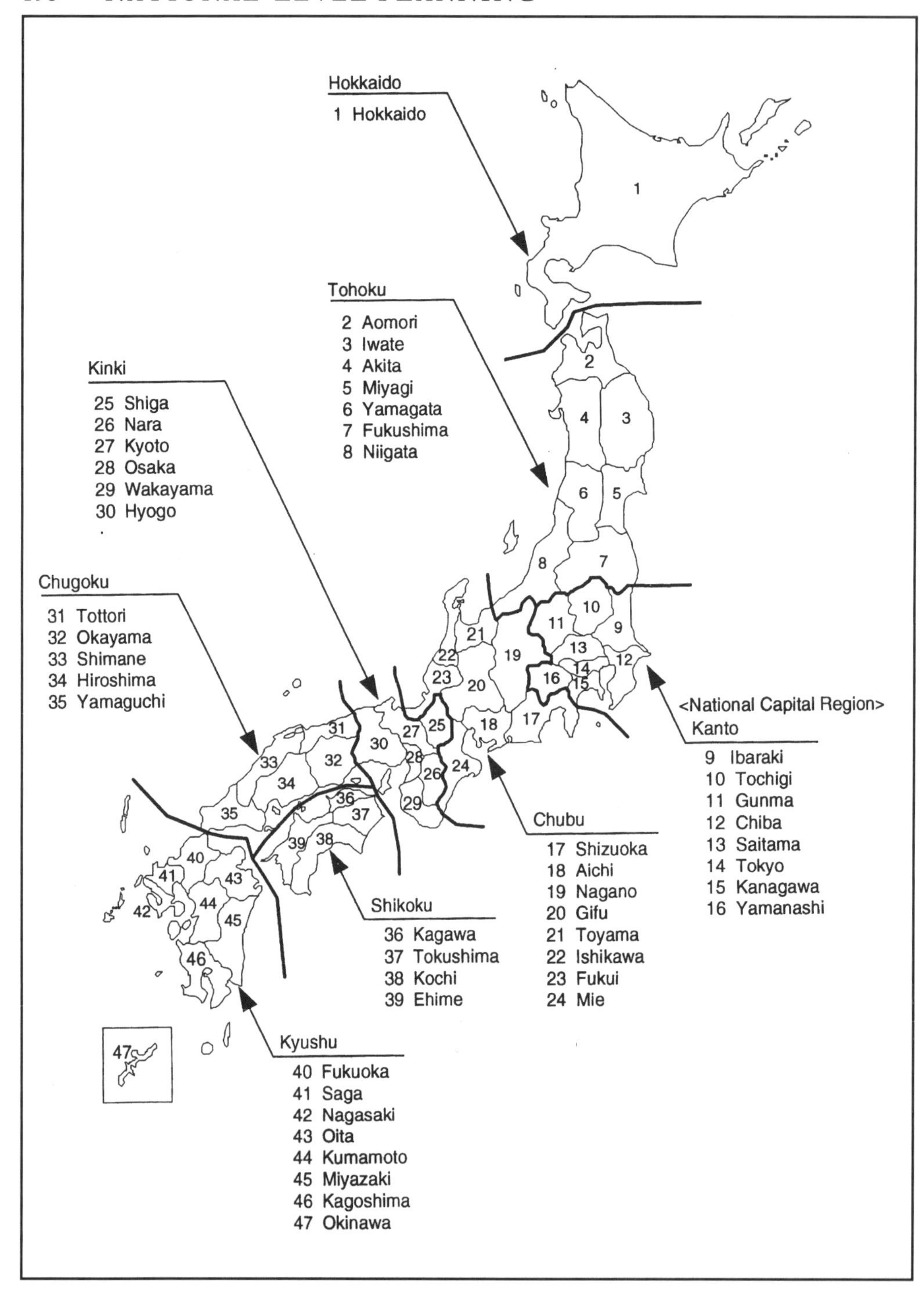

Figure 1 *Regions and prefectures in Japan*
Source: *National Land Agency, 1987.*

has taken place. Certainly, issues such as 'administrative deregulation of the economy' are currently on the agenda in Japan, yet there has been long-term consistency in the government's search for long-term goals and the development of targets to accomplish them (see, for example, Carlile, 1998).

The government also uses a number of coordinating mechanisms to great effect. Besides indicative planning machinery (considered in this chapter), Japan has evolved a set of official forums through which bureaucrats, business groups, academic exports and others are brought together to decide on all manner of public policies. These are the *shingikai* or 'deliberation councils' which contribute to smooth government–business relationships in Japan (Abe et al., 1994). They are conducted in private, shielded from interest groups or the press and under the control of Japan's elitist state bureaucracy (Johnson, 1995). Chalmers Johnson notes that 'by the time a proposed Japanese law gets to the Diet [Japanese parliament], it has already been thoroughly debated and a decision has been reached. All Japan's elected legislators do is rubber-stamp it' (Johnson, 1990, 49). In addition, a wide number of government and semi-government institutions supply domestic information, intelligence, publicity, and propaganda concerning commercial trends and new economic, technological and social challenges. This provides a framework for the corporate sector and facilitates interaction between public and private officials at a much more intense level than in most Western countries. Johnson (1990) argues that the level of awareness of even low public officials towards the problems of private sector issues is probably the highest in the world. Conversely, the influence of the labour unions, consumer groups, environmental groups, and citizens' groups generally is extremely weak when compared with Western countries.

The formal structure of national government agencies is shown in Figure 2. Japan has a parliamentary form of central government. The Diet, composed of two chambers, is the legislative organ. The cabinet is selected from among the members of the Diet as the executive organ. The court of law is the judicial organ independent of the Diet. Under the cabinet are 13 ministries, the most powerful of which are usually considered to be the Ministry of Finance (MOF) and the Ministry of International Trade and Industry (MITI). Together with the MITI, the Ministry of Construction (MOC), the Ministry of Agriculture, Forestry and Fisheries (MAFF) and the Ministry of Transport (MOT) have particular responsibilities for aspects of urban and regional infrastructure provision, as well as regulations relating to land and marine management. The Prime Minister's Office has 11 agencies: the Management and Coordination Agency, the Defence Agency, and six agencies involved in national-level planning requiring interministerial coordination. These six agencies can be grouped into two: one for the development of special regions and the other for the planning of special sectors, namely the environment, the economy, land use, and science and technology. The first group includes the Hokkaido Development Agency and Okinawa Development Agency. The latter group is composed of the Environmental Agency,[3] Economic Planning Agency, National Land Agency, and the Science and Technology Agency. Each of these government agencies and ministries derives its power from legislation passed by the national Diet.

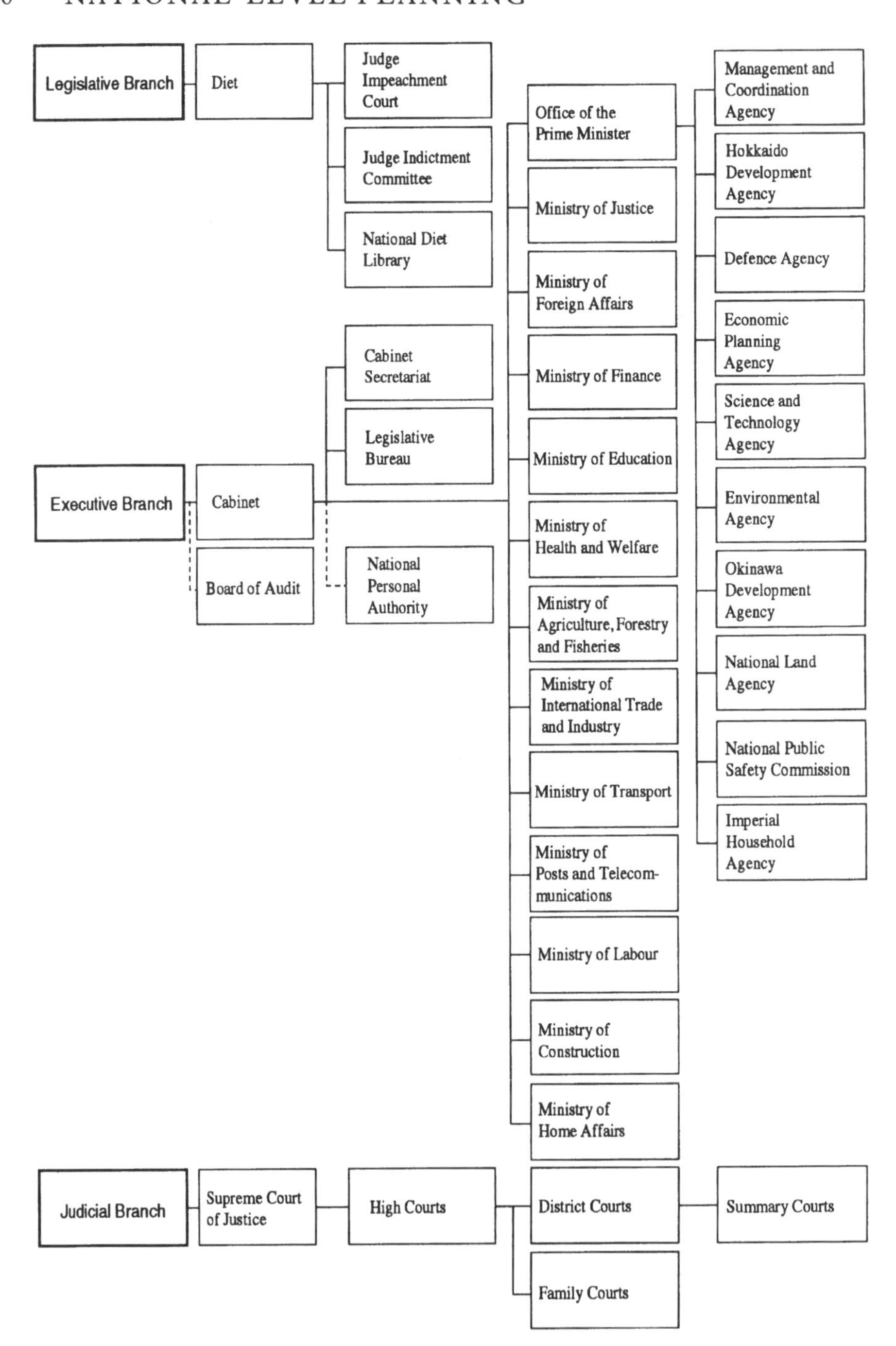

Figure 2 *National government organisation*
Source: Japan International Cooperation Agency/National Land Agency, 1995.

Japan's local government system is two-tier. The first tier comprises 47 prefectural areas (or *ken*), shown in Figure 1, each headed by a popularly elected governor. These are further divided into some 3000 minicipalities (designated as either city, town or village [*shi*, *cho*, or *mura*]). Each municipality has a local government council with directly elected members. The 47 prefectures are aggregated on a regional basis, depending on the nature of the planning issue involved. There is no fixed or standard definition of region in Japan for statistical purposes. For regional development plans, the 47 prefectures are usually aggregated into the ten regions shown in Figure 1, i.e. three metropolitan regions (the National Capital region [also called the Kanto region], and the Chubu and Kinki regions) and seven local or predominantly rural regions (Hokkaido, Tohoku, Hokuriku, Chugoku, Shikoku, Kyushu and Okinawa regions). As will be explained later, local government has an important role in implementing national-level plans and programmes.

A summary of Japan's planning system is shown in Figure 3. In the post-war period a number of national economic plans, national development plans, and national land-use plans have been established, along with complementary plans at a regional or local level. These are mutually influenced by each other, either through statutory or non-statutory links, and implemented mainly by local governments, as well as special government enterprises.[4]

Overall, Japan's planning system is highly centralised, and policy- and plan-making is often a struggle between powerful ministries with conflicting policy interests and directions (Barrett and Therivel, 1991; Johnson, 1989). The Economic Planning Agency (EPA) and the National Land Agency (NLA) are responsible for drawing up national and regional plans but have no separate budgetary powers. Consequently they must coordinate this separately with each ministry or agency charged with carrying out the proposed plan. As explained later, the annual budgetary process is highly politicised. Still, while sectionalism within the bureaucracy is rife, the strong links between ministries and levels of government

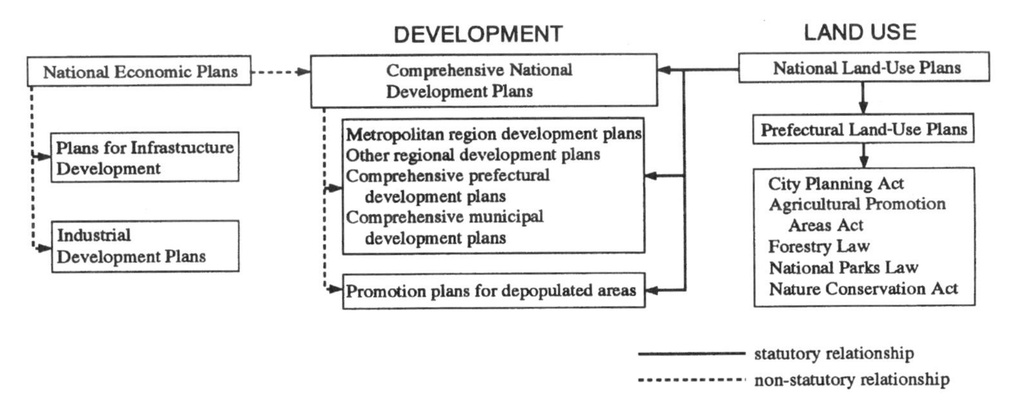

Figure 3 Summary of the planning system
Source: Barrett and Therivel, 1991.

are often seen as a major contributor to the post-war economic success (Johnson, 1982).

The remainder of this chapter explores the framework of Figure 3 by examining the planning work of the Economic Planning Agency and the National Land Agency, and then the implementation responsibility of local government. A final section comments on the validity of this system in meeting the challenges facing Japan in present and future years.

The Economic Planning Agency and national economic planning

Of the four planning agencies shown in Figure 2, the Economic Planning Agency and National Land Agency are most involved in long-term national planning. Comprehensive spatial policy and regional development planning in Japan are closely linked to national economic planning, as the Economic Planning Agency sets national economic targets in terms of annual growth rate of the gross national product and other economic indicators, which are in turn handed over to the National Land Agency for supporting land-use and infrastructure planning (Fig. 3). Consequently, this section discusses Japan's economic plans and how their role has changed over the post-war period.

The Economic Planning Agency is headed by the Minister of State responsible for Economic Planning and composed of six functional bureaux. They are the Director-General's Secretariat, Coordination Bureau, Social Policy Bureau, Price Bureau, Planning Bureau, and Research Bureau. The Economic Planning Agency is responsible for the formulation of the basic policy of steering the national economy as a whole and the annual economic programme and coordination of various basic economic policies concerning industries, labour, transportation, communications, foreign trade and exchange, and finance.

The EPA has created several long-term 'economic plans' since the end of the Second World War (Ito, 1992; Komiya, 1990; Nakamura, 1995; Sheridan, 1993). The word 'plan' might give an incorrect impression that Japan has rigid economic planning. In fact, the long-term economic plan is closer to a forecast than to a plan. Since the EPA does not have any regulatory power, it has less influence on the course of economic policies than the Ministry of Finance or the Ministry of International Trade and Industry. Between 1955, when the government first began formally to set up its own economic policy, and 1995, 13 plans have been adopted (see Table 1). Most were five-year plans, but the average lifespan of a plan has been about two-and-a-half years. Often new plans have been put together in an effort to build up a new Cabinet's image when there was a change of Prime Minister. Japan's post-war economic plans started as rationing plans with a socialist flavour, but soon there was a shift to 'indicative'-style plans that emphasised the market (Ito, 1992).

The post-war economic plans can be grouped broadly into three time periods (Nakamura, 1995). First, there are the plans concerned with recovery from the

Table 1 *National Economic Plans, 1955–2000*

Plan period	*Plan title*	*Average annual growth (%)* Plan	Actual
1955–60	Five-Year Self-Support Plan	4.9	8.8
1958–62	New Long-Range Economic Plan	6.5	9.7
1961–70	Ten-Year Income Doubling Plan	7.8	10.0
1964–68	Medium-Term Economic Plan	8.1	10.1
1967–71	Economic and Social Development Plan	8.2	9.8
1970–75	New Economic and Social Development Plan	10.6	5.1
1973–77	Basic Economic and Social Plan	9.4	3.5
1976–80	Economic Plan for the Latter Half of the 1970s	6.0	4.5
1979–85	Seven-Year Economic and Social Plan	5.5	3.9
1983–90	Economic and Social Guidelines for the 1980s	4.0	4.5
1988–92	Five-Year Plan for Economic Management	3.75	4.9
1992–96	Sharing a Better Quality of Life Around the Globe	3.5	0.6 (estimated)
1996–2000	Social and Economic Plan for Structural Reforms	3.0 (*1.75)	N/A

* Assumes structural reforms are not carried out
Source: Nakamura (1995); Economic Planning Agency, unpublished data.

damage of the Second World War and the achievement of a self-supporting economy. The Economic Rehabilitation Plan for fiscal year (FY) 1949–52 drafted in 1949 (but not approved by government) and the Five-Year Plan for Economic Self-Support for FY 1955–60 fall into this category. The plans in the second category can be called the plans for higher economic growth. Of this group, the Doubling National Income Plan for FY 1961–70 was the most well known and for the first time generally recognised the economy's capacity for rapid growth, and also carried a considerable propaganda effect. As examples of its influence, it prompted private firms to invest more boldly than before and led to a great expansion in plant and equipment investment, together with extensive wage increases obtained by the labour movement (over 13 per cent) in 1961. Throughout the period of high growth (from 1960 to 1973), actual economic growth kept exceeding the planned estimates, so that the targets and thrust of the plans were continually left behind by reality (Table 1).

The third category of plans (from the early 1970s to the present) can be characterised by an emphasis on balanced economic growth and social development. More than their predecessors, they have tended to present an 'outlook' for the economy rather than a hard forecast, and have set only approximate numerical targets for economic growth. Policy goals have also changed, from a fixation on high growth at any price to aims which stress 'achieving harmony with the rest of the world' and 'improving people's quality of life'. The 1992 Economic Plan in particular moved beyond merely indicating the desired direction of economic management and specified more concrete goals in areas such as reduced working hours, affordable housing, and social infrastructure improvements (Economic Planning Agency, 1992).[5] In the period since 1974, all plans (except the 1988 Plan) have overestimated economic growth compared with reality (see Table 1).

The procedure behind economic planning has become routine. Each Cabinet asks the Economic Council (an advisory committee reporting to the Prime Minister consisting of leaders of business, finance, trade unions, and other organisations, scholars, and journalists) to suggest the basic themes of the plan and to prepare a draft. The Economic Council sets up subcommittees and special study groups and invites temporary members of the Council to participate in its deliberations. The Economic Planning Agency serves as secretariat for the Council. The large number of members (often over 200 in total, including subcommittees) might give the impression that the Economic Council is collectively responsible for preparing the plan. In reality, however, the Council members' involvement in plan-making is rather minimal. Thus, most work is done at the subcommittee level, and these groups are dominated by corporate executives and ex-government officials. The Council itself usually only meets a few times. Besides, as the secretariat for the Council, the EPA's Bureau of Comprehensive Planning dominates the process of plan-making (Komiya, 1990). Employing econometric models, all plans make abundant use of refined techniques of economic forecasting and measurement of policy effects. The initial draft prepared by the EPA is circulated among the ministries concerned with its review and revision; at the same time, it is presented to the subcommittees for discussion. Thus, consensus formation occurs within the bureaucracy and corporate sector simultaneously. When the Economic Council presents its draft of the plan to the Cabinet, it is usually approved as the official plan without major alteration.

It is important to note that Japan's economic plans encompass only the general orientation of economic policy. They do not describe in detail how the plans will be implemented or which policy instruments will be utilised. Nor is the national economic plan binding, for at times important policy decisions are made without strict adherence to an existing plan and, in such cases, it is the plan itself that must be revised. In fact, Komiya characterises its growth predictions as a long-term forecast, 'as reliable or unreliable as a long-term weather forecast, and with some flavour of wishful thinking' (Komiya, 1990, 284). Interpreted in this way, it is not surprising that the government publishes a five-year plan as often as once every two or three years. Nonetheless, the very process of economic planning, with its emphasis on sharing information between ministries and between the private and public sectors, and with an emphasis on reaching a consensus, is typical of policy formation in Japan. The process fits well with the Japanese political model of 'patterned pluralism', with its wide scope of participants and deep contacts among the ministries and other organisations involved. On the other hand, consensus is often achieved by authorising the status quo. This approach often deprives economic policy of its efficacy in coping with change (Muramatsu and Krauss, 1987).

This aspect can be more clearly seen when we turn to question the effect of the national plans on economic growth. It is certain that the EPA's 'announcement effect' was more important in the early reconstruction days—especially the National Doubling of Income Plan of 1960. The current plan (Economic Planning Agency, 1995) reflects much higher levels of uncertainty in the 1990s, and today's unstable economic environment in which forecasts have to be made. The 1995 Plan is a significant departure from previous ones as it presents two opposing scenarios. One scenario, which emphasises the necessity of economic deregulation, is positive as it sees the economy growing steadily at an average rate of three per cent between 1996 and 2000, and the jobless rate dropping. The other is rather negative, as it is based on the assumption that economic reforms will not be carried out and the economy's high cost structure (due largely to regulations and official intervention) will stay in place (Table 1). In this case, the economy would continue to stagnate with the growth rate reaching only 1.75 per cent on average and the jobless rate rising to a new high of 3.75 per cent. It is quite unusual that a government economic plan should lay out a bad case with a good case, and may be taken as a tacit admission that it may be extremely difficult for the government to carry out the necessary structural reforms and make the economy more efficient and vigorous.

Since the 1970s, the EPA's plans have actively addressed problems of regional planning. For instance, the 1992 Plan states that efforts will be made to rectify the excessive concentration of the nation's economic, cultural and political life in Tokyo (the national capital), and this aim was continued into the 1995 Plan through a commitment to relocate government functions (i.e. the Diet and administration offices) out of the city.

As shown in Figure 3, the forecasts of the economic plans are taken by each ministry involved in public works in order to establish their own five-year plan. These investment plans cover sewerage systems, urban redevelopment, roads, urban parks and housing (MOC), airports, harbours and railways (MOT), as well as forest, marine and agricultural improvements (MAFF). However, the Ministry of Finance (MOF), possibly the most powerful government ministry in economic policy matters, has a tradition against committing to long-term investment plans. In addition, only in very special cases does the Diet allow the government to make capital expenditures beyond any particular fiscal year. Also of note is the five-year plan for industrial targeting produced by MITI, which controls industrial development and related locational decisions (Fig. 3). The latest MITI industrial vision for the 1990s aims at maintaining Japan's competitive structure, nurturing high-tech industries and fast-growing service sectors (such as medical and information technology), as well as avoiding a 'hollowing out' of industry (see MITI, 1990). Sectoral economic management (for example, MITI's policy of targeting particular industries through selective tax treatment, subsidies and loans) has often been regarded as a more important factor than macroeconomic management or economic plans in explaining Japan's economic performance (Komiya, 1990). However, given the poor macroeconomic performance of the 'post-bubble' economy (i.e. post-1990) and the subsequent depreciation of the yen, together with the Asian financial crisis, the power of sectoral management has declined considerably. A number of economic stimulus packages were announced by the government in 1997 and 1998, focusing in the main on public works investments and tax cuts (Morishita, 1998a).

The National Land Agency, Comprehensive National Development Plans and land-use plans

The National Land Agency, created in 1974, is headed by the Minister of State for the National Land Agency. There are six functional bureaux: the Water Resources Bureau, Planning and Coordination Bureau, Land Bureau, Metropolitan Development Bureau, Regional Development Bureau, and Disaster Prevention Bureau. As an external advisory board, there is the Land Development Council, composed of professionals and experts in the field. Preliminary versions of national plans are debated by this Council to reflect popular opinions (National Land Agency of Japan, 1993).

The role of the NLA is to coordinate the land-use policies of other ministries and to prevent land price speculation through the provisions of the National Planning Act of 1974 (National Land Agency, 1981). The NLA, not having an implementation budget of its own, cannot undertake projects at the local level and has to coordinate (which in effect means lobbying for) projects to be implemented through the five-year investment plans of other ministries and the annual budgetary process. Moreover, the staffing arrangements of the NLA ensure the other ministries are well represented, and often their voice dominates what might have otherwise been the more independent view of the NLA concerning national spatial issues.[6] In the post-war period, five Comprehensive National Development Plans have been formulated. The Agency takes the initiative in formulating each plan through conducting research and analysis of national land development patterns, drawing up the draft plan and consulting with related agencies (note that the Economic Planning Agency was responsible for the plan before 1974) (see Table 2 for a summary of these Comprehensive National Development Plans).

The relationship between the economic plan and the development plan began formally in 1962. In this year, the Comprehensive National Development Plan was decided by the Cabinet, and this plan was in effect the regional dimension of the Doubling of National Income Plan. Since then, the relationship has not been so close because of differences in the relative planning time periods. Nevertheless, as the following account shows, both plans have mutually influenced each other.

THE FIRST NATIONAL PLAN OF 1962

In 1946, the first national recovery plan was made by the Interior Ministry, for a target national population of 80 million. In 1950, the Comprehensive National Land Development Act was established, and in 1953, twenty drainage basins were

Table 2 *Summary of the five comprehensive national land development plans*

Plan	*First National Plan*	*Second National Plan*	*Third National Plan*	*Fourth National Plan*	*Fifth National Plan*
Adoption	1962	1969	1977	1987	1998
Planning horizon	1965–1970	1965–1985	1975–1985	1986–2000	Up to 2010–2015
Background	1. From post-war ruins to economic recovery 2. Regional disparities in demography and economic development 3. Income doubling plan 4. New industrial policy	1. Accelerated economic growth 2. Concentration of population and industries into large metropolitan centres 3. Regional income disparities 4. Efficient utilisation of natural resources	1. Stable economic growth 2. Decentralisation of industries 3. Regional social disparities 4. Limited natural resources 5. Awareness of quality of life	1. One-point concentration into the capital region 2. Decreasing jobs in provincial centres 3. Depopulation in remote prefectures 4. Technological innovation, ageing, global economy and industrial restructuring	1. Changing national ethos from quantity to quality 2. Global environmental problems 3. Globalisation of economic competition 4. Ageing and declining population 5. Advanced information technology
Goals	Economic growth and balanced development of national land	Remaking of infrastructure	Improvement of quality of life	Decentralisation through the development of advanced interaction network	Foundation of the grand design of national land for the twenty-first century
Objectives	1. Control of urban expansion and reduction of economic disparities 2. Efficient use of natural resources 3. Equitable allocation of capital, labour and technology 4. New industrial cities as growth poles	1. Preservation of nature 2. Expansion of development potential by improved infrastructure 3. Recognition of regional characteristics 4. Emphasis on safety, quality of life and culture	1. No more exploitation of natural resources 2. Appreciation of historical and traditional values 3. Balance between man and nature	1. No more concentration into Tokyo region 2. Strategic development of regional centres	1. Restructuring of the hierarchical urban system into the lateral lattice 2. Self-reliant and inter-linked regions
Planning instruments	Development of growth poles	Big national projects	Stable habitation zones	Multiple-pole decentralisation	Formulation of four developmental corridors
Major issues	1. Relocation of industries from metropolitan centres into remote regions 2. Medium-scale industrial parks 3. Development of new industrial centres for petrochemical, steel, and shipbuilding industries along coastlines	1. Interregional limited-access highways, super-express railways and information networks 2. Mega-scale industrial parks 3. Demarcation of settlement regions for public services planning	1. Preservation of historic environment 2. Integration of planning for jobs, housing and environment 3. Equalisation of availability of education, culture and health services	1. Promotion of interaction networks 2. Vitalisation of regional economies by innovation 3. Development of high-tech infrastructure	1. Creation of communities close to nature 2. Renovation of metropolitan regions 3. Promotion of regional interaction corridors 4. Development of regions with international interactions

designated for comprehensive development, taking the Tennessee Valley Authority (TVA) as a model. Leading industries at this time were labour-intensive manufacturing such as textiles and clothing. By 1960, post-war economic recovery was firmly under way and the first Comprehensive National Development Plan was adopted in 1962 following the National Income Doubling Plan. 'Growth pole' theory was the leading concept in industrial decentralisation, and this led to a plan for a number of new industrial centres, mostly along the Pacific coastline. The economic bases of these locations were to be 'smokestack industries' (e.g. petrochemical complexes, iron and steel production, and shipbuilding), adopting the most advanced production technology of the time. Raw materials were imported by huge ships directly into these industrial centres (also known as *kombinats* in Japan) and the produce exported directly to foreign markets, resulting in highly efficient industrial production. In later years, however, these first-generation industrial new towns became the symbol of 'economic growth at the expense of industrial pollution and environmental neglect'. Nonetheless, a total of 21 cities were planned as industrial growth poles and regional disparity in income was narrowed partly as a result.[7]

In the latter half of the 1960s, three issues manifested themselves as the unfavourable consequences of successful economic development through the heavy-industry growth pole strategy. First, industrial pollution occurred, such as heavy mercury poisoning in Minimata (in southern Kyushu), and asthma attacks caused by air pollution in Yokkaichi (Aichi prefecture) and Kawasaki (Kanegawa prefecture). Second, it became clear that the nation's infrastructure—such as highways, railways and trunk telephone lines—was inadequate to meet increasing demand created by high economic growth. Third, despite the larger number of industrial cities, rural depopulation increased in such a way as to make it difficult to maintain basic public services such as primary schools and outpatient clinics. Consequently, a review of the first national plan was initiated even before the end of the projected planning horizon. The growth pole strategy, although highly successful, was considered inadequate, and a structural change in national land policy was needed to realise further economic growth. At the same time, urban policy became a major political issue because of the increased importance of existing urban centres in the national economy.

THE SECOND NATIONAL PLAN OF 1969

In 1967, the Liberal Democratic Party established an Urban Policy Forum (headed by Prime Minister-to-be Kakuei Tanaka) that emphasised the need to formulate an integrated national policy on land use over and above the development of individual growth centres. This led to the second national plan, which was aimed at bringing structural changes to the communication network of the nation. Tanaka's concept was published as a book, *The Remaking of Japan*, which eventually became the year's bestseller in 1972. Its stated objective was to modernise railways and highways so that people from anywhere in Japan could conduct business in Tokyo in just a one-day trip. Soon after, the first super-express railway was built between

Tokyo and Osaka to connect the two cities in three hours, making it possible for a Tokyo business traveller to go to Osaka by train and come back by evening.

Following the improvement in transportation networks in the plan was the initiative to relocate large industrial establishments away from the major metropolitan centres. Laws were drawn up to prohibit new industries locating in urban areas and a number of industrial parks were developed to accommodate their relocation in regions away from Tokyo and Osaka. The thrust of industrial development shifted from 'heavy, thick, long and large' production to 'light, thin, short and small', or from material-based industries (e.g. steel and shipbuilding) to processing and assembly industries (e.g. sophisticated consumer and industrial electronics), adding more value with less pollution. Consequently, the location of these new industries favoured well-connected industrial parks along expressways rather than coastal sites.

THE THIRD NATIONAL PLAN OF 1977

The Third National Plan was placed on the drawing board in the autumn of 1973, about the same time as the crisis in the Middle East quadrupled the price of oil and ended Japan's era of 'high growth'. At the same time, citizens in Japan began to complain of a relatively poor quality of life in spite of their increase in nominal incomes. Thus social 'stabilisation' rather than migration to industrial cities for economic growth was considered necessary (National Land Agency, 1979). Atsushi Shimokobe, senior planner of the National Land Agency, advocated integrated development based on the drainage region as a unit to achieve balance between its water resource and water needed for agriculture and industry. The idea became known as the 'Integrated Settlement Zone' concept. About the same time, Masayoshi Ohira became Prime Minister and initiated the 'Garden City' concept. This idea was aimed at the balance between the countryside and urban areas, rather like the English Garden City movement, but it was presented as a conceptual model for a much larger urban area. These two ideas eventually became merged into the Third National Plan and some 300 drainage regions (or 'integrated settlement zones') were envisaged as the nation's planning unit; places where nature and human activities could be in a state of sustainable balance. This number may be compared with Japan's 47 prefectures, the first tier of the local government system, and some 3300 municipalities, the second tier.[8]

In the process of restructuring the national economy through two oil crises, the nation's industrial base was transformed from simple processing and assembly, such as televisions and home appliances, to technology-intensive products, such as cars and computers. Financial sectors of the economy gained much importance. Research and development of technology for industrial application became prominent. In short, the nation's 'knowledge industries' continued to concentrate in three large urban centres—Tokyo, Osaka and Nagoya, but overwhelmingly into Tokyo. The labour surplus in rural areas became almost drained to make any industrial development in the countryside unfeasible. A new concept for the national plan became necessary.

THE FOURTH NATIONAL PLAN OF 1987

The main target of the Fourth National Plan was to stop further concentration of the nation's activities in Tokyo. The problem became known as the 'one-pole concentration into Tokyo' at the expense of the viability of other urban centres. In spite of the policy to relocate industries and universities away from Tokyo since the implementation of the Third National Plan, the population in the Tokyo region continued to grow (National Land Agency, 1987).

At the same time, the planners' dream of making 300 self-sufficient drainage regions for sustainable settlement turned out not to be feasible. Japan's economic base very quickly shifted in the 1980s from agriculture and mass-production into sophisticated 'information-intensive' industries (e.g. telecommunications and computers) which require dense information infrastructure to be viable. Accordingly, the idea emerged of selecting a small number (about half a dozen or so) of urban centres which could compete internationally, independently (but complementarily) to Tokyo and Osaka, and so a 'multi-polar' approach was embedded in the Fourth Plan. While the early years of the plan (1987–1990) led to a short-lived but affluent 'bubble economy' (based on rising prices of land and stock assets), this eventually 'burst' in 1992. For instance, land prices in Tokyo increased three times between 1985 and 1991, but declined by half in the 'post-bubble' period up to 1995. Despite the slow-down in the economy, the yen continued to appreciate in that year to the historically high rate of US$1 to 80 yen. Now, even small and medium industries are leaving Japan to relocate in developing countries, resulting in an industrial 'hollowing out'. Japan's unemployment rate in 1995 was 3.5 per cent, which is another historical high for post-war Japan. Furthermore, the population is rapidly ageing, and people over 65 years of age, comprising 12 per cent of the population in 1996, will account for around 24 per cent in the year 2025. As a result, a fresh concept for a new national plan has been eagerly sought.

THE FIFTH NATIONAL PLAN OF 1998

The Fifth National Plan was finally approved by the Japanese Cabinet on 31 March 1998 with no political fanfare. It noted that the catching-up phase of Japanese economic development had ended and the people's aspirations had shifted from more material accumulation to more emphasis on the quality of life and harmony with nature. The plan also confirmed the ageing structure of Japan's population, noting that there were now more people aged 65 years or older than those 15 years or younger. The plan basically proposed to change the hierarchical urban system in Japan, with Tokyo at the summit, to a network type of urban system with emphasis on interdependence and collaboration. How such a mission was to be accomplished was not made clear, although the plan pointed to the expansion of information technology as a key. It calls for further development in Japan to take place in four decentralised 'belts', which would shift the focus of economic activity away from the most earthquake-prone regions in Japan along the Pacific coast. The plan, which is scheduled for completion between the years 2010 and 2015, also includes construction of regional hub airports for same-day travel to East Asia, and six major bridges across the nation's key straits of water. There is probably a good reason for the plan's failure to catch the public attention. Many commentators believe that, in light of the current economic crisis in Japan, priority should be given to restructuring the national government and its centralised decision-making system, rather than the redrawing of a national physical plan (Morishita, 1998b).

REGIONAL AND LOCAL DEVELOPMENT PLANS

As suggested by Figure 3, a number of regional development plans exist for each of Japan's eight regions (shown in Fig. 1). These were established under the National Capital Region Development Act of 1956, the Kinki Region Development Act of 1963, and the Chubu Region Development Act of 1966. These basic plans, as updated, provide a framework for population growth and land use, and set out policy for infrastructure provision. Development plans also exist for the Tohoku, Hokuriku, Chugoku, Shikoku and Kyushu regions (Fig. 1) and are formulated by the NLA in consultation with the prefectural governments. Hokkaido and Okinawa have their own land agencies to prepare development plans. At a lower level still, comprehensive prefectural development plans have been prepared by all the prefectural governments and about 90 per cent of municipal governments. Often these plans are merely statements of intent rather than definite plans with budgets and projects ready for implementation (Abe and Alden, 1988).

In addition to the Comprehensive Development Plan and various regional plans, MITI's industrial development plans promote industrial relocation throughout Japan, especially from the major metropolitan centres to the more peripheral regions, through a combination of both positive and negative incentives. This is done to redress regional disparities and control over-concentration of industry in large cities. In the last 15 years or so, MITI and other government ministries have promoted high-technology and information-based industries through a number of acts and programmes (see Edgington, 1994a).

Finally, as indicated in Figure 3, the NLA has responsibility for certain promotion plans for areas in Japan designated for special economic assistance, such as remote mountain villages which are experiencing rapid depopulation (National Land Agency, 1993). Since the 1950s, acts for these disadvantaged areas, which are scattered across the country, have been established one by one. These include the 'Act on the Extraordinary Measures for Disaster Prevention and Development in Areas with Special Soils' (1952), the 'Remote Islands Development Act' (1953), and the 'Act on Special Measures for Disaster Prevention in Areas Yearly Struck by Typhoons' (1958) of the 1950s. The 1960s saw the establishment of the 'Act on Special Measures for Promotion of Coal Producing Areas' (1961), the 'Act on Special Measures for Snowy Areas' (1962), the 'Mountainous Villages Development Act' (1965) and so on.

NATIONAL LAND-USE PLANNING

The National Land Use Planning Act of 1974 established a series of national, prefectural and local land-use plans to ensure coordinated control of land use and environmental conservation. Under this legislation, the NLA prepares a national land-use plan for five land-use categories: urban, agricultural, forest, natural parks, and nature conservation. This 'broad-brush' plan is spatially interpreted by each prefecture in prefectural land-use plans which broadly indicate on maps the location of the five land-use categories. These designations are only indicative, however, and have no legal value in controlling development. The statutory framework for detailed land-use control is administered by local governments under the City Planning Act of 1968 (as updated in 1992), the Agricultural Areas Promotion Act and the Nature Conservation Act (see Callies, 1994; Edgington, 1994b).

Local government and its implementation role

In the process of implementing plans, central government plays an overseeing role as fund supplier rather than acting as the direct instigator.[9] Consequently, one of the distinctive characteristics of public finance in Japan is that around 80 per cent of all government spending takes place at the local level (compared with a little over 50 per cent in the United States and the former West Germany) (Tabb, 1995). As local government is thus primarily responsible for implementing the various plans noted above, we should inquire what sort of central–local relations have been established in Japan (see Akizuki, 1995).

In essence, the legal structure of local government is highly centralised and characterised by strong administrative and financial control by the national government. As many researchers have pointed out, the relations between national and local government which characterised the pre-war period still persist, in spite of the democratic reforms following the Second World War, and in spite of rapid economic progress. It is often pointed out that local government has in practice '30 per cent autonomy' (*sanwari jichi*) (Horie, 1996; Nagata, 1996; Shindo, 1984).

Muto (1996) recognises several categories of difficulties. First, and perhaps the most well-known example of this top-down system of implementation, there are the so-called 'agency-assigned' functions, such as education or welfare, which the national government has delegated to local level. These services are implemented according to strict national-level standards, and according to a centrally determined financial formula. Tabb notes that this extends even to 'the selection of school superintendents, the contents of textbooks, and even the size of school gymnasiums' (Tabb, 1995, 196).

Second, there is the fact that local governments' own revenues (local taxes and charges), which they can spend freely, are on average only 30 per cent of total revenue. As for total tax revenue, central government collects about 60 per cent of taxes and local governments together collect roughly 40 per cent. Central government then gives about half of its tax revenue to local government. Thus, central government's share in the total tax becomes about 30 per cent, while local government's share is 70 per cent. Yet while this might seem generous, about half of the transferred money is in the form of specific grants—usually for particular public works such as roads, urban parks, rural drainage schemes and so on—the content of which is controlled by central government ministries. The grant rate, which means the percentage of grant in the total cost of any particular programme, ranges from 10 per cent to 100 per cent (local government must meet the rest of the cost of the programme or public works). Through these grants, central government thus steers local government expenditure, and it is evident that for public works spending, central government's subsidies form a large portion of any local government's total budget.[10]

Third, the central ministries often dispatch to prefectural governments an official of department director level who then acts as chief policy-maker for the prefectural government concerning a particular issue area (e.g. urban development and planning). Even though such a bureaucrat is assigned on a temporary basis from the corresponding central government ministry or agency, these 'visitors' usually act within the orders of their ministry or agency, rather than taking innovative or local approaches to regional development and planning.

The positive aspect of such strong central control is that the provision of services is highly uniform throughout Japan, with very little variation between rich (usually metropolitan) and poor (usually rural) prefectures (Horie, 1996). In fact, part of the income transfer from central to local government is specifically made to equalise the disparities in financial capabilities between local governments. This has in part led to the convergence of regional income levels over time, as noted earlier (see Abe and Alden, 1988).

The negative aspects of this system include the expensive and wasteful amounts of time and energy that local government expends in the pursuit of central government subsidies. Stories abound of subsidies that amounted to less than the cost of applying for them (e.g. through expensive journeys to Tokyo-based ministries by officials from peripheral prefectures, towns and villages [Abe et al., 1994]). Moreover, many agencies give small subsidies for similar projects independently of one another. As a result, when local government tries to cover a single project by adding together a number of separate subsidies from different agencies, its efforts are made far more difficult by the rivalries between ministries and even between bureaux in given ministries (Abe et al., 1994). Furthermore, the extreme 'vertical compartmentalisation' (*tatewari gyosei*) and strong relationships between individual local government departments and central ministries (e.g. the local planning control department is often strongly connected to the Ministry of Construction) tend to weaken local government's own institutional coordination (Muramatsu, 1986; Samuels, 1983).

Local government therefore finds it difficult to develop innovative policies under such a centralised system, and such strong centralisation produces passive local policy-makers. The quality of local government's regional policies is also often poor because the centralisation of power means that local government does not possess talented policy-makers. Consequently, the real democratic control of any local government by its citizens is severely reduced (Muto, 1996).

Evaluation

The distinctive feature of the national planning and decision-making process in Japan is that it is highly centralised but has a much greater input of information than other systems. While more comprehensive and centralised than most other countries, it has had far happier consequences than, say, its counterpart in the former Soviet Union. This is because it is carefully limited and pluralistic. Thus, rather than trying to micro-manage the entire economy, its emphasis has been on setting broad goals and objectives in conjunction with the private sector, and ensuring coordinated mechanisms for public sector implementation (mainly through public works schemes) at the regional and local levels. Such coordinated mechanisms have generated stability by reducing transaction costs and the risks that each participant faces.

Nevertheless, the stability which the Japanese planning system enjoys today does not guarantee stability tomorrow. Central and local government both have to deal with dynamic changes in society, such as concerns over the ageing population, internationalisation, and Tokyo's hyper-concentration and congestion, all of which require policy intervention (Hoshino, 1996; Ozawa, 1994). Most probably, how the government responds to this call for decentralisation will determine the long-term viability of Japan's planning system and economic development in the future. Accordingly, while positive changes to the system—such as the promotion of decentralisation—are currently being discussed (see, for example, Foreign Press Centre Japan, 1995), it is recommended that consideration be given to the following proposals.

Greater control over public works should be given to local government. With the exception of nationwide networks of expressways and international airports, public works and welfare, projects should in the main be left to local government. A smaller number of local government bodies (say 300 municipalities and a smaller number of prefectures) with their own sources of revenues (through changing tax allocation power to correspond with spending patterns) and upgraded human resources would be able to use their budget freely to meet residents' demands.

Moreover, strengthening the function of the Diet and the National Land Agency to keep a close eye on public works projects is also an urgent task. Much criticism of the present system concerns the collusion between politicians and the national construction industry, which benefits from the highly centralised planning system. Scholars such as McCormack (1995) and Woodall (1996) argue that while civil engineering is at the heart of Japan's political economy, it takes place beyond public scrutiny and with little serious assessment of environmental impacts (see also Barrett and Therivel, 1991). This may require refiguring the relationship between the budgetary and regulatory powers of the national ministries, including the powerful Ministry of Finance, and the coordinating requirements of the National Land Agency in favour of the latter. Clearly, the quality of regional development in the future depends on thorough decentralisation under the strong leadership of the Prime Minister and the National Land Agency, with strong support from the Diet.

Finally, the development of towns and cities in Japan has been led almost solely by government authorities or private developers. Citizens' groups need to be encouraged to promote the quality of life at the local level. Despite growing efforts to help such groups, many tax and legal regulations stand in their way, and these should now be changed.

REFERENCES

ABE, H. and J. D. ALDEN (1988), 'Regional Development Planning in Japan', *Regional Studies*, **22**, 429–38.

ABE, H., M. SHINDO and S. KAWATO (eds) (1994), *The Government and Politics of Japan*, Tokyo, University of Tokyo Press.

AKIZUKI, K. (1995), 'Institutionalizing the Local System: The Ministry of Home Affairs and Intergovernmental Relations in Japan', in Hyung-Ki Kim et al. (eds), *The Japanese Civil Service and Economic Development: Catalysts of Change*, Oxford, Clarendon, 337–66.

BARRETT, B. F. D. and R. THERIVEL (1991), *Environmental Policy and Impact Assessment in Japan*, London, Routledge.

CALLIES, D. L. (1994), 'Land Use Planning and Control in Japan', in P. Shapira, I. Maser and D. W. Edgington (eds), *Planning for Cities and Regions in Japan*, Liverpool, Liverpool University Press, 59–69.

CARLILE, L. E. (1998), 'Business and Government Relations in Japan and Canada: The Homestead and the Public Vessal', in K. Nagatani and D. W. Edgington (eds), *Japan and the West: The Perception Gap*, Aldershott, Ashgate, 109–32.

ECONOMIC PLANNING AGENCY (1992), *The Five-Year Economic Plan: Sharing a Better Quality of Life around the Globe*, Tokyo, Economic Planning Agency.

ECONOMIC PLANNING AGENCY (1995), *Social and Economic Plan for Structured Reforms: Towards a Vital Economy and Secure Life*, Tokyo, Economic Planning Agency.

EDGINGTON, D. W. (1994a), 'Planning for Technology Development and Information Systems in Japanese Cities and Regions', in P. Shapira, I. Masser and D. W. Edgington (eds), *Planning for Cities and Regions in Japan*, Liverpool, Liverpool University Press, 126–54.

EDGINGTON, D. W. (1994b), 'Chronology of Major Urban and Regional Planning Legislation in Japan', in P. Shapira, I. Masser and D. W. Edgington (eds), *Planning for Cities and Regions in Japan*, Liverpool, Liverpool University Press, 184–89.

ENVIRONMENT AGENCY OF JAPAN (1995), *The Basic Environment Plan—An Outline*, Tokyo, Environment Agency.

FALLOWS, J. (1994), *Looking at the Sun: The Rise of the New East Asian Economic and Political System*, New York, Vintage.

FOREIGN PRESS CENTRE JAPAN (1995), 'Government Struggles with Decentralization', *Press Guide*, **XX**, 1–3.

FOREIGN PRESS CENTRE JAPAN (1996), *Japan: A Pocket Guide*, Tokyo, Foreign Press Centre Japan.

HORIE, F. (1996), 'Intergovernmental Relations in Japan: Historical and Legal Patterns of Power Distribution between Central and Local Governments', in J. S. Jun and D. S.

Wright (eds), *Globalization and Decentralization: Institutional Contexts, Policy Issues and Intergovernmental Relations in Japan and the United States*, Washington, DC, Georgetown University Press, 48–67.

HOSHINO, S. (1996), 'Japanese Local Government in an Era of Global Economic Interdependency', in J. S. Jun and D. S. Wright (eds), *Globalization and Decentralization: Institutional Contexts, Policy Issues and Intergovernmental Relations in Japan and the United States*, Washington, DC, Georgetown University Press, 359–73.

HUBER, T. H. (1996), *Strategic Economic Planning in Japan*, Boulder, CO, Westview.

ITO, T. (1992), *The Japanese Economy*, Cambridge, MA, MIT Press.

JAPAN INTERNATIONAL COOPERATION AGENCY/NATIONAL LAND AGENCY (1995), *The National Administrative Organization in Japan*, Tokyo, JICA/NLA (mimeo).

JOHNSON, C. (1982), *MITI and the Japanese Miracle*, Stanford, CA, Stanford University Press.

JOHNSON, C. (1989), 'MITI, MPT, and the Telecom Wars: How Japan Makes Policy for High Technology', in C. Johnson, L. D'Andrea Tyson and J. Zysman (eds), *Politics and Productivity: The Real Story of Why Japan Works*, Cambridge, MA, Ballinger, 177–241.

JOHNSON, C. (1990), 'The Japanese Economy: A Different Kind of Capitalism', in S. N. Eisenstadt and Eyal Ben-Ari, *Japanese Models of Conflict Resolution*, London, Kegan Paul International, 39–59.

JOHNSON, C. (1995), *Japan: Who Governs?: The Rise of the Development State*, New York, W. W. Norton.

KOMIYA, R. (1990), *The Japanese Economy: Trade, Industry and Government*, Tokyo, University of Tokyo Press.

McCORMACK, G. (1995), *The Emptiness of Japanese Affluence*, Armonk, NJ, M. E. Sharpe.

McMILLAN, C. J. (1996), *The Japanese Industrial System* (third revised edition), Berlin, Walter de Gruyter.

MINISTRY OF CONSTRUCTION (1992), *Activities of the Ministry of Construction*, Tokyo, International Engineering Consultants Association (Japan).

MINISTRY OF INTERNATIONAL TRADE AND INDUSTRY (MITI) (1990), *International Trade and Industry Policy in the 1990s*, Tokyo, MITI.

MORISHITA, K. (1998a), 'Next Stimulus Plan Begins to Take Shape', *Nikkei Weekly*, 16 March, 1, 4.

MORISHITA, K. (1998b), 'National Plan Targets Quake-Safe Areas for Development', *Nikkei Weekly*, 6 April, 3.

MURAMATSU, M. (1986), 'Center-Local Political Relations in Japan: A Lateral Competition Model', *Journal of Japanese Studies*, **12**, 303–27.

MURAMATSU, M. and E. S. KRAUSS (1987), 'The Conservative Policy Line and the Development of Patterned Pluralism', in K. Yamamura and Y. Yasuba (eds), *The Political Economy of Japan: Vol. 1, The Domestic Transformation*, Stanford, CA, Stanford University Press, 516–54.

MUTO, H. (1996), 'Innovative Policies and Administrative Strategies for Intergovernmental Change in Japan', in J. S. Jun and D. S. Wright (eds), *Globalization and Decentralization: Institutional Contexts, Policy Issues and Intergovernmental Relations in Japan and the United States*, Washington, DC, Georgetown University Press, 68–73.

NAGATA, N. (1996), 'The Roles of Central Government and Local Government in Japan's Regional Development Policies', in J. S. Jun and D. S. Wright (eds), *Globalization and Decentralization: Institutional Contexts, Policy Issues and Intergovernmental Relations in Japan and the United States*, Washington, DC, Georgetown University Press, 157–75.

NAKAMURA, T. (1995), *The Postwar Japanese Economy: Its Development and Structure, 1937–1994*, (second edition), Tokyo, University of Tokyo Press.

NATIONAL LAND AGENCY OF JAPAN (1979), *The Third Comprehensive National Development Plan*, Tokyo, NLA.

NATIONAL LAND AGENCY OF JAPAN (1981), *The National Land Use Planning Act: The Comprehensive Land Development Act* (National Land Policy Series, No. 0-1), Tokyo, NLA.

NATIONAL LAND AGENCY OF JAPAN (1987), *The Fourth Comprehensive National Development Plan*, Tokyo, NLA.

NATIONAL LAND AGENCY OF JAPAN (1993), *The National Land Agency*, Tokyo, NLA.

OZAWA, I. (1994), *Blueprint for a New Japan*, Tokyo, Kodansha.

SAKAKIBARA, E. (1993), *Beyond Capitalism: The Japanese Model of Market Economics*, Lanham, MD, University Press of America.

SAMUELS, R. J. (1983), *The Politics of Regional Policy in Japan: Localities Incorporated?*, Princeton, NJ, Princeton University Press.

SHERIDAN, K. (1993), *Governing the Japanese Economy*, Cambridge, Polity.

SHINDO, M. (1984), 'Relations between National and Local Government', in K. Tsuji (ed.), *Public Administration in Japan*, Tokyo, University of Tokyo Press, 109–20.

TABB, W. K. (1995), *The Postwar Japanese System: Cultural Economy and Economic Transformation*, New York, Oxford University Press.

WOODALL, B. (1996), *Japan under Construction: Corruption, Politics and Public Works*, Berkeley, CA, University of California Press.

NOTES

1 Though not in the case of other East Asian countries (see Fallows, 1994).

2 The LDP was out of power in 1993–94 and formed a coalition with the Social Democratic Party in 1994–96. Following the election of October 1996, the LDP was returned to power after playing a lesser role for three years.

3 The Environment Agency of Japan has responsibility for implementing the 1995 Environment Plan (Environment Agency of Japan, 1995). This lays down environmental management policies such as waste reduction and recycling, protecting water environments, soil and land, and the atmosphere. However, in many cases the implementation of these broad policies has to be coordinated with other ministries and agencies (Barrett and Therivel, 1991).

4 Apart from the local government apparatus, there are several specially organised public corporations responsible for carrying out public infrastructure provision at the local level. Examples are the Housing and Urban Development Corporations, the Japan Highway Public Corporation, the Metropolitan Expressway Public Corporation, the Honshu Shikoku Bridge Authority, and the Housing Loan Corporations. Except for the last-mentioned, these institutions were established especially for the purpose of executing large-scale projects (Ministry of Construction, 1992).

5 These were never achieved during the plan's period (1992–96), mainly due to the

collapse of Japan's 'bubble economy' and the recession brought on by a sharp appreciation of the yen.

6 Thus, among the 200 or so employees of the NLA in 1995, 45 comprised administrative or technical officers seconded from the MOC, 30 from MAFF, 30 from MITI, and 20 from MOT, leaving just 10 full-time NLA staff and the remainder from the Office of the Prime Minister (interview with T. Kobayashi, Deputy Director, General Affairs Division, National Land Agency, Tokyo, January 1995).

7 In 1960, over 31 prefectures (out of 46) had incomes less than half of the income of the Tokyo capital region. In 1975, all prefectures had reported income more than the benchmark figure. This narrowing of income disparity was, of course, realised not only by industrial development in the regional growth poles, but also by massive population out-migration from the poorer provincial centres into the major urban centres along the Pacific industrial belt zone (e.g. Tokyo, Osaka and Nagoya) (Abe and Aden, 1988).

8 Incidentally, there were some 300 feudal states in the Edo period prior to Japan's Meiji restoration in 1868. The size of each feudal state reflected the importance of water management for rice production, the major economic activity of this period.

9 Although prefectures also instigate work programmes directly, they too often play a role similar to that of the state as fund supplier.

10 One international comparison of government expenditure carried out by Sakakibara (1993) has shown that Japanese public spending heavily emphasises public works. By comparison, Japan's expenditure on defence, medical services and welfare is much lower than other industrialised countries. This result would seem to confirm the general assessment of Japan as a 'construction state' economy, and the importance of state attention given to engineering and public works schemes, such as roads, sewerage, parks, agriculture, forestry and marine public works (see McCormack, 1995, Woodall, 1996).

ACKNOWLEDGEMENT

David W. Edgington wishes to acknowledge the funding received for his contribution by the Hampton Fund, University of British Columbia.

TEN

DUTCH NATIONAL PLANNING AT THE TURNING POINT: RETHINKING INSTITUTIONAL ARRANGEMENTS[1]

Hans (J. M.) Mastop

Introduction

The Dutch like the idea that foreigners visiting the Netherlands are more often than not impressed by the fact that the country looks tidy, clean and well organised. Needham (1989), speaks of a 'manicured environment'. Dutch spatial planners like to think that they have contributed to that manicured environment and that they have 'planned' its development. But Needham warns the reader: 'although there is massive public intervention with changing and maintaining the physical environment, and that massive public intervention indeed results in an overall good or even high quality, one would be wrong to include this under the planning system, "for it is not part of that system in the sense of statutory physical planning"' (Needham, 1989, 15). He argues that, in tandem with the virtues of orderliness and cleanliness and the egalitarianism of Dutch society, the active land policy of the municipalities and the highly effective work of executive agencies like those for water management, land consolidation, infrastructure, heritage conservation and forestry, contribute more to that 'manicured environment' than the planning system itself. Needham is willing to recognise that the Dutch physical environment is 'planned' only where he is convinced that the system has succeeded in coordinating public intervention.

With his rather paradoxical appreciation of the Dutch situation, Needham is really questioning the performance—or effectiveness—of the Dutch planning system. Its primary aim has been to secure that coordination of public intervention in the physical environment. In this paper we discuss whether or not it succeeds. We will limit ourselves to the institutional and legal-organisational arrangements that shape Dutch strategic planning at the national level and which have gained momentum since the 1950s. It is an important instrument in building the Dutch welfare state (Faludi, 1989). Here, we will examine the conditions within which the system has to work and ask whether it is functioning as it was intended.

To answer these questions, a broad outline of the context of strategic spatial planning in terms of some socio-economic statistics, the governmental system and the planning tradition will be given. This is followed by a description of the institutional setting of the planning system as it came into being in the mid-1960s. This has basically remained stable during the last three decades. The next section is devoted to how the system functioned and how it developed at the national level. This will show that within the same legal-organisational setting, strategic planning at the national level dramatically changed in a relatively short period of time. The final section concludes with an evaluation of the adjustments the Dutch planning system is making to the realities of the twenty-first century in general and to the European Union in particular.

The context

Whatever credit is due to national strategic planning in the Netherlands comes in the context of a small country, with a stable governmental system, a well-developed planning tradition, and a high level of continuity in overall policies.

SOME BASIC FEATURES

The Netherlands is a small country covering just over 41 500 square kilometres including nearly 500 square kilometers of the waters of the North Sea and the Ijsselmeer, the former Zuider Zee (see Fig. 1). The population amounts to well over 15 million inhabitants. Many areas in the north and west (more than 25 per cent of the country), as well as the polders of the former Zuider Zee (now the twelfth province, Flevoland), are below sea level. Sixty per cent of the population lives in this area.

The Dutch population is growing slowly. Between 1980 and 1992 it increased by eight per cent. It is expected to grow by almost one million, to 16.5 million, and to stabilise after 2015. The composition of the population is changing. One in three people will be over 55 years old in 2015. The number of single-person households grew from 22 per cent in 1980 to 30 per cent in 1992. In the bigger cities this proportion is 43 per cent (1988 figure). Ethnic minorities make up a small proportion of the population, but they are growing and are expected to increase to approximately eight per cent of the population by 2000.

Some 42 per cent of the Dutch population (6.5 million) lives in the *Randstad* area (average density 981 people per square kilometre, compared with 452 for the Netherlands as a whole). It covers about 16 per cent (6626 square kilometres) of the country, is located in the coastal region and is below sea level. Extensive waterworks (e.g. polders, canals) and permanent drainage are necessary. The area is a veritable node in the global ecological infrastructure (see Fig. 2).

The Randstad area provides about 50 per cent of the country's employment. This area is the national economic and business core of the Netherlands. In the urbanised areas (about 60 square kilometres) we find about 70 per cent of the head offices of the 100 biggest Dutch companies, 65 per cent of research and development activities and high-tech companies, well over 90 per cent of foreign services and 80

Figure 1 *Map of the Netherlands*
Source: Faludi, 1989 (Originally newsletter on 'Urban and Regional Research in the Netherlands')

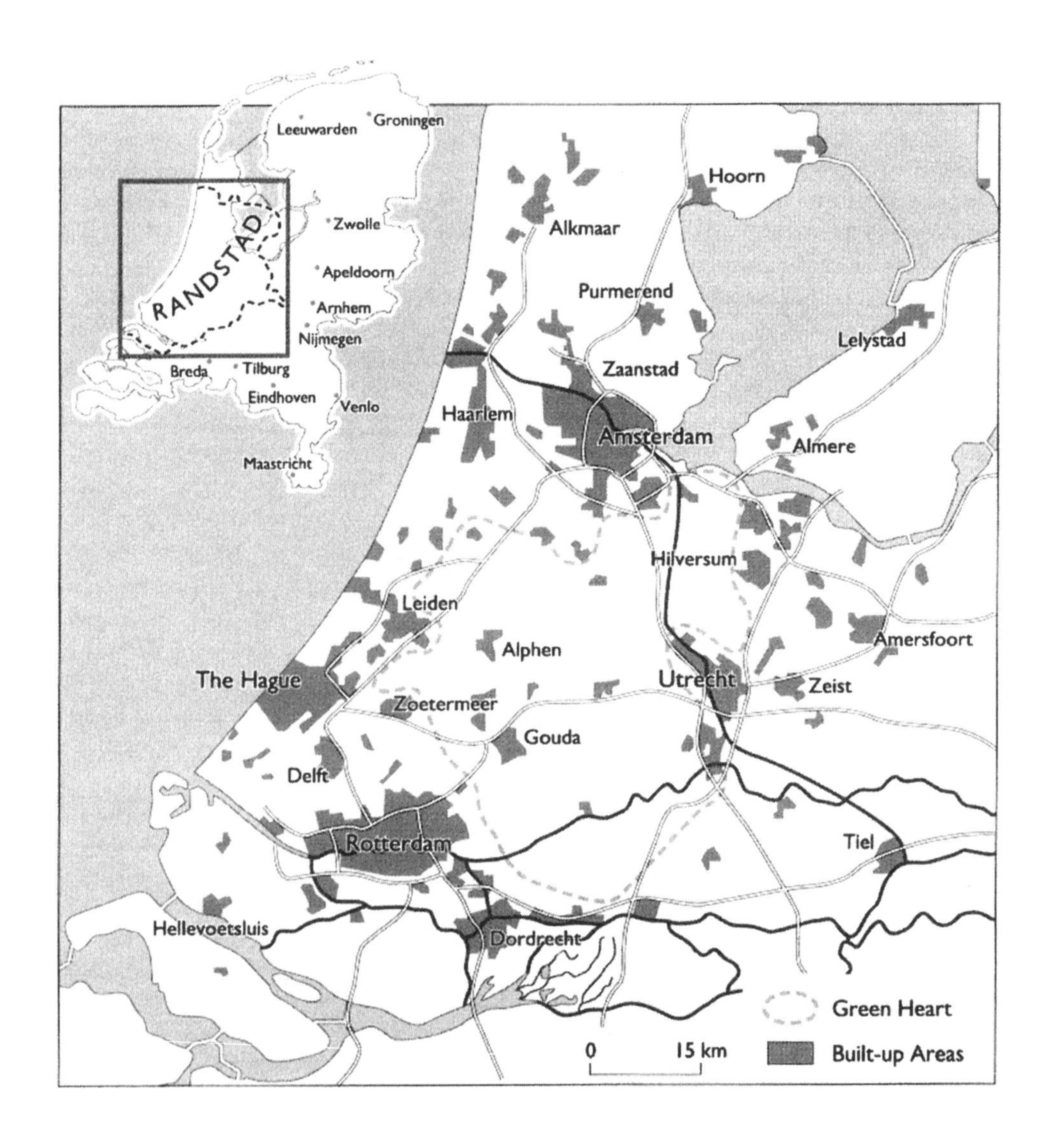

Figure 2 *The Randstad*
Source: Faludi and Van der Valk, 1994

Table 1 *Population and primary functions of the main cities of the Randstad area, January 1994*

City	*Population*	*Primary functions*
Amsterdam	725 000 (city region about 1 100 000 inhabitants)	Financial and air transport centre
Rotterdam	599 000 (city region about 1 100 000 inhabitants)	World port, centre of the petrochemical industry and trans-shipment activities
The Hague	445 000 (city region about 700 000 inhabitants)	Seat of government
Utrecht	234 000 (city region about 550 000 inhabitants)	Centre of rail network and national focal point for services

per cent of foreign commercial enterprises. The area accounts for more than 50 per cent of the country's gross national product. Two of the main Western European ports are located in the Randstad area—the port of Rotterdam (Europort) and Amsterdam airport (Schiphol). More than a quarter of all sea cargo destined for Europe is shipped through Rotterdam's harbour. In general, the Dutch economy is highly international in orientation. Trade and transport are two of its most important sectors.

The Randstad has a polynuclear structure of medium-sized cities and towns interlinked by a highly advanced infrastructure and grouped around a central open space—the 'Green Heart' which comprises mainly agricultural land, countryside and areas for leisure activities. The main cities of the Randstad area and their population figures (as of January 1994) are shown in Table 1.

THE GOVERNMENT SYSTEM[2]

The Netherlands is a constitutional monarchy. It has a three-tier governmental system—national government, provincial government (twelve provinces) and local government (in 1996 there were 647 municipalities). The political system is based on a party system, on the principle of direct elections and on proportional representation. At all levels, coalition governments are the norm. The most prominent parties in the political spectrum are the Labour Party, the Liberal Party (both right and left of centre) and the Christian-Democrat Party. Although changes in coalitions do influence government policy, there is a general level of continuity.

The Netherlands is a decentralised unitary state. Lower tiers of government are guaranteed constitutional autonomy, and higher-tier bodies have powers to prevent lower-tier bodies from intervening in their policies.[3] At all three levels of government, powers are divided between elected representative bodies and executive bodies. Like the members of the Cabinet at the national level, the Queen's commissioner and burgomaster are appointed executives at the provincial and local level. Whereas at the provincial and local levels, principles of unitary government and collective responsibility (used to) prevail, the national level has a departmental organisational structure and is based on the principle of ministerial responsibility.

The administrative structure is under revision. In addition to continuous municipal amalgamations to deal with problems of scale, there is an ongoing process of decentralisation and regionalisation. One of the main reasons for pursuing regional or cooperative municipal authorities is that spatial planning problems are becoming more regional in character. Local government is increasingly a matter of cooperation between municipalities, particularly in the urban areas. This cooperation is found on subjects like housing development, traffic, industrial sites and the environment. Cooperation between municipalities may take the form of a regional authority with its own powers. In July 1994, interim legislation came into force introducing a form of regional government in the main city regions which would also deal with spatial planning. It is not yet clear how these regions will fit in with the Dutch system of public administration. Recent opposition to these proposals from the provinces, municipalities and the public in general makes implementation as yet uncertain.

THE PLANNING TRADITION

As in many other Western countries, the Netherlands' spatial planning system developed 'bottom-up'. The roots of spatial planning as a governmental activity date back to nineteenth century town expansion schemes. There is evidence, however, of planned local intervention in towns and rural areas in earlier centuries through extension schemes and precursors of the land-use regulations we know today. These existed in seventeenth-century Amsterdam. The Dutch are also well known for their tradition of planning water works. This earlier planning was basically a matter of civil engineering, architecture, urban and landscape design and legal disciplines. These planning traditions prevailed even when, in the 1920s, the system developed to include the provincial levels of government and, during the Second World War, the national levels of government.

In the post-war period, spatial planning really gained momentum and became a leading area of government policy. The post-war reconstruction schemes—in housing and regional development—contributed to this, as did the development of the social, geographical, planning and administrative sciences. Professionals from these disciplines contributed much to the growth of planning departments and the planning profession, and planning flourished.

The Dutch Scientific Council for Government Policy (Den Hoed et al., 1983; see also Mastop, 1989) has shown how in the 1970s systematic, synoptic, rational and coordinative spatial planning led societal planning. The 1980s, however, saw much criticism of the system and its performance, and in the 1990s belief in the planning style of the 1970s faded. Planning has become more market-oriented, selective and open-ended. Implementation is a matter of negotiation rather than scheduling, and communicating 'appealing plans' and mediation strategies have become a principal method of gaining societal support. Environmental planning is rapidly broadening its domain as a result of increased environmental concern.

Although these developments are in no way unique to the Netherlands, they differ from those in Britain and the USA where strategic planning practically vanished. In the Netherlands, it readjusted and attuned itself to changing societal conditions. To many writers this is accepted as evidence for the positive attitude of Dutch society to a fair amount of government planning and planned intervention (see, for example, Needham, 1989; Faludi and van der Valk, 1994; Dutt and Costa, 985; Mastop, 1997).

NATIONAL PLANNING ISSUES

Strategic national planning in the Netherlands is closely linked to planning for the Randstad. The idea of the Randstad dates back to the 1930s, but it took until the mid-1950s for the Randstad (including the Green Heart) to become a planning concept (Faludi and van der Valk, 1994). The impetus for central government to develop a planning strategy was threefold—first, the reconstruction works in the aftermath of the Second World War; second, increased population pressure in the western parts of the country; and third, the need for coordination of the investment programmes of the Ministries of Economic Affairs, Agriculture, Transport and Public Works with the urban planning activities of the major cities.

At the time, a major concern was the lack of space for activities like port industries, head offices, airports and new road infrastructure which were thought essential for the economic development of the west of the Netherlands. A second source of concern was the loss of agricultural land desperately needed to promote exports like horticulture and dairy products. A third concern was the western part of the country, which was felt to need extra care because of its strategic vulnerability due to its location below sea level. In 1953 there was a huge flood disaster in the south-west. In addition there was an awareness of the social dangers of 'big cities', and an overall feeling of scarcity of land. In the 1950s and 1960s, many people migrated to 'the new world'.

Once central government took on responsibility for overall strategic planning, it developed rather quickly. Within a decade, an overall spatial planning strategy for the Netherlands as a whole had been developed.[4] The 1958 development scheme set the tune. Its general features included:

- Retention of the urban ring of the Randstad;
- Green buffers between cities and city regions;
- Conservation of the central agricultural area as a large-scale open space;
- Urbanisation in an outward direction, for the overspill population of the cities, business and industry to the north, the north-east and the south;
- A new towns programme to contain the projected overspill from the major cities at a distance of 20 to 30 kilometres from them while ensuring good public transportation connections to the central areas; and
- Policies to ensure a more just and even geographic distribution of prosperity and well-being throughout the country.

The first national policies emphasised economic growth, the accommodation of a strong population increase in the Randstad area and regional-economic development in other parts of the country. Within the limits of the ever-changing population forecasts (1962: 18–20 million by the year 2000, of which nine million in the Randstad area; 1974: 15.4 million by the year 2000) and an ever-growing amount of space per capita, these features were repeated in the first and second report and again in the third report in the mid-1970s. These reports, too, clearly demonstrate how Randstad policies became national policies and how national government succeeded in developing an overall comprehensive spatial planning strategy for the whole of the country.

There were changes in emphasis, too. Until the Second Report (1966), it was important to make the case for systematic, plan-led, national spatial policy. Since then, and especially in the Third Report, this has changed to a strong emphasis on implementation matters.

In the 1960s, suburbanisation threatened Randstad policies. Growth occurred in the smaller towns and villages on the ring and in the central open space. So policies changed to *clustered deconcentration*, which set out to establish a hierarchy of regional centres. At the same time, the smaller villages were prevented from expanding, and limited to an increase in their housing stock of one per cent per annum.

In the 1970s, the programme of specified *growth centres* and *new towns* was launched, 13 of them within the vicinity of the Randstad. The 1970s also saw the introduction of the concept of the *city region*, which was to become the building block of overall Dutch urbanisation policy. Policies for population dispersal and economic development in the north and south of the country were abandoned because of poor performance. By the end of the 1970s, urban policies had shifted to the concept of the *compact city* and to an overall concern for the economic development and international position of the Randstad area.

This latter shift coincided with the preparation of the Fourth Note (1988) and the Fourth Note Extra (1990). The main issues in these notes are the sustainability of development, especially taking account of the Netherlands' open, and therefore competitive, economy, the need for more housing, the maintenance of spatial diversity, dealing with the crisis in the countryside, and finally, the desire to play an important role in Europe. The overall policies are concerned with:

- Strengthening the economic base of the Randstad area and of the city regions with a vast housing programme, including between 600 000 and one million dwellings in the Randstad; the development of office space in the four main cities; the development of industrial and business sites; and the funding of key urban innovation projects;
- The promotion of the Randstad area as a metropolitan region on a European and global scale;
- The acceptance of the fact that the economic core area of the Netherlands had 'grown' to embrace cities in the east and south of the country—the *central ring of cities*;

- The promotion of the Netherlands as an important European core area in terms of distribution, infrastructure and logistics backed by the spatial and economic development of the main ports of Rotterdam and Amsterdam airport, the development of main infrastructure links with the Ruhr area and the development of a high-speed rail system (TGV);
- The promotion of a system of urban nodes as centres of innovation and services at the regional level and linked to the European network of cities; and
- Active landscaping, especially in the Green Heart and the central river areas of the country, in close cooperation with the provinces (e.g. the *Randstad-Green-structure*).

The system

DEVELOPMENT OF THE SYSTEM

Local autonomy, rooted as it was in the constitution which had been drawn up in the nineteenth century, was of vital importance to the development of spatial planning. Local authorities still have a considerable amount of autonomy and are relatively free to regulate land use within their own boundaries.

The Housing Act of 1901 introduced the local extension plan, the forerunner of the present land-use plan (*bestemmingsplan*). The extension plan defined the basic layout for future extensions and was a necessary by-product of local housing policies. It was legally binding. In successive amendments of the Housing Act, the extension plan developed to become a more or less 'modern' land-use plan in the 1930s.

In the 1920s and 1930s the provinces became more involved in spatial planning. At first they supervised local land-use planning. Later, the increased overspill effects of local activities and a growing concern for conserving agricultural land and areas of natural beauty led some provinces to develop master plans for the areas under their jurisdiction. This development was 'legalised' by the introduction of the (legally binding) provincial structure plan in the 1930s.

Next came the national level. By the end of the 1930s, Dutch planners pleaded for the introduction of, again, a legally binding land-use plan at the national level—the national master plan. During the Second World War this request was met when the German occupation forces introduced a strict and centralised hierarchical system of legally binding land-use plans on the national, provincial and local levels of government.

In the post-war period, this system was upheld[5] in the Housing Act and the accompanying Act of the National Plan and Provincial Structure Plans. It was augmented with, again, rather centralised land-use planning regulations for reconstruction areas laid down in the Reconstruction Act. These three acts served as the legal framework for Dutch spatial planning until 1965, when a new Act on Spatial Planning came into force. However, apart from the practice of extension and reconstruction plans at the local level, it never was a framework for a spatial planning policy. Work on a national plan hardly left the national planning agency and only a few provincial structure plans were drawn up. Strategic spatial policy in the 1950s emerged outside the statutory system.

In the mid-1950s, this paradoxical situation led to a rethinking of the legal framework for spatial planning in the Netherlands. Experience showed that a hierarchical system did not work. Therefore, a system was developed that maintained the legally binding land-use plan at the local level, but introduced a system of non-binding indicative strategic planning documents at the local and provincial levels. A national plan was not thought necessary as, at the national level, spatial planning policies would be given the form of separate facet plans or policies, which were to be integrated—'translated' into actual claims for land use—in the structure plan at the provincial level.

THE KEY FEATURES

This system came into force in the 1965 Act on Spatial Planning, and although the act was substantially amended in 1985 and again in 1994, the basic principles still hold. The key features are (see Figs 3 and 4):

- A plan-led system of statutory planning at all three levels of government;
- The local land-use plan is the only one that has direct legally binding power for all actors—including higher tiers of government. The plan is mandatory for the non-built-up areas of the municipalities. Procedures give ample opportunities for appeal to all concerned;
- The local structure plan, the provincial structure plan and national plans are in the 'indicative planning' mode and therefore have only indirect legal effects, if any;[6]
- Extensive surveys and administrative and public consultation in every planning procedure;
- Plans are laid down by the representative bodies in democratic procedures;
- Local land-use plans require the approval of higher levels of government, while local or provincial structure plans need to be brought to the attention of these levels of government; and
- Higher levels of government can intervene in the planning activities of lower levels.

The 1985 amendment dealt mainly with the municipal level of the system and introduced much deregulation, for example, regarding the need for surveying and consultation required by the national level. It included the following features:

- The introduction of the procedure for key national planning decisions in a variety of interlocking national planning documents (sketches, schemes, notes and special projects—see below). This legalised a practice begun in the early 1970s; and

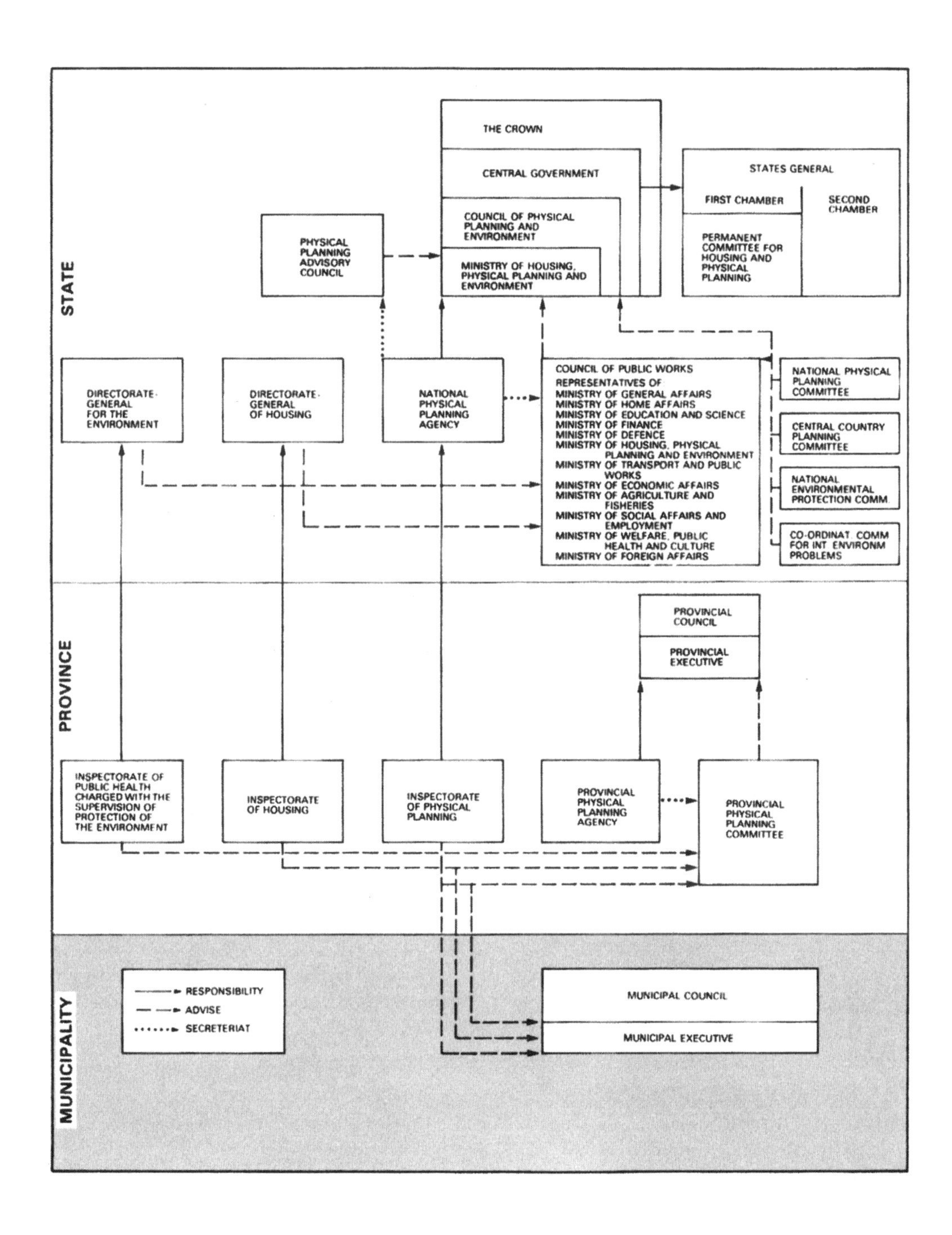

Figure 3 *Bodies in statutory spatial ('physical') planning*
Source: Brussaard, 1987.

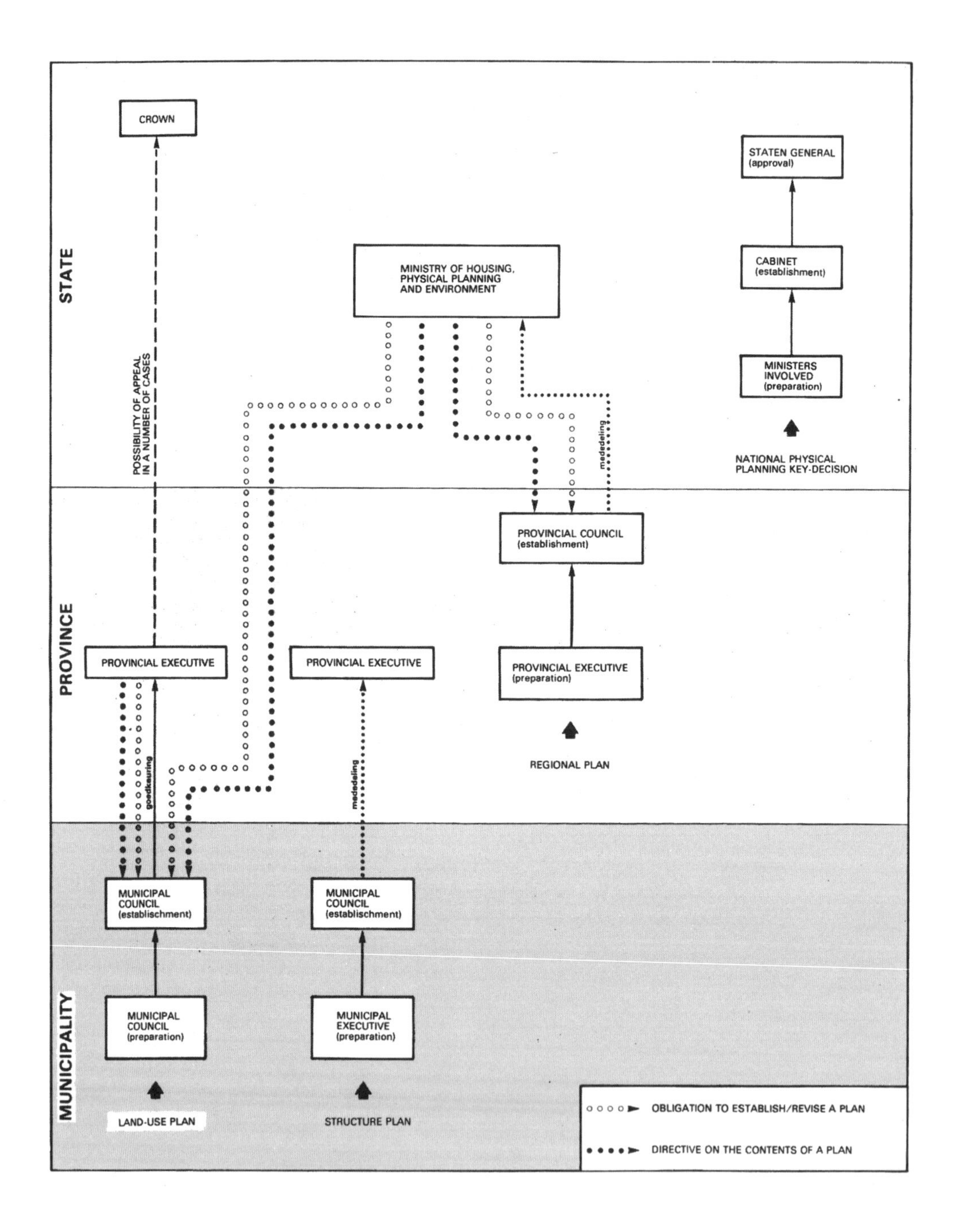

Figure 4 *Instruments of statutory spatial ('physical') planning*
Source: Brussaard, 1987.

- Procedures for direct intervention by the national government in local land-use planning, where the 1965 Act had introduced a step-by-step procedure with the provincial structure plan as link-pin.

The 1994 amendment related to the national government. It included:

- Even more direct powers enabling the national and provincial governments to intervene in local land-use planning. This amendment became known as the NIMBY Act; and
- A strategy to accelerate the procedures for key national planning decisions in large projects of national importance.

THE IMPLICATIONS

An account of some significant implications of the system will allow a better understanding of how it works and has developed since 1965. First, the 1965 system reemphasised the formal separation of statutory spatial planning from all kinds of public servicing of the physical environment like infrastructure, housing, waterworks, health care, reallotment and redevelopment of agricultural land. Moreover, under the Dutch legal system, the principle of '*détournement de pouvoir*' forbids the use of legal powers for any other purposes than the ones for which they were granted. For example, zoning cannot be used for regulating non-spatial aspects of environmental policy. Thus, the formal scope of spatial planning is carefully circumscribed. It will be evident that the Dutch system is therefore heavily burdened by extensive coordination procedures.[7]

Second, this need for extensive coordination is especially acute for national-level planning because many of the powers for dealing with areas such as public services are not formally held by the minister responsible for national planning, but by ministers responsible for the spending departments.[8]

Third, and finally, in theoretical terms the Dutch system is a good example of Faludi's 'proto-planning-theory B' (Faludi, 1987, ch. 12). These systems are built around mandatory and legally binding plans, which have the character of imperative zoning. Deviations from these plans are regulated and subject to more or less objective norms. Legal certainty for citizens is supposed to coincide with the moment of the formal adoption of the plans and is essentially material in character—'by reading the plan one knows one's rights'. Implementation, then, boils down to upholding the plans. Permits are bound by the content of the plan. Even exemptions, variances or amendments to existing plans are supposed to be guided by the legal norms. Clearly, this theory and the systems based on it are akin to the classic ideas of blueprint and end-state planning. According to the 1965 Act, only the local land-use plan would perfectly fit the 'theory', but discussions on the role, function and power of the framing non-binding plans in the system were often influenced by these premises.

The experience

THE 'LIBERALISATION' OF STRATEGIC PLANNING

The growth of national planning coincided with the preparation of the 1965 Act on Spatial Planning. Discussions in Parliament on this Act in 1962 coincided with those on the (first) Note on Spatial Planning. The Second Note on Spatial Planning was discussed in Parliament in 1966, just one year after the Act on Spatial Planning came into force. It opened new horizons, especially because planning would no longer be bound by strict legal regulations. The planning documents on the provincial and national levels of government would no longer be legally binding but would be strategic documents and framing devices. Spatial planning became a major instrument for overall socio-economic development planning, albeit translated into spatial terms.

Spatial planning at the national level, in the 1960s, was still of the type of long-term blueprint or end-state planning so typical of that period. By contrast, at the provincial level planning experiments flourished. The province was seen as the first level of government where integrated spatial planning was possible. This became a vehicle for more general socio-economic development planning on a regional scale. During the following years there was tremendous planning activity in all the provinces. Their planning departments grew in personnel and influence and they employed a new generation of planners whose education was greatly influenced by international thought on planning and by the development of various planning schools.[9] Dutch planning thought was especially influenced by the Anglo-Saxon theoretical debate. The rational synoptic approaches which were so typical for US metropolitan planning, as well as the sub-regional planning studies and the subsequent structure plan practice of the UK, were brought to bear on the Dutch situation. In political terms, being in control of spatial planning, notwithstanding the 'unity of government' at the provincial level, was important for elections.

Within a period of just over a decade, the whole of the Netherlands had been covered with modern provincial structure plans and every province had developed the practice of systematic structure planning on a regular basis, with an average of between four and eight years for plan reviews. The provincial level, then, proved to be of the utmost importance for the growth of modern strategic spatial planning (Mastop, 1987; 1989) and its development at the national level.

SYSTEMATISING THE NATIONAL LEVEL

With the adoption of the Second Note on Spatial Planning (1966), the national government embarked on a programme of long-term development studies on topics like sea ports, the housing programme and outdoor recreational facilities. These were still thought of as separate studies or plans which were to be integrated into the provincial structure plans. Things changed when, in 1968, the national government allowed the development of an industrial estate for Shell, the national oil refinery industry, in Moerdijk, just east of the Rotterdam harbours. This decision was in conflict with the existing provincial structure plan and led to serious debate in

Parliament. The right of the national government to interfere so deeply with provincial policies was questioned. Without prior notice, provincial policies, laid down in a structure plan which had been adopted according to the 'best' democratic procedures, were set aside. The debate did not alter the outcome—'Moerdijk' was built—but it did give the impetus for serious thought on the organisation and institutionalisation of strategic spatial planning at the national level. The conceptualisation of national strategic planning as an interlocking system of sector and facet planning, and proposals for the establishment of democratic planning procedures at the national level, were especially important.

The report of the De Wolff Commission (1970) conceptualised planning at the national level as 'sector' and 'facet' planning[10] and linked it with the activities of the various departments and ministerial offices. According to the Commission, sector planning relates to the provision of public services within the responsibility of individual ministers and could best be organised as a departmental activity. Facet planning, on the other hand, relates to elements which the different sectors have in common. Since there is extensive competition among many sectors of society for land and financial resources, these claims need to be coordinated in order to achieve the best possible future outcomes. That, then, is what spatial planning is all about. It is basically coordinative.

This conceptualisation of national planning activities as an interlocking system of sector and facet planning was the planners' 'Holy Scripture', as Faludi and Needham (1994, 31) call it. This is particularly so since the De Wolff Commission hinted at the idea that facet planning, because of its coordinative character, should overarch sectoral planning. This conceptualisation, then, would bring spatial planners to the forefront of government planning, as the Act on Spatial Planning at that time was the only legal arrangement for such types of facet planning.

Two years later, proposals for new democratic and open planning procedures at the national level were discussed in Parliament. The procedures for these key national planning decisions involve interdepartmental consultation and coordination at the national level, consultation with lower tiers of government, consultation with the public in general and final discussion in Parliament. Together, the sector-facet system and the procedure for key national planning decisions provided the framework for the programme of the Third Note.

THE PROGRAMME OF THE THIRD NOTE

The programme of the Third Note involved the national government in a tremendous effort to establish a fully coordinated, interlocking system of long- and medium-term strategic and programmatic spatial plans dealing with nearly every aspect of the physical environment. The equally massive exercise in systematic structure planning at the provincial and local levels of government gave extra momentum to the Third Note programme. The programme was based on modern theoretical views on planning which included systematic, rational approaches, a process-mode approach, extensive monitoring and systematic review, as well as open and democratic procedures. When the programme was started, the intention was to combine it with an equally systematic programme for the development and pursuit of regional economic policy. This approach was, however, abandoned in the late 1970s.

Starting with the Orientation Report in 1973, which gave the overall framework both in terms of goals and objectives and in terms of the programme itself, before the end of the decade well over a dozen key national planning decision-making procedures had been instituted. The 1976 Urbanisation Report and the 1977 Report on the Rural Areas together outlined the overall spatial policy for the Netherlands until the end of the century. Both reports were accompanied by structural outline sketches—on the urban areas and on the rural areas—which set the policy programmes for the next 10 years and the targets for such topics as population dispersal, housing, town growth development, inner-city development, redevelopment of agricultural areas, conservation of areas of natural beauty and non-urban population policies. The reports and sketches were the kind of spatial facet planning the De Wolff Committee had proposed. Both specified general spatial policies.

Together with these, a host of so-called structural outline schemes were issued. These schemes represented the kind of sector planning the De Wolff Committee had suggested. Schemes specified the programme of various departments responsible for public development in so far as this was spatially relevant and land was required for the proposed development. Schemes were issued on housing, traffic and transport, military areas, general airports, land redevelopment, pipelines, and the like (see also De Lange et al., 1997). The schemes, which were developed in close cooperation with the spatial planning department, really were an innovation in national planning for, until then, most of the departments concerned had had little experience with long(er)-term planning. The sum total of all reports, sketches and schemes together was nothing less than a sophisticated national spatial plan for the Netherlands, albeit not legally binding.

INSTITUTIONALISING COORDINATION AND THE PRINCIPLE OF 'BILATERALITY'

The programme of the Third Note also entailed the huge task of building the coordinating machinery. Although it did not establish a hierarchy of plans, it was assumed that the system of the Act on Spatial Planning would ensure the *vertical* coordination, elaboration and application of national policies in the 'facet line' (the line of 'classic' spatial planning). Together with planning at three levels and consultation, mandatory exchange of information and procedures for approval and correction, the Act prescribed planning agencies and committees at the national and provincial levels as well as regionalised inspectorates for spatial planning. The agencies were to ensure the professionalisation of planning. The inspectorates were to be 'the eyes and ears' of the Minister of Spatial Planning at the regional level and oversee both provincial and municipal spatial policy. The Act rules that the inspectorate be informed on all proposals for land use. The provincial planning committees are advisory committees for the provincial executive and have a central role with regard to both provincial structure planning and the approval of local land-use plans by the province. The committee consists of representatives from the various national departments (housing, infrastructure and waterworks, agriculture and the like, including the inspector for spatial planning), representatives from municipalities and from private organisations. Consultation with the committee is mandatory on all plans and proposals that have a bearing on land use.

The structural outline schemes are a kind of link-pin between statutory spatial planning on the one hand and all kinds of spatial decisions outside that framework on the other. The schemes play a vital role in *horizontal* coordination at the national level. The schemes were intended to link the 'facet line' with the 'sector line' of elaboration and implementation. The De Wolff Committee had proposed that facet planning would overarch sector planning, but, over the years, it has become clear that the two are of equal weight and significance. Thus, spatially-relevant matters can be the subject of planning and decision-making within the frameworks of the spatial facet policy and within the sector concerned. This principle is called *bilaterality*. It is the responsibility of the government of the various tiers to ensure the spatial coherence of different developments, and specifically, of its own measures and decisions. Finally, all have to be specified in the legally binding local land-use plan.

At the national level, the mechanisms for horizontal coordination—and indirectly vertical coordination—are manifold. The Minister for Spatial Planning—who is responsible for Housing and the Environment too—has a coordinating role in all matters of national spatial policy. In addition to his or her main responsibility for preparing facet plans, he or she is the official co-author of all relevant sectoral planning documents and his or her consent is mandatory for many sectoral planning decisions. The Directorate for Spatial Planning (the National Spatial Planning Agency) gives professional support. Professional interdepartmental coordination is the responsibility of the National Spatial Planning Committee, one of the official 'ante-chambers' of the Cabinet. The committee advises the Minister for Spatial Planning on spatial planning matters and, if requested, advises other ministers as well. The ministers responsible for adopting measures and making plans which are of significance for government policy on spatial planning are expected to consult with the committee. This makes the committee the most important body, at the national level, for the official coordination and preparation of administrative synchronisation in the sphere of physical planning.

The Cabinet coordinates the administration of spatial policy, which is officially the responsibility of the whole Cabinet. Spatial planning decisions are prepared in the Council for Spatial Planning and Environment, a cabinet sub-committee. Like every sub-committee, the Council is presided over by the Prime Minister. The Council forms an important link between the administrative decision-making in the Cabinet and the preparatory work in interdepartmental coordination bodies.

In addition to the mainly officially-composed National Spatial Planning Committee, the Act also provided for a Spatial Planning Advisory Council. This Council was created to facilitate citizen interaction with government in the field of spatial planning. It acts as a channel of communication between government and people. To this end, the Council is composed of representatives of social organisations such as employers, employees, building cooperatives, recreation and nature conservation bodies, agricultural boards and transport organisations, as well as experts. Some of these experts are proficient in the administrative and technical aspects of spatial planning while others are proficient in provincial and municipal administration. The Council has up to now been charged with mobilising and guiding societal support for national spatial planning[11] and with handling public inquiries. These many mechanisms for vertical and horizontal coordination make Dutch spatial policies at the various levels of government fairly coherent.

FEELINGS OF CRISIS

However impressive the programme of the Third Note was in terms of systematic forward planning, in coordination and in mobilising societal support, it did not prevent the general feelings of crisis which pervaded the early 1980s. The sources of these feelings were substantive, institutional and political.

As for the substance of the policies, the Third Note programme was basically a continuation and a refinement of the population and urbanisation policies set out in the Second Note (1966) and which were based on assumptions of continued overall economic prosperity. It was closely linked to the housing and urban renewal programme and growth town development policies on the one hand, and to ensuring equally systematic overall planning for the non-urban areas on the other.[12] The housing programme was one of the most important political thermometers for judging the success or failure of the policies. However, the coordinated spatial and regional economic planning effort which had been envisaged in the early 1970s did not materialise.

The signs of the oncoming crisis in the early 1970s were not immediately recognised by the planning community. These signs included an economic crisis, growing concern for the environment, the weakening financial position of the national government and a loss of population in urban and rural areas. It took until the end of the 1970s for the spatial planning community[13] to appreciate their significance. They were so encapsulated in their desire to build the national planning system through spatial planning that they were unable the read the signs, indicating that its context was drastically changing.

In institutional terms, it became clear that the Act on Spatial Planning and the mechanisms for coordination were not as powerful as had been intended. Although the programme really had ensured close coordination between the national and provincial levels of spatial policy, the adoption of local land-use plans—the legal backbone of coordination in the facet plans—lagged behind. They did so because their approval was held up by various legal restrictions on flexibility and the practice of much detailing. In addition, this legally binding plan (the only legally binding plan for the physical environment!) became an important weapon in the hands of environmental conservationists and the emerging anti-growth lobby. Thus it took from four to six years for a local land-use plan to gain legal authority. Higher tiers of government, both provincial and national, were unable, or rather unwilling, to use their corrective powers to speed up local land-use planning and to ensure close coordination. In fact much (if not most) of the actual changes in land use were not, as the system had intended, plan-led, but were realised by exemption procedures. The Article 19 procedure, which allowed exemption from existing land-use plans while anticipating future changes in policy, had become notorious.[14] The system of plan-led coordination proved too cumbersome to ensure coordinated decision-making within a reasonable period of time. This was especially true for issues of national or regional importance, such as important infrastructure, large-scale industrial sites or waste disposal sites. By the mid-1970s, studies had been started and proposals developed to ensure swifter decision-making procedures for these kinds of projects. However, it took until the 1990s before these could be implemented. In addition, the sectoral departments, impeded by the experience of the structural outline schemes, started to develop their own departmental planning procedures, indicating that, while still adhering to the principle of bilaterality, the balance was changing in their favour. Another significant development was the emergence of environmental facet planning which drew together the rapidly growing number of specific and rather fragmented environmental policy fields.

Politically, too, the coming of the 1980s proved disruptive. The national government changed. During the 1970s, coalitions of Christian Democrats and Labour held power. They were replaced by a coalition of the Christian Democrats and Liberals. As the financial position of the national government deteriorated, unemployment figures of well over 10 per cent became common, and concern for the environment grew. The national government engaged in policies of deregulation, privatisation and decentralisation, stimulated by the apparent successes of 'Reaganomics' and 'Thatcherism'. The programme of the welfare state was weakened. The 'no-nonsense cabinets' of Prime-Minister Lubbers set the tune. In terms of processes of societal mobilisation and political importance, environmental issues were taking the lead over spatial planning issues. The spatial planning programme lost momentum, and it was not long before the system of plan-led spatial planning of the 1970s was called a paper tiger, a system that failed to deliver the goods. And while the preparation, adoption and revision of sketches and schemes continued within central government, the first ideas for a new programme were being developed.

THE PROGRAMME OF THE FOURTH NOTE

Work on the Fourth Note started as a programme for updating the policies of the Third Note. As the overall economic situation changed, environmental concern grew and the internationalisation of the Dutch economy led to a rethinking of its position, especially within the European context. The National Spatial Planning Agency initiated studies on these emerging issues.

In June 1985, the Minister and Cabinet were urged to start preparing a Fourth Note on Spatial Planning, which would not only update existing policies but also rethink the planning programme of the Third Note. The Minister was explicitly asked to prepare one single planning document covering all relevant general spatial policies for the next 25 years. This document was to replace both the Urbanisation Report and the Report on the Rural Areas, and to develop a programme for reducing the number of structural outline schemes and sketches, since the various sectoral departments had established their own strategic planning procedures. The goal was to establish a leaner and more transparent set of national planning documents.

In June 1986, just before elections for Parliament, the department issued a non-political *Report on Spatial Perspectives* for the twenty-first century. The report stressed the need for the Netherlands, with its open economy, to prepare itself for the coming century. This entailed working on the international repositioning of the Randstad and other parts of the country, ensuring the continued main port functions of the country for Europe as a whole, and rethinking the fostering of economic growth while paying attention to the environment. It received a warm welcome.

The Fourth Note on Spatial Planning was issued in 1988, and existing policies of the Third Note were updated and presented in the Structural Outline Sketch for the Urban and Rural Areas. The Fourth Note was presented as the national planning document, valid to the year 2015. Together with new strategic targets and policies (see above) the Note introduced a rather dramatic change in national spatial policy. It included the following features:

- National policies could become much more selective;
- A firm commitment to spending departments at the national level for investment;
- More market-oriented than its forerunners, the Note was attuned to working with private organisations;
- Instead of a systematic hierarchical plan-led approach, it introduced a kind of communicative project-led type of planning, attuned to market forces and able to seize opportunities as they appeared;
- It made a serious effort to re-establish spatial and environmental quality, i.e. the quality of living conditions, as a prime goal for planning decisions; and
- It aimed to reposition the Randstad and the Netherlands in an international context.

The Fourth note was prepared in close cooperation with the revision of the Structural Outline Scheme on Traffic and Transport and the first National Plan for the Environment. The link with the revision of the structural outline scheme was necessary, as the Fourth Note incorporated many proposals for investment in the Dutch infrastructure, not only in terms of the main ports of Rotterdam and Schiphol, but also to improve the connections with the European hinterland. The link with the National Plan for the Environment was intended to ensure that spatial and environmental policies would be closely coordinated and complementary to one another, since environmental planning had developed its own system with strategic integrating planning documents at the provincial levels, together with the structure plans. The link between these two planning systems, both dealing with conditions of the physical environment, became even closer when, after the elections and a change of Cabinet (Christian Democrats and Labour again), it was decided to give even more priority to environmental goals in spatial planning. So, 1990 saw the updates of the two national strategic planning documents—the Fourth Note Extra and the National Plan for Environment Plus. Both were formally adopted in 1993.

The Fourth Note triggered a host of new strategic planning activities and many public–public and public–private partnerships. These are listed below:

- A programme of strategic plans for selected regions which are of special importance for national spatial development. These include the Randstad and the Green Heart areas and regions close to the German border, which were expected to thrive in the impending European integration (1992);
- A programme of operational project planning for selected city regions to deal with the housing programme (600 000 to one million). Large new housing estates would be built close to the existing cities;
- A programme of integrated planning for a number of spatial and environmental target areas. These areas are crucial for national development. It was felt that 'normal procedures' would consume too much time. The targeted areas like the Rotterdam harbour area and certain agricultural regions suffer from severe environmental pollution or are areas like the Amsterdam Airport region where new spatial and infrastructural developments could conflict with environmental protection policies. In addition to dealing with acute problems, the programme was also intended to experiment with intensified forms of cooperation between spatial and environmental planning bodies;
- A programme for the development of large-scale infrastructure projects. These include the high-speed rail connections (TGV) of the Amsterdam Airport region with the south (Brussels, Paris), and rail connections with the Ruhr area;
- A programme of key investment projects in urban innovation in the Hague, Rotterdam and Amsterdam. This would promote the international metropolitan nature of the Randstad; and
- Developing the Randstad and the Netherlands to fit in with the European perspective.

The various sectoral departments intensified and renewed their strategic planning. The Department of Infrastructure issued the Second Structural Outline Scheme on Traffic and Transport. The Department of Housing, Spatial Planning and the Environment issued the National Plan for the Environment. The Department of Agriculture followed with a new Structural Outline Scheme for Green Space (1992). This scheme integrated all the departments' separate policy documents into one overall strategic plan for the non-built-up areas.

More recently, the Departments of Economic Affairs, Agriculture and (again) Infrastructure each presented a general strategic planning document. These documents all contain policy proposals and options—and better even, spatial concepts and development perspectives that impinge upon future spatial developments in the Netherlands. The Department of Economic Affairs has developed a spatial arrangement for additional land claims for industrial areas in the Randstad area. The Department of Agriculture has developed the concept of town landscapes, which addresses the omnipresent problem of how to deal with urban fringe areas. The Department of Infrastructure has developed a new, corridor-like concept for dealing with the location of new urban extensions and new towns from the point of view of spatial mobility.

Together, these documents have set the tone for a new Fifth Note on Spatial Planning. The formal decision on the need for a Fifth Note was taken in 1998, but preparation for it started in 1996 when the Department of Planning launched the Netherlands 2030 project. This project, which aims to mobilise societal debate on the future spatial arrangements of the Netherlands, offered four competing perspectives for Dutch society and its spatial arrangements. One concept stresses the Netherlands as a country of cities; another stresses future infrastructure as the case for future developments; a third sees the Netherlands as a country of park-like landscapes; and a fourth—called 'pallet'—stresses regional differences and a future development that is predominantly led by private-actor decisions. Each of these perspectives builds on different assumptions on topics such as expected population growth, economic development, social cohesion, the role of government, and governance questions in general.

It is quite easy to link each of these four perspectives with the position taken by one of the four national departments most concerned with spatial planning—planning, infrastructure, agriculture, and economic affairs. All four departments are thus positioning themselves to play a role in the future debate. Together, these policy documents and the perspectives of the Netherlands 2030 project are, in any case, a clear sign that the leadership in discussions on the future spatial development of the Netherlands is no longer the automatic prerogative of the Minister and Directorate for Spatial Planning. These documents are, perhaps, a sign that the balance of power has definitely shifted towards the 'sectoral' departments.

INSTITUTIONAL CHANGES

The Fourth Note led to the revival of strategic planning and, concomitantly, to innovations in the institutional settings. First and foremost, the principle of bilaterality was expanded to include the principle of *diagonal coordination*. This principle holds that coordination procedures cut through existing statutory systems in order to ensure close cooperation between the departments, tiers of government and, if necessary, private partners in the planning and implementation of complex and urgent strategic projects. Diagonal coordination has become the mechanism by which most of the strategic planning programmes generated by the principles enunciated in the Fourth Note are implemented. It is interesting to note that most of these plans and programmes are extra-legal, which means that, in the end, a way has to be found to link the results to the statutory system, especially to the land-use plan.

In 1993 the Act on Environmental Management came into force. This set the formal legal framework for mandatory integrated environmental planning at the national and provincial levels. Municipalities are free to prepare their own environmental plans. Compared with the spatial planning system, environmental planning is much more centralised. Goals, norms, standards and budgets are fixed at the national level. This makes environmental planning both strategic and operational in character. The planning system not only intended to establish a second facet line of planning for the physical environment, but also called for mandatory mutual adjustments between the two strategic planning documents, at the national and the provincial levels.

Still in 1993, the Act on Infrastructure Projections was passed by Parliament. It introduced a planning procedure for the projection of nationally important infrastructure projects. This procedure integrates all the necessary planning and consultation procedures at the various levels of government through diagonal coordination. The Act is intended to accelerate decision-making as well as to ensure democratic procedures and legal certainty when decisive action is taken. Citizens' legal recourse for objecting to proposed measures has been curtailed because the different procedures dealing with separate aspects of the same decision have been abolished, and these can therefore no longer be appealed individually. Undoubtedly, the Act has brought more central control to the spatial planning system.

We have already seen that in 1994 the Act on Spatial Planning was amended to introduce swifter and leaner measures for the implementation of national government directives (the so-called NIMBY Act) in local land-use planning. It also aimed to accelerate the procedures for implementing key national planning decisions and carrying out large nationally important projects. Both acts were intended to accelerate decision-making in situations of horizontal and vertical coordination and to overcome possible deadlocks in the procedures. Like the Act on Infrastructure Projections, these acts also introduced more central control into the system and reduced the number of separate instances whereby citizens could exercise their right to object and their right of appeal. The sum total of all these developments is that national spatial planning today is still plan-led, but is increasingly becoming project-led too. In a no-nonsense political climate, spatial planning has to deliver the goods.

Preserving the old and adjusting to the new[15]

Despite the fact that there is much continuity in Dutch spatial planning at the national strategic level,[16] the analyses thus far have shown that in institutional and legal-organisational terms, the past four decades have been rather turbulent. The system has always been committed to the following basic objectives:

- Ensuring timely and adequate decision-making on the future use of land;
- Ensuring balanced procedures (public and democratic); and
- Ensuring reasonable transaction costs and reasonable time-spans.

Once national government really had a role in spatial planning, the characteristics of the system posed severe problems. The delicate balance between central control and local democracy entailed time-consuming procedures, a multiplicity of regulations, a fragmentation of power and a need for extensive coordination. The regulations were required to ensure citizen input and legal recourse. The *de facto* and the *de jure* powers were fragmented between and within the tiers of government which represented Dutch democracy's attempt to achieve a system of checks and balances. Coordination was a *sine qua non* of the system and its objectives. However, this need has been extensively debated since the mid-1970s.

Only if we are ready to accept that spatial planning is a very intricate and sophisticated institutional arrangement can we give a positive, albeit conditional, answer to Needham's pressing question on the degree to which the Dutch physical environment is 'planned'. It *is* planned, though less straightforwardly than the statutory planning system wants us to believe. Theory and practice, as often, are not in harmony; many auxiliary mechanisms had to be invented, and new issues which emerge force us to rethink the present institutional arrangement.

THEORY AND PRACTICE

Firmly rooted in a proto-planning theory B conception, the Dutch spatial planning system adheres to the primacy of the legally binding plan. This even holds for those plans which are not formally legally binding.

In a review of the research on the administrative aspects of the spatial planning system, Bos (1995), a former assistant inspector for spatial planning, compared the system with those of other countries (see Table 2). His conclusions are critical of both the 'theory' of the Dutch planning system and the extensive coordination procedures which, from a legal and institutional point of view, can be understood as auxiliary devices to ensure concerted action at and between the various levels. According to Bos, in practice there is hardly any sign of the primacy of the plan. Legal security is not on the side of those who initiate spatial development, but instead on the side of existing rights and of those who oppose development. Horizontal and vertical coordination hardly ensure internal consistency, and the system proves itself to be inefficient and inflexible. In short, Bos judges the statutory system to be inert and not to work in the way that was intended. Spatial planning at the higher levels of government—national and provincial, which together are central to the coordination machinery—is very vulnerable because of weaknesses at the local level. But Bos is too hard on the system. It is definitely changing. In theoretical terms, Dutch legislation has introduced elements of Faludi's proto-planning theory A into the system, particularly at the lower level. The 1985 amendment not only led to less time-consuming procedures for adopting local land-use plans; it also introduced new options for less detailed planning documents, thus offering flexibility. It also added new rules for exemption, deviation, elaboration and 'bypass' procedures (van Damme and Verdaas, 1996). In practice, while still adhering to the primacy of the binding plan, Dutch planning is moving closer to the UK procedures of development planning. In addition, on national procedures the

Table 2 *Comparison of systems of statutory planning in selected countries*

	The Netherlands	*Belgium (Flanders)*	*France*	*Germany*	*England before 1991*	*England after 1991*	*Ireland after 1993*
Basic principle	Blueprint	Blueprint	Blueprint	Blueprint	Process	Process	Process
Internal consistency	+	+	–	+	+	–	–
Assignment of tasks						+	–
Vertical coordination	+	+	+	+	+	+	–
Horizontal coordination				+	+	–	–
Systematic planning procedures	+	+	–	+	+	+	+
Democratic procedures							
Societal mobilisation	+	+	+	±	–	–	–
Legal security initiator	±	±	±	+	+	+	+
Legal security 'neighbour'	+	+	+	–	–	–	–
Efficiency	±	±	–	+	+	+	+
Primacy of the plan in theory	+	+	+	+	–	–	–
Primacy of the plan in practice	+	+	+	–	?	+	–
Flexibility to the civilian	–	–	–	+	?	+	+
Perceptibility of existing rights							

+ Adequately dealt with/ensured by the system.
– Not adequately dealt with/ensured by the system.
± More or less dealt with/ensured by the system.
? Uncertain.
Source: Bos (1995).

Dutch are looking at the experiences of France, the UK and Switzerland in order to find ways of introducing more central control and direct intervention. Diagonal planning procedures, the NIMBY (amendment) Act and the Act on Infrastructure Projections are examples of this.

PROJECT-LED PLANNING

Undoubtedly, the most significant change in Dutch national planning is the move to a type of project-led planning, based on the principle of diagonal coordination. The programme of the Fourth note is based on these approaches, as are most of the more recent changes in planning legislation. This development means that the institutions in spatial planning have to adapt to new approaches and cannot simply rely on the rather bureaucratic procedures of the statutory system. Most of these new approaches use communicative and negotiation types of planning and are often extra-legal. Much of national policy-making today comprises up-to-date concepts such as networking, looking for 'win-win' situations, negotiating the concerted use of legal powers and ensuring financial back-up from the various levels of government and important change agents from the private sector. Planning along these lines also means that principles of administrative law are increasingly being supplemented by general principles from civil law to ensure the implementation of policies. Today, private contracts between government agencies and private sector partners are common.[17]

Such procedures undoubtedly ensure that the development of the physical environment is plan-led, and indeed the system does work. However, more often than not, the statutory system poses a severe problem. When, for instance, the Integral Policy for Amsterdam Airport and its Surroundings was agreed upon and adopted in 1990 by all the government and semi-private agencies involved, it turned out that all the agreed policies had to follow separate statutory procedures before any action could be taken. The agencies involved included the government Departments for Spatial Planning, the Environment and Traffic and Transport, the province of North Holland, the municipalities of Amsterdam, Amstelveen and others and (semi-)private organisations like those of Amsterdam Airport and the Royal Dutch Airlines. The procedures included those reserved for key national planning decisions, an environmental assessment procedure, a revision of the provincial structure plan and revisions of relevant local land-use plans. All these procedures in their turn entail mandatory consultation of the public, possible appeals, and so on. So, while time was won on the one hand, it was lost again on the other.

The Scientific Council for Government Policy (WRR, 1994) recently suggested that these kinds of paradoxical situations, which are expected to increase, could be dealt with by the adoption of a general Act on National Projects. Basing itself on the Act on Infrastructure Projections and on international experience, the Council suggests a step-by-step procedure. First, the national government should propose that a project of national importance be implemented, giving information on the what, when and how. This proposal should be opened to public consultation. The national government should then decide in principle on the project's implementation. A special committee would give advice on the spatial, environmental and economic/financial implications of the project. In the second stage, the implementation should be planned and all necessary procedures combined. At this stage, lower-tier government would no longer be able to oppose the project, and compliance would become mandatory. In the third stage, the development process must be efficiently structured. The Council expects that such a procedure not only would be less time-consuming, but also that government proposals would gain more societal support because there is public consultation on the decisions in principle.

Ensuring less time-consuming procedures might well be the outcome of such an Act, as the decision in principle (first stage) actually is 'the point of no return'. But organising public consultation on rather general proposals, without being able to clarify what will happen in 'one's backyard', might turn out to be an illusion. It would be far better to look at the problem the other way around, and to concentrate such public consultation on the later stages of decision-making when the planning policies and the legal interests and positions are clearer and more concrete. If the Council's suggestions are followed, the decision-making process will become more centralised and will be even more legalistic than it is today.

In its most recent advice (WRR, 1998), the Scientific Council even goes one step further. After analysing present-day problems with the existing spatial planning legislation and institutions, it comes to the conclusion that the so-called 'infrastructure' or 'spatial investments' model of (project-led and multi-actor) decision-making is much more promising for a proactive kind of policy-making for spatial development than the existing system. According to the Council, the Netherlands should invest in a thorough debate on how to innovate existing institutions and to build new types of political and administrative arrangements. In the Council's view, this would inevitably lead to a new division of powers and responsibilities between the various levels of government (actually, selective centralisation).

STRATEGIC PLANNING IS BECOMING MORE LEGALISTIC

The primacy of the legally binding plan and its ensuing plan-led legal security for all sides means that these plans are perceived as if they were legal contracts. This is understandable at the local level, where direct measures which affect privately held land are taken. Citizens have to know their rights. One might wonder, however, whether 'knowing one's rights' can also apply to strategic policy-making for future development, especially because of the legalistic and litigious tendencies in Dutch planning legislation.

Over the years, because the national government increased its role in directing overall policies and national policy documents became more prominent, these documents gained a rather hybrid legal character. On the one hand, they are indicative and non-binding, which means that they would become binding only if additional decisions are taken. On the other hand, the need for more specific policy statements has reinforced the tendency to seek greater certainty through legally binding means. When the General Act on Administrative Legislation came into force (1994), this tendency was intensified. The Act grants a general right of appeal against administrative decisions that directly influence existing rights, and also grants the right to financial compensation if these rights are affected. There is evidence that governments at all levels are anxious not to interfere with those rights, so they adapt their plans, are less specific in their policies, and adjust their procedures. The growing dependency of strategic planning on legal procedures thus might counteract the current preference for swifter and leaner procedures and more direct intervention and control by the national government. However, the danger of this dependency on legal procedures is that the system will become even less flexible than it already is. The proposals of the Scientific Council described above have the same pitfalls.

The legal status of Dutch strategic planning has been debated since its inception. The only way to deal with this problem is to make strategic planning and planning documents non-binding in a legal, not extra-legal sense, and to secure existing rights by a zoning ordinance, which would only state the current situation. Such a solution would integrate the two aspects of any spatial plan—the description of the current situation and the recognition of existing rights on the one hand, and the proposals for future developments on the other. The first would be legally dealt with in the zoning ordinance. The second would be brought to bear on existing rights when decisive measures need to be taken. Public control over the development of policies would be confined to the political system. To be sure, such a solution is in severe conflict with Dutch legal tradition. But institutionalised spatial planning is not a matter of legal rulings alone.

INTEGRATED STRATEGIC PLANNING

The programmes of the Third and Fourth Note have recognised that statutory spatial planning is only one of many planning and policy mechanisms that deal with the development of the physical environment. Coordinating these policy mechanisms can be difficult, not just because the different plans are complementary and can to some extent be competitive, but also because each has its own procedures. Thus the national and provincial tiers of government need to find ways of coordinating, for example, the environmental planning system, the spatial planning system and the system of water management planning. Encouraged by the experiments on integrated planning for spatial and environmental target areas, several provinces have developed formats for combining the strategic plans for spatial development, the environment and water management into one single plan for the physical environment. This approach would make a tremendous contribution towards simplifying the system and making it more transparent. Were this approach to combine the various legal procedures as well, it would solve the problem of '*dètournement du pouvoir*'. In fact, at the operational level, steps have already been taken in this direction. Experiments on 'integral zoning', based on environmental legislation and including spatial planning aspects, are cases in point. The Minister for Housing, Spatial Planning and the Environment has recently stated that if a next (fifth!) National Strategic Planning Note were to be prepared, it would most certainly be combined with a new Note on environmental planning.[18] Such integrated strategic planning for the environment might well be a blueprint for the future of Dutch planning.

DECENTRALISATION AND REGIONALISATION

The programme of the Fourth Note championed more selective national spatial planning and stressed projects of national importance. It also brought about considerable decentralisation in areas like housing, landscape development, regional economy and daily living conditions. It was motivated by a need to cut back on national expenses. From a national point of view then, regional and city region planning is increasing in importance. Whereas in the 1970s and 1980s this intermediate level of government was sometimes seen as a mechanism to hand national policies on to the local level, it is now increasingly being appreciated as a level of integrated planning in its own right. The 1994 proposals for administrative reform, introducing the city region as a separate governmental level with powers to develop mandatory structure plans, are a case in point. These structure plans would to some extent pre-empt local land-use planning for the area in question. Diagonal coordination, in which a 'regional' authority often plays a central role, strengthens this tendency towards the regionalisation of planning.

This process of the decentralisation of powers and the regionalisation of planning is firmly under way. Not only are city regions being created and municipalities amalgamated, but the traditional role and position of the provinces are being rethought. The impetus for this rethinking has come from the EC. As European policies more often than not aim at regional development, and as Europe is increasingly conceptualised as a Europe of the Regions, it is of crucial importance to position these regions in such a way that they count at the European level. So, given the new reality of the national government and Europe, the provinces are cooperating in discussions for their reorganisation. There are also many cross-border planning initiatives. In addition, the existing supranational cooperation between the Netherlands, Belgium and Luxembourg (Benelux) has resumed work on strategic spatial planning, which led to a Second Structure Plan for the area. This Benelux plan also stresses the need for regional planning. These recommendations for strengthening the regional level are based on approaches akin to diagonal coordination, with regional authorities as the link-pin.

In many respects, institutionalised spatial planning in the Netherlands is changing. However, these changes do not always point in the same direction.

REFERENCES

BOS, J. P. (1995), 'Streek en bestemmingsplannen: Samenvatting', The Hague, Werkprogramma Planningstelsel in bestuurlijk perspectief, Rijksplanologische Dienst (unpublished memorandum).

BRUSSAARD, W. (1987), *The Rules of Physical Planning 1986*, The Hague, Ministry of Housing, Physical Planning and Environment.

DAMME, L. VAN and J. C. VERDAAS (1996), 'Bestemmingsplannen als instrument van beleidsvoering' (PhD thesis), Nijmegen, University of Nijmegen.

DE LANGE, M., H. MASTOP and T. SPIT (1997), 'Performance of National Policies', *Environment and Planning B*, **24**, 845–58.

DE WOLFF COMMISSION (1970), 'Commissie voorbereiding onderzoek toekomstige maatschappijstructuur', Second Chamber 1970–71, 10914.

DREXHAGE, E. C. and M. H. B. PEN-SOETERMEER (1995), *Het ruimtelijk planningstelsel: Een bestuurlijk perspectief*, The Hague, Rijksplanologische Dienst.

DUTT, A. K. and F. J. COSTA (eds) (1985), *Public Planning in the Netherlands*, Oxford, Oxford University Press.

FALUDI, A. (1987), *A Decision Centred View on Environmental Planning*, Oxford, Pergamon.

FALUDI, A. (ed.) (1989), 'Keeping the Netherlands in Shape', *Built Environment*, **15**, special issue.

FALUDI, A. and B. NEEDHAM (1994), 'Development Plans and Development Planning in the Netherlands', in P. Healey (ed.), 'Trends in Development Plan-Making in European Planning System' (Working Paper 42), Newcastle upon Tyne, CREUE, Department of Town and Country Planning, 19–37.

FALUDI, A. and A. J. VAN DER VALK (1994), *Rule and Order: Dutch Planning Doctrine in the Twentieth Century*, Dordrecht, Kluwer Academic.

HOED, P. DEN, W. G. M. SALET and H. VAN DER SLUŸS (1983), *Planning als onderneming* V34, The Hague, Staatsuitgeverij.

MASTOP, J. M. (1987, first published 1984), *Besluitvorming, handelen en normeren, Planologische Studies*, 4, Universiteit van Amsterdam.

MASTOP, J. M. (1989), 'The Case of Provincial Structure Planning', in Faludi (1989), 49–56.

MASTOP, J. M. (1991), 'Over plan, beleid en de vrijstellingsroute', in P. J. J. van Buuren et al. (eds), *25 Jaar WRO*, Deventer, Kluwer.

MASTOP, J. M. (1997), 'Performance in Dutch Strategic Planning: An Introduction', *Environment and Planning B*, **24**, 807–13.

MINISTRY OF FOREIGN AFFAIRS, FOREIGN INFORMATION SERVICE (1994), 'The Netherlands in Brief', The Hague.

MINISTRY OF HOUSING, SPATIAL PLANNING AND THE ENVIRONMENT (1989), *Fourth Report (EXTRA) on Physical Planning in the Netherlands: Comprehensive Summary*, The Hague.

MINISTRY OF HOUSING, SPATIAL PLANNING AND THE ENVIRONMENT (1991), *Perspectives on Europe*, The Hague.

Nederland 2020. Discussienota (1996), The Hague, SDU Uitgeverij.

NEEDHAM, D. B. (1989), 'Strategic Planning and the Shape of the Netherlands through Foreign Eyes: But Do Appearances Deceive?', in Faludi (1989), 11–16.

WRR (1994), *Besluiten over grote projecten*, The Hague, SDU Uitgeverij.

WRR (1998), *Ruimtelijke Ontwikkelingspolitiek*, The Hague, SDU Uitgeverij.

NOTES

1 This paper was originally written in 1996. It has been updated for formal legislation and statistics.

2 See also Brussaard (1987).

3 These powers are often of a corrective character, there is no formal hierarchical relation between the levels of government, and each level has the autonomy to deal with its own affairs. Only if matters are clearly beyond the (territorial) scope of lower bodies are they dealt with by higher levels. There are many formal regulations for ensuring that the different governmental bodies do not pursue contradictory policies (the principle of the unitary state). Some of these allow a higher authority to intervene correctively to supervise a lower authority (an example is the general financial supervision by a higher authority over a lower). Another is the fact that the local land-use plan has to be approved by the provincial executive. There are in addition many formal mechanisms for coordination between the levels and there is a tradition of much informal consultation between the levels. This is strengthened by the practice whereby top civil servants and cabinet ministers are often recruited from the top politicians of the big cities, and vice versa.

4 1956: *The West and the Rest of the Netherlands* (brochure); 1958: *Development Scheme for the West of the Country* (National Working Committee); 1960: *First Report on Spatial Planning*; 1996: *Second Report on Spatial Planning*. These were to be followed by the *Third* (1973–1982, issued in separate volumes), *Fourth* (1988) and its addendum, the *Fourth Report Extra on Spatial Planning* (1990, formally adopted in 1993). The latter states the current policies.

5 The existing laws were scrutinised for 'non-Dutch' (i.e. German) elements, which were subsequently removed or changed.

6 Plans in the 'indicative' mode present projections and guidelines for private and public actors, and are not expected to be legally binding.

7 Next to all kinds of legislation for specific 'works' in the physical environment, the key legislative basis for spatial planning is provided by the Act on Spatial Planning (*Wet op de ruimtelijke ordening*) with its accompanying Decree (*Besluit op de ruimtelijke ordening*) and in combination with the Housing Act (*Woningwet*), supplemented by the Act on Urban and Village Renewal (*Wet op de stads- en dorpsvernieuwing*).

8 One way of securing the close coordination between planning and implementation of course is through ministerial responsibility. The 1950s already saw combinations of responsibility for housing and planning within one department and under one minister's responsibility. In the 1970s this department and the minister became responsible for housing, planning and the environment.

9 In 1962, planning—or *planology*—was recognised as an academic discipline which led to the founding of the first two chairs in planning in the Netherlands—in Amsterdam and Nijmegen.

10 The Commission also conceptualised 'integral planning' as the planning level overarching 'sector' and 'facet' planning. As this kind of integral planning never really got off the ground, it is not discussed here.

11 Due to an overall reduction in the number of advisory boards at the national level and a revision of their tasks, the role of the Council was recently made less prominent.

12 The policies of the Second Note had a clear focus on urbanisation policies and on ensuring an equal spread of economic prosperity over the country, but lacked any sophisticated general policies for the non-urban areas.

13 See Faludi and Van der Valk (1994) on this topic.

14 Research on this topic revealed that this Article 19 procedure had become a generally accepted procedure for spatial planning at the local level, notwithstanding the fact that the local land-use plan was mandatory and that Article 19 was meant only for exemption in individual cases of minor importance (see Mastop, 1991). In practice, many, if not most,

town expansion schemes were realised using Article 19 in an 'unauthorised manner', to be 'legalised' afterwards by the formal adoption of a land-use plan.

15 This section is partly based on the final report on a project of the National Spatial Planning Agency on administrative aspects of the spatial planning system (see Drexhage and Pen-Soetermeer, 1995).

16 Faludi and Van der Valk (1994) argue that this is probably due to a strong existing planning doctrine.

17 In the 1960s and 1970s when, due to welfare state policies, public intervention grew at a fast rate, there was much debate on whether government was bound to administrative law to ensure its goals or whether it could and should also make use of civil law. This discussion on 'the magic line' led to the conviction that administrative law had to prevail. The spatial planning system was based on this principle. But especially when environmental legislation grew in importance and it became apparent that in order to reach environmental goals, industry had to comply with new products, civil law was 'rediscovered'.

18 On the occasion of the fiftieth anniversary of the Ministry of Housing, Spatial Planning and the Environment, 21 December 1995.

Appendix A. Current policies

The current policy for spatial planning in the Netherlands is set out in the Fourth Report (Extra) on Physical Planning in the Netherlands (abbreviated as VINEX, formally adopted in 1993). This policy has been chosen as a response to the following issues:

- Sustainable development
  The Netherlands is a small and densely populated country, with few natural resources, and great problems of polluted soil, air and water. It wished to use land efficiently, to reduce the emission of pollutants, to restrict the use of raw materials and energy, and to improve the organisation of waste processing.
- Competitive economy
  The economy of the Netherlands is very open, especially to the other countries of the EU, and it must be able to compete effectively. Otherwise the country will become poorer and unemployment will rise even higher. This has important consequences for transport policy.
- The need for more housing
  In 1989, a report by the Minister of Housing said that there was no quantitative housing need, i.e. no housing shortage, and that housing policy should be directed at improving the quality and use of the housing stock. That proved to be a demographic or political error, and the targets for house building have had to be revised upwards. Added to this are the housing needs of political asylum seekers.
- Threats to spatial diversity
  The spatial quality in certain urban and rural areas is worsening. In urban areas this contributes to social problems and the retreat of the better-off, and in rural areas, to ecological impoverishment. Moreover, the result is a reduction in

diversity—all town centres are beginning to look alike. The same can be said for residential areas. Landscapes are losing their distinctive features.

- The crisis in the countryside
  Agriculture is employing fewer people and is withdrawing production from certain areas. This is threatening the viability of some rural areas. Also, agriculture has in some areas become a notorious polluter. This retraction is therefore not entirely unwelcome. But what function does the countryside have then? Is it to become a playground for urbanites and a reserve for wildlife? Can that provide the necessary economic basis for the countryside?
- The desire to play a central role in Europe
  Some of the above issues (e.g. increasing international economic competition, the crisis in the countryside) arise out of, or are exacerbated by, membership of the EU. Nevertheless, the Netherlands wants to continue to play a central role, and it is clear that it is in the national interest of the Netherlands to cooperate with its neighbours by, for example, cross-border planning and with its near-neighbours by, for example, developing transport axes linking the Netherlands to France and to Denmark.

The content of the policy can be divided into two hearings:

- The daily living environment; and
- The national spatial development scenario.

Responsibility for realising both is divided between the national, provincial and municipal governments, with the national government providing the necessary changes in legal powers, some of the funding, examples of good practice and the initiative for selective policy experiments.

THE DAILY LIVING ENVIRONMENT

This set of policies must secure the quality of the spatial environment by ensuring that it is well maintained, clean, safe, and offers choice and diversity.

Urban functions are to be concentrated into urban regions, since this reduces the need to travel and creates a larger base for amenities and public transport. This is a continuation of the 'compact city policy'. Within these urban regions, sites for development have been chosen on the basis of certain criteria. Residential, work and recreational areas should be as near as possible to the centre. They should be accessible to public transport and bicycles. They should not be near sources of environmental nuisance. Urban renewal, then, has the task of using the potential of the existing urban area to the full and improving its quality. Extensive networks of green open spaces are to be developed within the urban regions.

Car traffic in cities and urban regions must be reduced, while ensuring accessibility for good and necessary car traffic. This is to be achieved by a location policy that keeps distances and the number of trips to a minimum, providing superior amenities for slow traffic (pedestrians and cycles) and public transport and encouraging the use of public transport by, for example, a strict parking policy.

Water must be managed to combat drying out of the ground and eutrophication of surface waters, and to prevent pollution. That means protecting vulnerable and important groundwater stocks. The approach is based on the hydrological system where the inflow of water alien to the area has had a polluting effect. This will be restricted by land-use measures and engineering works. Clean water belonging to the area will be retained as long as possible. Natural water treatment methods will be promoted.

Rural areas have been put into one of four categories, each with its own planning policy. These are: the green course, where ecological qualities will guide development; the yellow course, where agricultural production will be concentrated in regional complexes; the blue course, where functions will be closely integrated using the specific regional qualities, for example combinations of agriculture, recreation, nature, forestry, landscape management and water management; and the brown course, where agriculture is the main activity but is integrated with other activities. The measures taken to combat pollution, for mineral extraction, or for military exercise grounds, will depend on the 'course' into which the land has been placed.

THE NATIONAL SPATIAL DEVELOPMENT SCENARIO

This set of policies must lead to a disposition of activities whereby the Dutch economy will be strong enough to grow and hold its own against international competition, without endangering the spatial diversity of the country.

Schiphol Amsterdam Airport and the port of Rotterdam must be allowed to grow as international main ports, and certain transport axes (road, rail and water) will be developed for the international distribution of goods. The economic core of the Netherlands consists of the Randstad and large parts of the provinces of Gelderland and Noord Brabant. This is the Central Netherlands Urban Ring, and it is of vital importance to the country's national and international economic potential. This must be strengthened, and the country's 'Green Heart' within the Randstad must be retained and improved.

A number of 'urban nodes' have been selected which fulfil a central function for the region and which are well placed to compete nationally and internationally with other cities. These nodes (there are 13 of them) receive preference when funds for higher-order facilities are being allocated. For three of these nodes—Amsterdam, Rotterdam and the Hague—measures are to be taken to improve them as locations for commercial and business services, in order that they can compete better internationally.

The Netherlands is a 'waterland'. The connection between the functions of water supply, nature, tourism, recreation and transport will be strengthened. The major stretches of water will be better linked. Both nature development and nature conservation will receive attention, and a linked waterway network for pleasure craft will be created.

The regions must develop on the strength of their own assets. Each region must seek a balance between exploiting its own individual qualities, orienting itself towards the economic core of the Netherlands, and connecting to developments in adjacent frontier regions and economic core areas beyond the national frontiers. (Put another way, the national government is going to give hardly any help to less developed regions in the country!)

Spatial planning and environmental policy must be integrated in the different areas. This is because the environmental quality of some areas is such that the desired spatial development is being jeopardised. Eleven areas have been selected where, on an experimental basis, spatial planning and environmental policy are being integrated.

Appendix B. Relevant actors in spatial planning

The Department of Housing, Spatial Planning and the Environment, and especially its Directorate General for Spatial Planning, play a central coordinative role in the development and elaboration of the Dutch national planning strategy. But other agencies play decisive roles too, both in policy development and in implementation. Other relevant ministries, departments or agencies of the public administration are as follows:

- **Directorate General for Housing** (*Directoraat Generaal voor de Volkshuisvesting*)
  A department of the Ministry of Housing, Spatial Planning and Environment, which has regional agencies.
  Relevant activities: Implements national government policy for housing, including the huge subsidies available. These are financed by the national government and transferred to municipalities, which pay them to housing associations and housing tenants.
- **Council of Spatial Planning and Environment** (*Raad van de Ruimtelijke ordening en Milieubeheer*)
  Relevant activities: Nationally, the minister VROM is responsible for government policy on spatial planning. Coordination with the portfolios of other ministers takes place in the Cabinet. The Council of Spatial Planning and Environment is a sub-committee of the Cabinet and prepares the decisions on spatial planning. It is presided over by the Minister President and attended by all ministers and junior ministers with responsibilities affecting spatial planning.
- **Council of State** (*Raad van staten*)
  Relevant activities: This is the supreme administrative court and therefore the last court of appeal against decisions (*besluiten*) of a body of the public administration.
- **Department for Conservation** (*Rijksdienst voor de Monumentenzorg*)
  Under the Ministry of Welfare, Public Health and Culture (*Welzijn, Volksgezondheid en Cultuur*).

Relevant activities: Advises on the listing of monuments and the paying of subsidies for the restoration and management of monuments.

- **Directorate General for Environmental Protection** (*Directoraat Generaal voor het Milieubeheer*)
  Relevant activities: The preparation and design of environmental policy. Responsibility for supervising the observance of legal regulations.
- **Government Service for Land and Water Use** (*Landinrichtingsdienst*)
  A department of the Ministry of Agriculture, Nature Management and Fisheries (*Ministerie Landbouw, Natuurbeheer en Visserij*).
  Relevant activities: Most of the rural areas have been subjected at least once to land-consolidation schemes (*landinrichting*). Landownerships are pooled, water management projects carried out, farm units reformed. This is carried out by local agencies of this service.
- **Ministry of Agriculture, Nature Management and Fisheries** (*Ministerie van Landbouw, Natuurbeheer en Visserij*)
  Relevant activities: Agricultural policy (which has an enormous influence on rural land use). The protection of nature (which includes nature reserves, national parks, etc.).
- **Ministry of Defence** (*Ministerie van Defensie*)
  Relevant activities: In connection with military exercise grounds, the defence of the realm.
- **National Agency for Transport and Public Works** (*Rijkswaterstaat*)
  Relevant activities: A large and powerful agency which started with water defence and land reclamation works and is now responsible for most large-scale public works.
- **National Forestry Agency** (*Staatsbosbeheer*)
  Relevant activities: Managing the forests, woodlands and nature reserves owned by the state.
- **National Property Agency** (*Rijksgebouwendienst*)
  Relevant activities: Owns and manages property used by the civil service and many other public services (e.g. the law courts).
- **Water Control Boards** (*Waterschappen*)
  Relevant activities: Water management at a regional scale (e.g. maintaining drainage and flood protection works); also in many cases the provision and management of water purification (sewage) works.

Private associations or trusts recognised by the public administration as fulfilling a public function are:

- **Agricultural Board** (*Landbouwschap*)
  Relevant activities: To promote good and profitable conditions for living and working for farmers and employees in the agricultural sector.
- **Chambers of Trade** (*Kamer van Koophandel*)
  Relevant activities: Representing local businesses, also registering them on behalf of the government.

- **Coordinative Committee for Urban Renewal** (*Coördinatiecommissie Stadsvernieuwing*)
  Relevant activities: To achieve better coordination in national policy for urban renewal. It advises the relevant government ministers on urban renewal policy and instruments.
- **Dutch Railways** (*Nederlandse Spoorwegen*)
  Relevant activities: Managing the railway system.
- **Local Housing Associations** (*woningcorporaties*) and their national representation (*koepelorganisaties*)
  Relevant activities: The local housing associations build and manage most of the subsidised housing. To this end they work very closely with the municipalities. The national bodies are very powerful politically.
- **National Advisory Council for Housing** (*Raad voor de Volkshuisvesting*)
  Relevant activities: To advise the national government on legislation and other policy matters concerning housing. Installed by the Housing Act, Article 89.

Private interest groups include:

- **Agricultural Pressure Groups** (*boerenbonden etc.*)
  Relevant activities: Representing farming interests, especially important in rural areas.
- **Association of Dutch Municipalities** (*Vereniging van Nederlandse Gemeenten*)
  Relevant activities: Representing the interests of municipalities, the Association has great political influence.
- **Association of Dutch Town Planners** (*Bond van Nederlandse Stedebouwkundigen*)
  Relevant activities: Maintains a register of recognised urban designers. The title of urban designer (*stedebouwkundige*) is protected by law.
- **Association of the Provinces of the Netherlands** (*Interprovinciaal overleg*)
  Relevant activities: Representing the interests of provinces.
- **Commercial Transport Interests** (*Nederland distributieland*)
  Relevant activities: Representing the interests of the private hauliers by road and water. Freight transport is economically very important for the Netherlands, so this group is a powerful political lobby.
- **Dutch Association for the Conservation of Natural Monuments** (*Vereniging natuurmonumenten*)
  Relevant activities: Owns and manages large nature areas (often including historic buildings and estates). Also an action group for a better environmental policy.
- **Dutch Association of Town Planners** (*Bond van Nederlandse Planologen*)
  Relevant activities: Maintains a register of recognised town planners. However, the title of town planner (*planoloog*) is not protected by law.
- **Dutch Property Federation** (*Raad van Advies voor Onroerende Zaken*)
  Relevant activities: Representing the common interests of its members, who have commercial interests in property.

- **Netherlands Association of Property Developers and Investors** (*NEPROM*)
  Relevant activities: Representing the interests of developers of and investors in property.
- **Netherlands Institute for Spatial Planning and Housing** (*NIROV*)
  Relevant activities: Providing a forum for all those engaged and interested in questions of spatial planning and housing.
- **Netherlands Society for Nature and the Environment** (*Stichting Natuur en Milieu*)
  Relevant activities: Works towards a better environment and protection of nature. It is one of the most important private bodies doing this work.
- **Royal Institute of Dutch Architects** (*Bond van Nederlandse Architecten*)
  Relevant activities: Maintains a register of recognised architects and protects their interests. The title of architect is protected by law.

THE ROLE OF CONSULTANTS IN THE PRIVATE SECTOR

All the policy institutions in the public administration may engage consultants to assist them in the execution of their tasks. This can include carrying out research, providing advice on policy, providing advice on applications for a building permit and making spatial plans. The smaller municipalities in particular make much use of consultants and usually have no planning department of their own.

NOTE

Appendixes A and B are based on the work of B. Needham and others in *The Compendium of Spatial Planning Systems and Policies of the Netherlands*. This work is part of *The EU Compendium of Spatial Planning Systems and Policies*, European Commission, Regional Development Studies 28, 1997.

ELEVEN

NATIONAL-LEVEL PLANNING IN ISRAEL: WALKING THE TIGHTROPE BETWEEN GOVERNMENT CONTROL AND PRIVATISATION

Rachelle Alterman

Compared with most other Western countries, Israel has maintained a very high dosage of national-level planning institutions and powers. This is not surprising. Given Israel's unique constraints and national goals, it should be a 'natural' for national-level planning. Yet, as our story will show, these institutions have not always functioned to the same degree or held the same status.

The exposition of national-level planning in Israel begins with an introduction to Israel's 'vital statistics' and built-up form. Next comes a section that introduces the key national urban and regional policies in order to give the reader a feel for the context. We then move to the general constitutional and institutional setting for policy-making by national and local government. Then, we focus on the major national-level agencies charged with a comprehensive view and with setting overall policy. This leads to an analysis of the relatively large degree of state involvement in sectoral planning and implementation, and is followed by a detailed presentation of the statutory land-use planning system and of national statutory plans. A separate section recounts how leading planners cleverly utilised the national crisis brought about by mass immigration from the former USSR in the early 1990s, to raise national-level planning to a new plateau through initiatives like the 'Israel 2020' project. To conclude, I sketch my view of the future role for national-level planning in Israel, shaped as it is by the conflicting forces of centralisation on the one hand, and decentralisation and privatisation on the other.

Some background geographic and demographic statistics

Israel has a population of six million people, 80 per cent of whom are Jewish and 20 per cent Arab. (Here and throughout this chapter, unless stated otherwise, I am referring to Israel in its international borders, without the still occupied parts of the West Bank and the Gaza strip.[1]) Israel's population is 92 per cent urban—among the

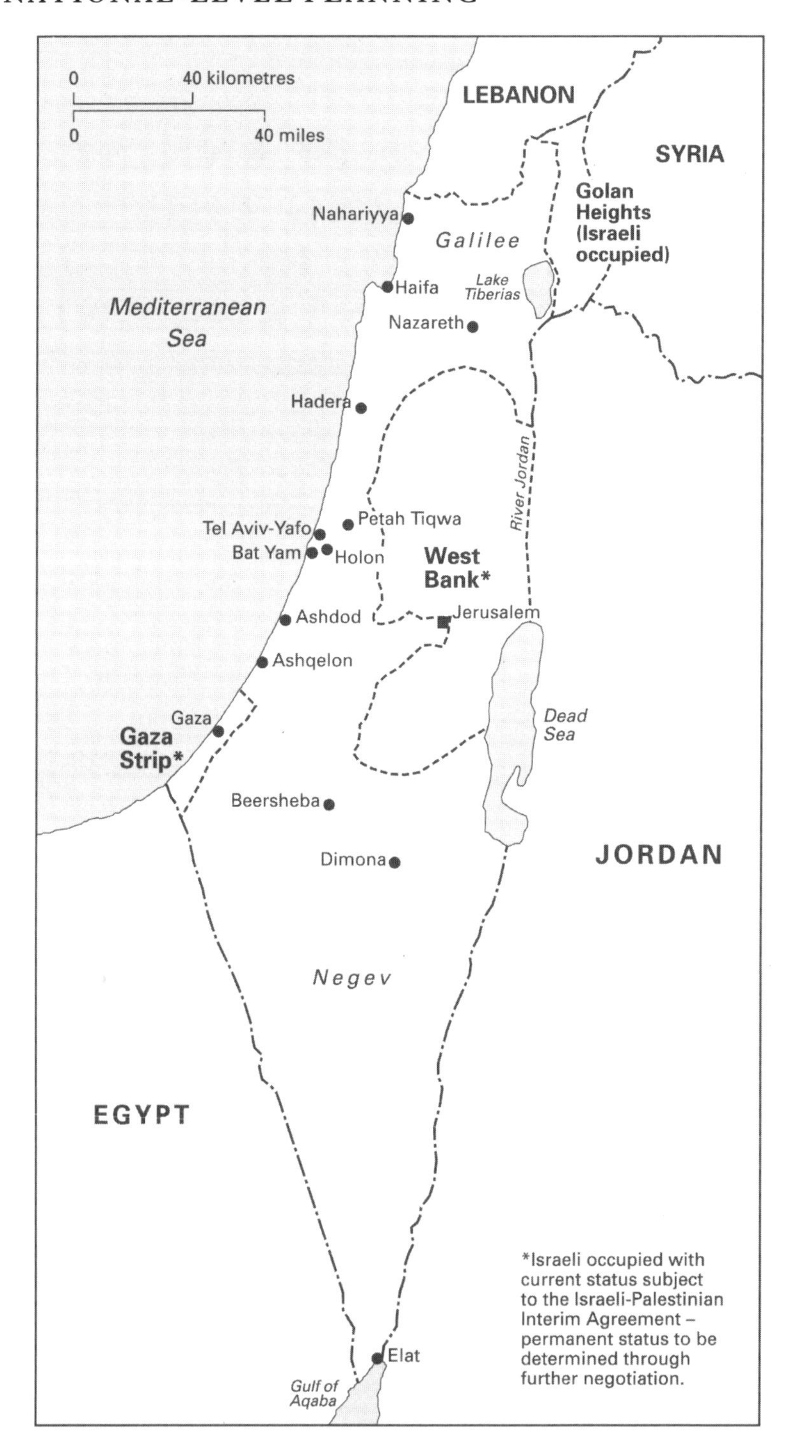

Figure 1 *Map of Israel showing neighbouring countries and occupied areas*

highest urban populations in the world (74 per cent of the population in the USA is urban, 77 per cent in Canada, 89 per cent in Britain, 84 per cent in Sweden and 89 per cent in the Netherlands) (United Nations, 1996). Israel's land area is approximately 21 000 square kilometres (see Fig. 1). The population density is 300 persons per square kilometre. Furthermore, since over 50 per cent of Israel's land area is in the inhospitable southern desert, the effect density is much higher. Although this density level is not the highest among this book's sample of advanced-economy countries—the Netherlands and Japan are still higher—Israel is unique among Western countries in having a high positive natural growth rate (births over deaths). Before long this will make it the most densely inhabited country among the advanced economies.

Israel is also the only country in the West that is ideologically committed to taking in mass immigration (of Jews and family members). Since 1990 it has taken in about a million former Soviet bloc citizens. It thus holds the Western world's record for immigration absorption, proportionate to population size.[2] Furthermore, since its establishment in 1948, Israel's economy has grown steadily from a level of per capita GDP characteristic of developing countries, to a level typical of many advanced-economy countries, albeit on the lower rungs of that ladder with a per capita GDP of $18 000 in 2000. This remarkable economic growth has been expressed through a steep rise in demand for land and built-up space.

To this cocktail of needs and constraints, one should add that Israel is the only country among our sample which, for most of its history, has been in a state of war with all its neighbours. After the peace treaty with Egypt in 1978, with Jordan in 1994, the Oslo peace accords with the Palestinians in 1993 and its continuation with the Wye and Camp David accords in 1998 and 2000, Israel is still in a state of war with the remaining neighbouring countries and under military threat from several directions. This combination of factors—the country's small geographic size, its demographically growing population, its policy favouring mass immigration, its accelerated economic growth, and its geopolitical and security needs—has made it a natural candidate for strong national planning powers and actions. It was reasonable for Israeli policy-makers to wish to harness urban, regional, land and infrastructure policies for the achievement of national goals.

At the same time, Israel's economic and socio-political development—the march to an open society and an open economy—has rendered Israeli public administration and policy more and more akin to other advanced-economy democracies. As most of the other countries have shifted to greater decentralisation, increased deregulation and more privatisation, so has Israel; but its starting point has been different (Shefer, 1996).

An introduction to national urban and regional policies

Israeli cities and towns are quite compact, and resemble cities and towns in Europe. Most Israelis live in medium- and occasionally high-density apartment buildings (mostly condominiums). Until the 1980s, there was almost no construction in urban areas of land-consuming 'ground-attached' housing (Israeli professional jargon for single- or double-family low-rise houses). This mode of living was reserved for rural areas. However, since the mid-1980s, consumer demand on the upmarket side has shifted new construction of this type to towns and cities, so that today, it holds a hefty 40 per cent share of annual housing starts.[3] The 1990s also saw the proliferation of shopping malls on the outskirts of urban areas.

Readers might find this puzzling. In a small country with a growing economy, a strong natural growth rate, and a policy of 'open gates' towards potential immigrants, one would expect a policy of careful stewardship of land reserves, rather than a policy of allowing new land-consuming uses. Yet until the 1990 mass immigration crisis, the concern with land as a depletable resource was not very strong (Mazor, 1993). The reasons for this lie with the planning goals into which Israeli planners and policy-makers were locked during most of Israel's history (Shachar, 1993).

Israel's national-planning goals were rooted in the agenda of nation-building and territorial stabilisation (Brutzkus, 1988). This is understandable in a country which, in the years following its 1948 War of Independence, was seeking to establish its legitimate standing within its international borders, some of which were (and still are) officially only 'armistice lines' in international law.[4] This geopolitical agenda yielded a strong focus on 'population distribution', which led to a policy of constructing many new towns and new villages, distributed as widely as possible along Israel's borders and throughout, so as to create a 'Jewish presence' in most areas of the country (Yiftachel, 1992, 95–98). The population distribution goal was reinforced by the traditional ideological emphasis on rural development as a utopian form of living, to symbolise the return of the Jewish people to the Holy Land and to agriculture (Cohen, 1970). This double ideology led to the establishment of several hundred cooperative and communal rural settlements, distributed as widely as possible within Israel. These were the focus of considerable attention from planners and politicians, and they received generous land, water and budget allocations. The pro-rural policy was, however, at odds with the reality that the vast majority of residents at all times have preferred to live in urban areas (Alterman and Hill, 1986).

The imperative of housing masses of immigrant refugees in the 1950s and 1960s led planners to shift some of their attention to urban areas. The resultant policy was the establishment of some 30 new ('development') towns in various parts of the country, many in outlying areas. The neighbourhoods of the 1950s and 1960s were, for the most part, constructed by government or other public agencies on public, national land. As in many other countries (Alterman and Cars, 1991), the housing of that time was characterised by uniform blocks of apartments designed by central government architects, with little regard for consumer diversity, and with little attention to the differing landscapes. A new town in the green hills of Galilee might be planned at a density similar to a neighbourhood in Tel Aviv. But preservation of open spaces *per se* was not yet part of the politicians' agenda. The fact that the major national parks were—luckily—established in the 1960s and 1970s is a feat of the leadership and conviction of a few senior planners, rather than a reflection of ideological priorities (Brutzkus, 1988).

Israel's population distribution policy has been implemented in many ways, not only through the location of rural and urban units, but also through subsidies to industries wishing to locate in peripheral areas. Households of young couples, new immigrants or needy families are offered preferred mortgage terms. Employees are offered reduced income tax.

Despite the deep-rooted changes in Israeli demographics and geopolitics, the population distribution doctrine was not challenged until the mass-immigration crisis of the early 1990s. Leading national-level planners used the crisis to challenge this 'sacred cow', and offered a doctrine more befitting Israel's current objective conditions. The new doctrine revolves around the scarcity of land and the need to preserve enough open spaces for future generations. The new awareness of the need to use land more wisely is now shared by many decision-makers across most of the political spectrum, and since the late 1990s, they have promoted multi-use, higher urban densities, and urban recycling.

But the consensus is only on those aspects that relate to land use in existing cities, while aspects pertaining to open space preservation (i.e. contrary to the population distribution policy) are in dispute. Some government agencies still promote contentious land-gobbling projects: the Cross-Israel Highway is being built despite considerable criticism (Alexander, 1998), and the Ministry of Housing still proposes new towns. But now that the population distribution policy has been untied from the ropes of consensus, these projects are subject to intensive debate. This debate is, to some extent, being conducted along partisan lines: hawks tend to promote population distribution for perceived security and land control reasons,[5] while others debate the issue as a 'normal' domestic planning issue, raising universal dilemmas such as distributive and inter-generational justice, and freedom of consumer choice (there is much demand for 'ground-attached' homes, malls and ex-urban communities).

The constitutional context and central–local government relations

THE PARLIAMENT AND CENTRAL GOVERNMENT

Israel is a unitary state with a parliamentary political system. The parliament—the Knesset—is composed of 120 members, in a single chamber. Aside from its legislative role, the parliament has no direct involvement in approving plans or policies. During most of Israel's history, the Knesset has shown very little interest in urban and regional matters since much of Israel's political agenda has been preoccupied with issues of war, peace, the future of the occupied territories and religious controversies. Until the early 1990s, only a handful of Knesset members (MKs) would show up to meetings of the Knesset Committee for Interior and Environmental Affairs, where bills related to land use or environmental planning are prepared for legislation.[6] This has changed to some degree in the 1990s. As Knesset committees have become more open to interest groups, slightly more MKs have become active in planning affairs.

Although the country is divided into six statutory districts, these do not hold any intrinsic powers, and Israel does not have a federal structure. Powers held by the district officers—oversight of local government and land-use planning—have been assigned to them by legislation. There are also administrative districts that some ministries have set up for administrative convenience, but these have no statutory powers and often their boundaries do not coincide with each other. Israel's constitutional structure vests within central government all residual executive powers not specifically assigned by law to local government or to a specific agency within central government or outside it.

Given the highly centralised structure of decisions, it is not surprising that central government is closely involved in many aspects of spatial planning. Yet, despite the absence of formal decentralisation, the national-level policy-making process and the content of policies have changed significantly. These changes reflect the growing local government assertiveness noted below, the accelerating trends of privatisation of public services, and the reduced government involvement in housing supply.

Many national-level agencies make decisions which have a direct bearing on urban and regional development. These include the Cabinet itself, most government ministries, several statutory bodies directly entrusted with urban and regional issues, and non-governmental bodies that are unique to Israel's historic context.

LOCAL GOVERNMENT AND ITS RELATIONS WITH CENTRAL GOVERNMENT

In 2000 there were approximately 250 local government bodies, of which 71 were in the Arab sector (most of the latter are villages that have become urbanised). These bodies are of three types: city-status local authorities (usually with a population larger than 25 000—approximately 60), town-status local authorities (approximately 135), and regional authorities (56), which include not only agricultural land but also ex-urban housing, and commercial and industrial units. Central government regards this number of local authorities as too large and fiscally wasteful and has promoted a policy of merging local authorities. But despite the recommendations of the several public commissions that the government has set up over the past 20 years, few mergers have yet been successfully completed because of local government resistance (Razin, 1996; 1997; Razin and Hazan, 1995).

Looking at legal and administrative powers only, it is not immediately obvious why there should be such a fervent defence of local government independence. On paper, the legal powers of local authorities are weak and their financial powers are severely constrained by central government. Local government is burdened by a whole gamut of responsibilities. Israeli local government bodies are legally and financially weaker than their counterparts in most Western countries, with the possible exception of the United Kingdom. Most major budgetary decisions and spending require central government approval. All but the most prosperous local authorities have a weak tax base, and are dependent on hefty central government transfers. At the same time, they have little leeway in adding or raising taxes. As we shall see, land-use and development control powers are more highly centralised than in most democratic countries.

Yet, much of the rapid rise in living standards and provision of public services is due to local authority initiatives which have probably been stimulated by the dynamics of electoral politics. Since the 1980s, mayors have been elected directly rather than as heads of a party slate. It seems there is nothing like political competition among candidates to bring out the best of initiatives and creative action. Such action has characterised not only the better-off local government bodies, but also those in the peripheral areas or with poorer neighbourhoods. Backed up by their electorates (who, the mayor will remind national politicians, also vote in the national elections), proactive mayors have taken two main routes to jack up their resources. They lobby central government and their own political party for more resources; and they use creative ways of getting developers to participate in the upgrading of public services to the dismay of central government (Margalit and Alterman, 1998).

Israel is one of the few advanced-economy countries where no major decentralisation and devolution of powers have (officially) taken place. Central government legally still retains most of the powers it possessed when Israel was in its formative stages. Yet, since the 1980s, various incremental trends towards decentralisation have been occurring, most without a legislative stamp. Despite this, the span of topics that are still dealt with at the national level in Israel is astounding. So what remains for local jurisdictions? This picture is, however, somewhat misleading. Despite the rather hefty central government presence in many sectors, much of the day-to-day development policy and initiatives—the things that affect consumers most—are dealt with at the local level. Local authorities have learned all too well how to negotiate with central government to stave off 'locally unwanted land uses' (LULUs) and to increase their *de facto* autonomy despite central government powers of oversight.

Agencies with a potentially comprehensive view

Four agencies have the authority to take a comprehensive, multi-sectoral, long-range view of spatial issues, define priorities, and coordinate the plethora of sectoral agencies—the Cabinet, the Ministry of the Interior, the Ministry of Finance and the Prime Minister's Office. The defunct Ministry of Economics and Planning also had this authority. Yehezkiel Dror, a leading public policy expert, contends that this function is essential for Israel because of its special circumstances and needs (Dror, 1989; 1998). Although the institutional structure has not changed much since the state's establishment in 1948, there have been many shifts in the roles played by these bodies, reflecting changing politics, policy emphases, and attitudes to planning.

The problem is that most of these agencies have never fulfilled their comprehensive planning role—as many planning and public policy theorists would have predicted. All but the Ministry of the Interior, which is in charge of statutory planning, are usually involved in ad hoc issues that the politicians view as 'burning'. Yet, the story of each agency shows clearly that its involvement in urban and regional issues has increased in recent years.

THE CABINET

According to the 'Basic Law: the Government', the Cabinet is comprised of 18 ministers.[7] The Cabinet has only one statutory duty in land-use planning—it is the agency that gives final approval to national land-use plans. But according to Israel's constitutional structure, the Cabinet holds residual powers for any issues not allocated by law to a particular minister, and it has general authority to coordinate the policies of all the ministries. To the best of my knowledge, never—certainly in recent decades—has the Cabinet devoted a meeting to a comprehensive discussion of urban and regional policies and priorities. The Cabinet has not even used its statutory duty in approving national-level plans as an opportunity to engender a discussion of overall spatial policy priorities. For example, when the most comprehensive statutory plan approved to date—National Plan number 31—was brought to the Cabinet for its approval in 1993, as part of the immigration crisis alignment, only the Minister of the Interior was present. Planners joke that the plan was approved *unanimously*. All attempts to present the conclusions of the most comprehensive multi-sectoral non-statutory plan—the 'Israel 2020' plan—to the full Cabinet have, to date, also failed.

But the involvement of the Cabinet in ad hoc urban and regional policies seems to be increasing.[8] For example, since 1995, fearing a further hike in housing prices, the Cabinet has become concerned about the inadequate number of annual housing starts and the insufficient number of approved plans for housing that could serve as a reserve in case of another crisis. A concomitant concern has been the length of time it takes for statutory plans and building permits to be approved. So the Cabinet periodically instructs the Ministers of Housing and Interior, respectively, to take implementation actions, including changing the Planning and Building Law to streamline procedures.[9]

Another urban and regional planning issue that has entered the Cabinet's agenda since 1996 is the scarcity of developable land and the need for planning policies that intensify land utilisation and recycle under-utilised urban fabric.[10] Distinctive 'green' policies have not yet entered the Cabinet's agenda, so that its decisions on the land intensification policy are not yet motivated by an open space conservation policy. As in the past, these issues are being furthered by the planners who have now been joined by green interest groups.

THE MINISTRY OF THE INTERIOR

The Minister of the Interior is responsible for implementing the Planning and Building Law of 1965. He stands at the top of the administrative hierarchy of the planning pyramid and is directly responsible for appointing many of the members of the national and district planning boards.[11] The Ministry of the Interior runs the professional support units for the national and district planning bodies—

the Planning Administration[12]—and is also in charge of all aspects of local government oversight—administrative, political and financial. Unfortunately it has not capitalised on the coincidence (in international comparative terms) that a single ministry is responsible for both land-use planning and local government. Had it done so, it might have made better use of its local government powers as an incentive for implementing national planning policies.

Until the late 1980s, the Ministry was regarded as a weak bureau, and was usually allocated to one of the religious orthodox parties. The Planning Administration was not regarded as the centrepiece within the Ministry. In fact, its budgets and staffing were so ludicrously low that statutory planning seemed doomed to sink into further inaction and disrepute. But during the 1990s, the Ministry's position in the political pecking order had risen significantly. This reflected the transformation in the image of land-use planning and the growing realisation among politicians and stakeholders that the person who controls land use holds the keys to public policy implementation.

Since all land use and development falls under the jurisdiction of the planning law, one can argue that the Ministry of the Interior should be responsible for ensuring a comprehensive view of spatial planning, even though this is only implementable through land-use and development controls. But until the early 1990s, no comprehensive national plan had been prepared, only sectoral ones. The immigration crisis brought about an impressive strengthening of comprehensive land-use planning, along with an increase in the budgets allocated to national and district plans.

THE MINISTRY OF FINANCE

The Ministry of Finance is responsible for budgetary oversight of all policy areas, including urban and regional, housing and land policies. In that capacity, it should advise the Cabinet on the economic impacts of policies. It is not expected to be a policy-making body in areas covered by particular government bureaux. In practice, however, the Ministry has been gradually increasing its direct involvement in making urban and regional policies. Since the early 1990s, representatives of the Ministry of Finance have been involved (to some extent) in all new legislation regarding planning and related matters. For example, in 1990, representatives of the Ministry worked side by side with the Ministry of the Interior to initiate crisis-time legislation to streamline the planning and building procedures. The Ministry of Finance was clearly overstepping its normal jurisdiction since the bodies responsible for planning legislation are clearly specified in law—the Ministry of the Interior in consultation with the National Planning and Building Board.

In the late 1990s, officers from the Ministry of Finance began to develop new urban policies that would normally be under the jurisdiction of the Ministries of Housing or the Interior. In 1995, a committee headed by the Director General of Finance determined central government allocations for local public facilities; and, in 1998, a committee under the head of the Budget Branch proposed policies for urban regeneration. The Cabinet has been increasingly allocating leadership roles in urban and regional matters to the Ministry of Finance. This situation all but usurps the clear statutory duties and distorts a normal administrative division of labour. Officers of Finance—all economists[13]—have become regular and active members on the steering committees of all major national statutory plans and some regional plans.[14]

This emerging division of responsibilities is not just a case of 'he who pays calls the shots'. I surmise that Finance's interest in planning stems from two directions, one bearing 'good news' for planning, the other not. The good news reflects an impressive rise in the saliency of urban and regional planning as perceived by senior government officers. Since much larger budgets are now involved and there is more public exposure, the Ministry of Finance wants to 'get into the act' so as to steer policies in the direction it desires.[15] The bad news is that this encroachment also reflects the still prevalent weakness of the Ministry of the Interior and of statutory planning, despite the positive strides taken in the 1990s.

THE PRIME MINISTER'S OFFICE

Some public policy experts have seen the Prime Minister's Office—a bureau with a rather small staff—as an appropriate locus for a national strategic policy and planning unit focused on security and other issues (Dror, 1989; 1998). This has never materialised. But the 1990s and the crisis-time alignment did intensify the hitherto low level of involvement of the Prime Minister's Office in urban and regional policies.[16] As in the case of the Ministry of Finance, current involvement of the Prime Minister's Office in urban and regional policy testifies to the growing importance of land and of planning. In the early 1990s, a small professional planning unit was set up. Initial tasks included monitoring the outputs of the statutory planning bodies in approving plans for housing units (remember the Cabinet's concern about insufficient land reserves for housing), and it developed the first computerised system for that purpose. Since then, the unit has increased its span of involvement in promoting the implementation of the Cabinet's ad hoc decisions on urban and regional matters, on housing and on land policy.

THE (NOW PHASED OUT) MINISTRY OF ECONOMICS AND PLANNING

Without being overly blunt, one can say that the Ministry of Economics and Planning was established as a lip-service expression of the Cabinet's awareness of the need for a coordinating body to promote planning modes of policy-making among government agencies. The Economics and National Planning Authority was also established and was to be guided by a council of that name. But very little was ever heard from the Council,[17] and the Authority had only a minor impact on a few policy areas. The Ministry was regarded as a minor portfolio and was not coveted by most Cabinet members. It therefore changed ministers frequently. The Authority's research unit produced some interesting analyses with an economics emphasis, but it was too weak to compete with the powerful Ministry of Finance, or with the highly professional Bank of Israel and its large macroeconomics research unit.

Potentially, the Ministry could have played a role in promoting planning modes in government. In 1994–95 (during Labour's term of office), the minister initiated a draft bill to require all agencies with state budgets to submit a middle-range budgetary plan in addition to their annual budget. Had the law been enacted and implemented, it might have strengthened planning in government agencies in general, and urban and regional planning in particular. But the bill went nowhere.

For a brief while, the Ministry of Economics and Planning came close to playing a direct role in promoting urban and regional planning. In 1995, Dr Yossi Beilin—a political scientist and co-architect of the Oslo Peace Pact, who took over the Ministry—was approached by the Israel 2020 Master Plan team in its search for an appropriate agency with a macro perspective that could promote the implementation of this multi-sectoral, non-governmental planning initiative.[18] Beilin was personally very supportive of the project, but was very sceptical of his bureau's future. In January 1996, he initiated its final closure.[19]

Sectoral national-level planning and the agencies involved

Israel has a plethora of agencies with responsibilities for national-level sectoral planning. By sectoral planning I mean initiatives taken by a government agency to prepare plans that further the interests of a particular sector. Usually, such plans are submitted for approval to the national statutory planning bodies. This section describes the major sectors that have distinctive spatial impacts and indicates which agencies play important roles in each sector. The obvious question and challenge is the degree of coordination among them. As we have seen, there is no agency which systematically takes a multi-sectoral, long-range view of public policy. Later in this paper I shall ask whether the statutory planning system can fulfil this role.

HOUSING: PLANNING AND CONSTRUCTION

Housing has always played a key role in national spatial planning. During Israel's formative decades, the Ministry of Housing used to be *the* land-use and development planning agency, shaping the form of the country almost at its whim (Law Yone and Wilkansky, 1984). Planners in the Ministry of Housing, not of the Interior, made the renowned national development plans of the 1950s (there were no statutory national plans then). Even a decade or more after the Planning and Building Law was enacted in 1965, the Ministry of Housing was reluctant to let go of its national planning role. It had its own 'national population distribution plan', and continued to prepare plans for new towns and regions even though these were poorly coordinated with national statutory planning.

Until the early 1970s, the Ministry of Housing was responsible for 60–70 per cent of housing starts, and had constructed over 30 new towns and many new neighbourhoods in existing cities. It was also responsible for the construction of much of the infrastructure and public facilities. Its rural development department was jointly responsible (with the Jewish Agency) for housing and infrastructure in the cooperative and communal villages. But by the late 1980s, privatisation trends had reduced the share of public sector housing to 18 per cent, and most of this housing, called 'public programme housing', was given incentives by the state but financed and constructed by the private sector. The developer had to conform to a particular mix of size and types of housing, and in exchange could tender for public land. Most indirect subsidies have also been phased out, except in peripheral regions. Today, infrastructure duties in all cities are the responsibility of local government, which sometimes passes them on to developers. For a short while in the early 1990s, during the mass immigration crisis, the Ministry of Housing returned to its earlier mode, but when the crisis ebbed away, it resumed its former down-sized role. In the 1980s, the Ministry of Housing was the co-partner with the Jewish Agency—an arm of the Jewish people worldwide which has quasi-government functions within Israel——in initiating Israel's ambitious and successful project renewal in 90 urban neighbourhoods (Alterman, 1991). This project still exists, but on a modest level.

Today, the Ministry of Housing's role in national planning is more akin to that of other sectoral agencies. It remains a powerful and sought-after portfolio because of its large budgets and political importance. This holds even though, since the mid-1990s, about half of the responsibilities for public sector housing have been passed to the Lands Administration, as part of an ad hoc political calibration of Cabinet portfolios. The Ministry of Housing or the Lands Administration are responsible for the land-use planning of public programme housing areas throughout the country, and submit these plans for approval to the statutory bodies. In addition, the architects in the Ministry of Housing or the Lands Administration oversee the actual design of the housing sites. Both agencies are active in pushing for a new national statutory plan that would ensure enough land reserves for a large housing stock for the future.

What I have described is, indeed, a very high degree of centralisation which might surprise readers from most Western countries, where social housing initiatives—if they still exist—are usually undertaken by local government. In Israel, however, local government has never been significantly involved in social housing supply, because of its legal and financial structure. Thus, in Israel, as the third millennium sets in, despite the significant privatisation trends, central government is still highly involved in the planning and even the design of social housing.

LAND POLICY AND MANAGEMENT

With 93 per cent of the land area owned or managed by the State,[20] the Israel Lands Administration Law of 1960 assigns to the Israel Lands Council and the Israel Lands Administration the roles of making policies regarding public land (the Council) and implementing them (the Administration). The degree of government intervention in the land market is, however, much less onerous than it may seem, because of the strong quasi-privatisation trends that have gradually rendered public land, once released, almost tantamount to private land.[21]

The responsibilities of the Administration and the Council might suggest a national planning system independent of the regulatory system. That was indeed true until 1965. Until that time, the Ministry of Housing and the predecessors of the Lands Administration both planned and implemented whatever they wished. But the 1965 law expanded the decision-making fora and regulated all development—including private development on public land, and state-initiated development on any land. Until the 1990s, the Lands Administration played a relatively minor role in initiating land-use plans, leaving most planning to the various sectoral agencies or to the private sector. Although it always did have special standing in its capacity as the surrogate of the landowner (the state), the Administration confined itself mostly to its major statutory role—to manage public land by determining the contractual conditions under which land would be released for development or agricultural use. The Administration played a direct planning role only in areas allocated to single-family homes in urban areas.

Ironically, despite the overwhelming trends towards the *de facto* privatisation of public land, the Administration has not reduced its participation in national-level spatial planning, but rather, has significantly expanded its direct involvement in initiating plans and even in implementing them. As we have seen, about half the public sector housing has been assigned (for Cabinet turf-division reasons) to the Lands Administration. The Administration prepares and oversees the implementation (through outsourcing) of a new type of programme—urban regeneration and recycling. And, most importantly, it plans and releases for development the country's major land reserves and future development initiatives—the transformation of agricultural land in cooperative or communal villages to urban development. Thus, although the Administration does not initiate national sectoral plans on its own, in recent years it has become an active participant in major national planning initiatives. These changes once again demonstrate the pivotal role played by real-estate development, and, therefore, the growing recognition of the importance of land-use planning. Everyone wants to 'get in on the act'.

THE RURAL SECTOR

National-level agricultural planning and management has been extremely strong in the Jewish rural sector—even stronger than in the urban areas. All cooperative and communal villages—some 600, and these constitute most of the Jewish rural sector—were sited, planned and built by a national-level agency. Until the 1970s, agricultural production itself was nationally planned and regulated. Today it is mostly left to economic competition.

Three major agencies have played a role in rural sectoral planning: The Rural Planning Authority—a joint unit of the Ministry of Agriculture and the Jewish Agency—allocated the agricultural 'land square' for each village (on public land), based on land-allocation norms per household that were deemed egalitarian. The Jewish Agency was responsible for planning and co-financing the construction of housing and infrastructure in most of the cooperative and communal villages. In some villages, the Agency was the lessee of the land from the Lands Administration, which it sub-leased to the residents. The Agency was also responsible for Israel's exemplary agricultural training and economic development. As the co-partner of the Rural Planning Authority, the Jewish Agency has always played a major planning role in initiating new rural villages (into the 1980s), and has been extremely influential in regional planning and economic-development planning. Until the late 1980s, the Agency's planning department was probably the largest in the country. In the 1990s, the Agency's planning role declined because of financial constraints. The third major body involved in rural planning is the Rural Department of the Ministry of Housing. It has worked jointly with the two other bodies in the design of rural villages in Israel, but has shifted much of its focus to West Bank Jewish settlements.[22]

Today, the rural economy has declined and the rural sector has become the fastest-changing sector. Due to Israel's small size, most agricultural land is potentially a real-estate bonanza for urban or ex-urban development. The questions surrounding the future of agricultural land and of the rural 'villages' are among the most controversial planning issues. They are vehemently contested between development-oriented agencies or interest groups on the one hand, and 'green' and long-range planning interests on the other. Since the 1990s, conversions of agricultural land into urban or ex-urban land uses have been authorised by the Lands Council for the first time. Such conversions are being energetically implemented. The Rural Planning Authority and its policies have become a pivotal clearing-house for this conversion process. In 1998, recognising that agricultural land *per se* is no longer the exclusive focus, the Ministry changed its name to Agriculture and the Rural Milieu, and, for the first time, articulated its goals to include an active role in the preservation of open space *per se*.[23]

Another important body concerned with agriculture—the Commission for the Protection of Agricultural Land—is not dealt with here because it is part of the regulative statutory land-use planning system discussed below. In the past, this Commission was dominated by agricultural interests. These are much weaker today.

TRANSPORTATION AND OTHER INFRASTRUCTURE

The responsibility for sectoral planning in this area has historically been rather fragmented, and it is growing more so, probably as a reflection of the traditionally politically unattractive status of most infrastructure Cabinet portfolios. The Public Works Department is an exception because it is a large spending agency, and has thus tended to be moved from one coveting ministry to another. In the distant past, the Public Works Department was in the Ministry of Labour. Later it passed to the Ministry of Housing, which then gained the Construction part of its name. In 1996, when a Likud-led coalition came to power, the Department was moved to the Ministry of National Infrastructure. The latter was created from bureaux carved out of other ministries in order to produce a suitable portfolio for Mr Ariel Sharon. The Public Works Department is responsible for initiating and implementing one of the most important national sectoral statutory plans—highways and roads. The Ministry of Transportation, a politically weak bureau, is responsible for policies concerning metropolitan-area roads, the regulation of public transport, the Airports Authority, and general transportation policy. It has initiated a national statutory plan for airports.

The rationale for the National Infrastructure Ministry was that it would co-ordinate infrastructure policy—not only transportation, but also rail, ports, electricity, gas and oil pipelines, water, telecommunications, etc. These have been chronically uncoordinated in their locations and strategies, despite the powers of national statutory planning. Lately, Infrastructure has indeed taken initial steps to coordinate several national sectoral plans in these areas. But the continued existence of the Ministry of National Infrastructure intact depends on the convenience of portfolio redistribution after every national election. The shuffling and reshuffling process may have exacerbated coordination among some infrastructure sectors rather than improving it. Some areas of infrastructure have remained outside the new office—like the Ministry of Energy, which has been left only with responsibility for power plant siting (also a national sectoral plan). Another relevant ministry is Telecommunications. In conjunction with the Bezek Israel Telecommunications Company and the cellular corporations, it is in the process of initiating a statutory plan to regulate telecommunications facilities. In Israel, due to extreme land shortage, if poor coordination among infrastructure services continues, a major crisis or partial collapse can be anticipated.

WATER

Because of water scarcity in the Middle East, water resources have been subject to exemplary national-level planning and regulation since Israel's early days— a story beyond the scope of this paper. All water resources—streams, the only sweet-water lake, and all aquifers—are nationally owned, planned and rationed, and there is a battery of laws and regulations to protect them. The Water Commissioner's office—a pivotal position for national land-use and development policy—was transferred from the Ministry of Agriculture and the Rural Milieu, where it has been for decades, to the new Infrastructure bureau (perhaps symbolising the decline in power of the agricultural sector). It will be interesting—and telling—to see which Cabinet portfolio receives this key responsibility after the next election.

ENVIRONMENT AND OPEN SPACE

Compared with most advanced-economy countries, environmental policy and planning in Israel has occupied a rather low position in national politics, on the Cabinet's agenda, and in budget allocations, but it is definitely on the ascent. In the mid-1970s, the Environmental Protection Agency was established within the Ministry of the Interior and later became the Ministry for Environmental Quality. With a highly motivated and professional staff, this bureau has consistently increased its presence and importance in all environmental planning areas and in legislation for regulating pollution and other environmental hazards.

Together with other agencies, the Ministry of the Environment plays an important role in initiating statutory plans for areas such as streams, coastlines, solid and toxic waste sites, and sewage purification facilities. In all these areas, the national level has an important policy-making and planning role. Under the Law for National Parks and Nature Reserves, the Ministry of the Environment has ministerial responsibility for two of the three major national open space authorities: the National Parks Authority and the Nature Reserves Authority (recently amalgamated). They initiated the important national statutory plan for parks and nature reserves. The third body active in this area is the Jewish National Fund

(JNF)—the second non-state body of the Jewish people which, like the Jewish Agency, carries out quasi-state functions. The JNF plans and manages the country's major forests and has initiated a national statutory plan for that purpose.

INDUSTRY AND TOURISM

The ministries associated with economic planning also have a sectoral national-level planning function. Israel's long-term population distribution policy has been implemented not only through housing and other construction, but also through a battery of economic incentives. These are offered through the Law for the Encouragement of Capital Investments, and implemented by the Ministry of Commerce and Industry which gives hefty loans, write-offs and tax perks to industrialists willing to locate plans in peripheral areas. But unlike some other countries represented in this book, central government in Israel is not directly involved in planning the location of industrial plants. A similar subsidy policy is implemented by the Ministry of Tourism, which has also been active in initiating a statutory national plan for tourism. This plan has important land-use implications for coastline, open space, and other issues of preservation and development.

The land-use planning system

At some point in the process, most sectoral agencies described in the previous section require the approval of the statutory planning bodies for their policies regarding land use or their initiatives regarding development or conservation. Much of the give-and-take among the national-level sectors—as well as between central and local government and private interests—occurs through the mediation of the statutory land-use planning system.

Israel has a rather centralised land-use planning system that combines top-down planning with bottom-up initiative. Central government is involved, first, through its extensive powers to oversee local-level planning decisions, and second, through its power to make binding national land-use plans. National involvement is channelled, at least on paper, through the hierarchy of plans, from national plans, through district plans, down to local plans.

NATIONAL-LEVEL OVERSIGHT OF LOCAL AND DISTRICT DECISIONS

The Israel Planning and Building Law of 1965 controls all planning and development. This law replaced the legislation introduced by the British in 1922 and 1936 during their Mandate over Palestine, and that remained in force after the establishment of Israel in 1948 (see Alexander et al., 1983).[24] British Mandate legislation did not have a national planning body or national statutory plans. The 1965 legislation added both.

Until 1965, planning controls did not apply to central government bodies. The 1965 law required all government jurisdictions—central, district or local (defence-related uses have special procedures in the law)—to submit a plan for approval and obtain a building permit just like any private developer. The institutional hierarchy under the Planning and Building Law has remained more or less the same since 1965, with some incremental amendments, the major one in 1995 (see Fig. 2). Although the figure shows the Cabinet at the top, it actually has only one direct role in the statutory system—to approve national plans.

The National Planning and Building Board has 31 members. In 1965 the Knesset took a rather progressive view and included not only representatives of all relevant government ministries (11), but also representatives of the various levels of

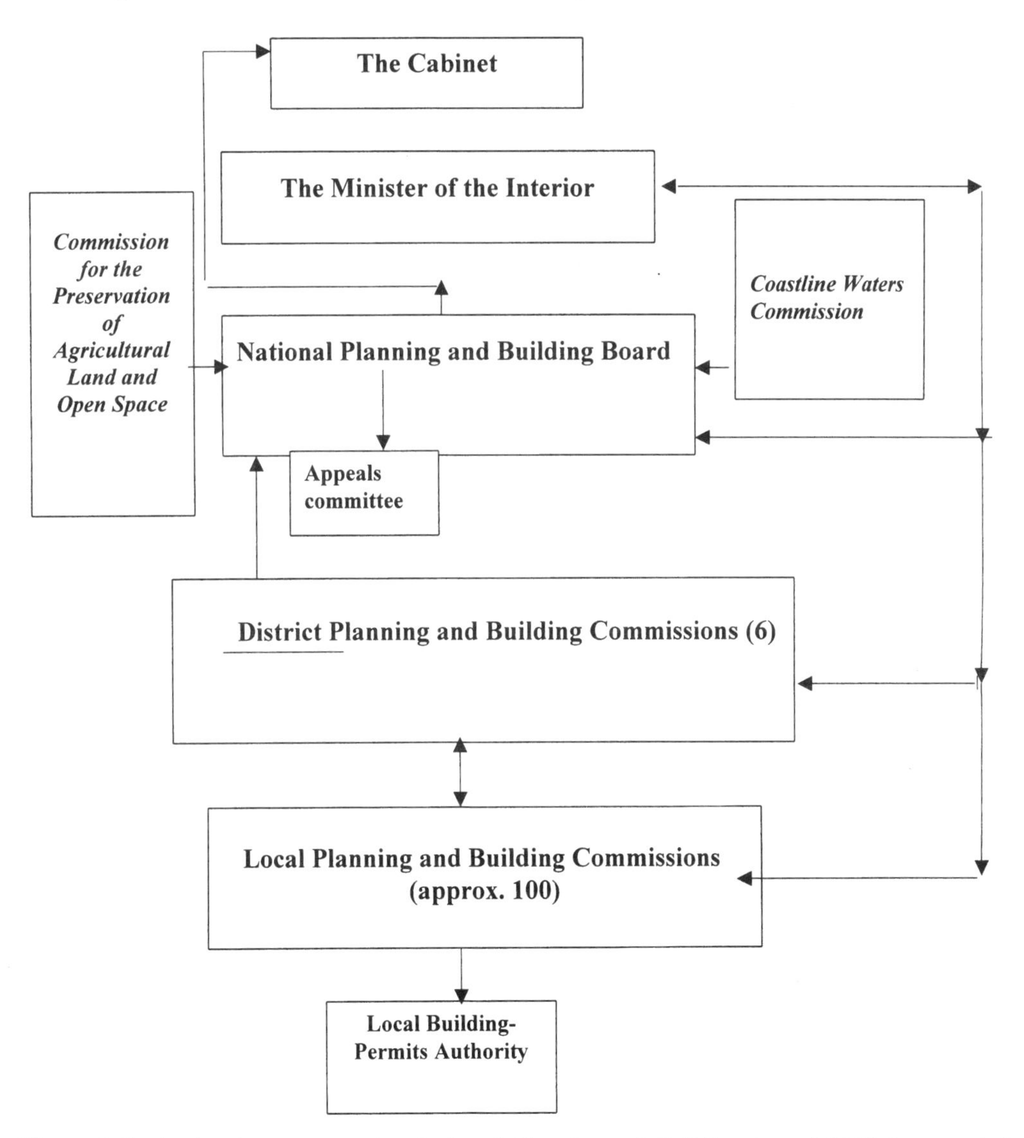

Figure 2 *Institutional structure under the Israel Planning and Building Law of 1965 (as amended 1995)*

local government, non-governmental environmental bodies, rural interests, the planning and building professions, women's organisations, a sociologist, a representative of academe (the Technion), and even a representative of 'the younger generation'. Hard to believe, but in 1998, the then Minister of the Interior convinced the Cabinet of the need to reduce the number of members to 21 by ousting the representatives of the 'green' and other interest groups. A bill to that effect was prepared but, happily, not floored.[25] The Minister's rationale was that a large forum makes deliberations cumbersome and that government bureaux represent the 'greens' and other interests well enough!

Another statutory planning body 'alongside the National Board' is the Commission for the Preservation of Agricultural Land (CPAL). The CPAL was in the past extremely powerful. It has no counterpart in any country (Alterman, 1997a). Legally, it stands above the National Board and even above the Cabinet because every plan that encroaches on declared agricultural land—a local, district or national plan—cannot be approved until the CPAL has reviewed it and decided whether to change the classification temporarily or permanently. If one considers that in 1968 almost all vacant land—whether suitable for actual agricultural use or not—was declared to be agricultural land through the Agricultural Land Declaration, one realises the power and importance of the CPAL. Another way of looking at it is to say that the Agricultural Land Declaration is like a super, overriding national plan (although its legal status is quite different). The CPAL, which used to be the bane of local authorities and developers, began to lose power in 1990, when it was perceived as a barrier to rapid development during the mass immigration crisis. Its legal powers, however, have remained intact. In October 1998, the Minister of the Interior aired the idea of transferring the CPAL's powers to the districts. If carried out, this will effectively mean the demise of the CPAL—but such action is unlikely.

The district planning commissions are composed mainly of representatives of central government bureaux—almost all the ministries relevant to planning. This body was probably conceived as a forum to encourage coordination between national and local planning policies, and among sectoral policies. Local planning and building commissions are, in the majority of cases,[26] composed solely of the local government-elected council. Central government representatives have onlooker status only. The local planning commissions are the first and usually crucial clearing-house for local policies and most development initiatives.

Central government oversight is carried out in several ways. Most of the local planning commissions' decisions require the approval of the district commission, so central government in effect controls the heart of the planning system. The Minister of the Interior, who is authorised to call in any local plan for his additional approval, provides added oversight. A third, indirect means of oversight is through appeals—though rare—to the National Planning Board.

A small degree of decentralisation did occur in the planning system in 1995. This was expressed in two modest ways. First, whereas in the past the Minister of the Interior's signature was required for all local and district plans, now he is authorised to exempt any plan from his signature and he does so in most cases. Furthermore, after a period of time, approval is by default. Second, the 1995 amendment allowed certain types of local plans to receive final approval at the local level—those that do not alter basic land use but only make certain minor changes to a pre-existing plan. That same amendment, however, also introduced a modicum of greater centralisation in allowing developers who are unhappy with the local commission to go directly to the district level. Central government thus has almost all conceivable instruments to oversee local planning initiatives. In theory at least, these institutions could serve to ensure coordinated action to implement national planning policy—provided such a policy existed.

On the eve of the mass immigration crisis in 1989–90, the Planning and Building Law and the planning system came under severe criticism. Despite its grossly understaffed planning administration, the statutory planning system held back-burner priority in public interest and in budgets. The multi-layered approval process was regarded as chronically lethargic, and the planning system began to be seen as a bureaucratic, unnecessary impediment to economic development. Consumers often took their grievances to the High Court. When the crisis broke, the Planning and Building Law became the target for legislative change.

As a response to the crisis, the Planning and Building Procedures (Interim Law) was enacted in June 1990 (see Alterman, 2000). Although intended to self-terminate in two years, it was extended three times, until 1995. The crisis-time law greatly shortened procedures by creating even greater centralisation. With a 'one-stop shopping' philosophy, plan approval would bypass the local elected councils altogether. A new type of planning body—the Commission for the Construction of Housing and Industry (*valal* in its Hebrew acronym)—was set up in parallel to each of the district commissions. These small, compact bodies were composed of a few key central government bureaux, while the local authority was represented on a 'warm chair' basis only—it would join the commission for the discussion of a plan pertaining to its own city. The *valals* were authorised to decide on any local plan pertaining to 200 housing units or more, or to the expansion of industrial sites. The CPAL's authority was also decentralised, and was vested in the new commissions, with the CPAL's representative as a member of the committee with call-in powers (that were rarely exercised).

The streamlined procedures came not only at the expense of local political debate, but also at the cost of insufficient time for professional review, for public perusal, and for participation. The *valal* law did, however, succeed in speeding up the approval process of hundreds of plans for hundreds of thousands of housing units (Alterman, 2000). When the crisis law finally ended in 1995, the Knesset enacted what was supposed to be a partial substitute—an amendment to the regular Planning and Building Law which introduced some streamlining measures, but turned out to create even more litigiousness than the earlier law. The *valals* still have avid followers in central government and among developers, who recall crisis-time speed with nostalgia. Periodically, Cabinet decisions concerned with speeding up plan approval ask the Minister of the Interior to propose amendments or new legislation similar to the *valal* law. So far that has been only a threat.

THE FOUR TYPES OF STATUTORY PLAN

The 1965 law added national planning over the two tiers that had existed previously—the local level and the district level. The result is a three-tier edifice of plans. Lower-level plans must be strictly consistent with all higher-level plans. Since every action of construction or demolition, whether big or small, requires a building permit, national land-use and development rules should 'seep down' all the way to the building permits issued by local government planners (Fig. 3).

The top tier consists of national plans prepared by the National Planning Board, discussed in detail below. Occupying the second tier are the district plans prepared for each of Israel's six statutory administrative districts and approved by the National Board. The function of these plans is to translate national plans to the district level, to coordinate among local plans, and to propose regional plans. But district plans, tightly sandwiched between national and local plans, have always been of less importance than either of the other two. In most cases, district plans were prepared much later than the law's 1971 mandated date and have had little visibility

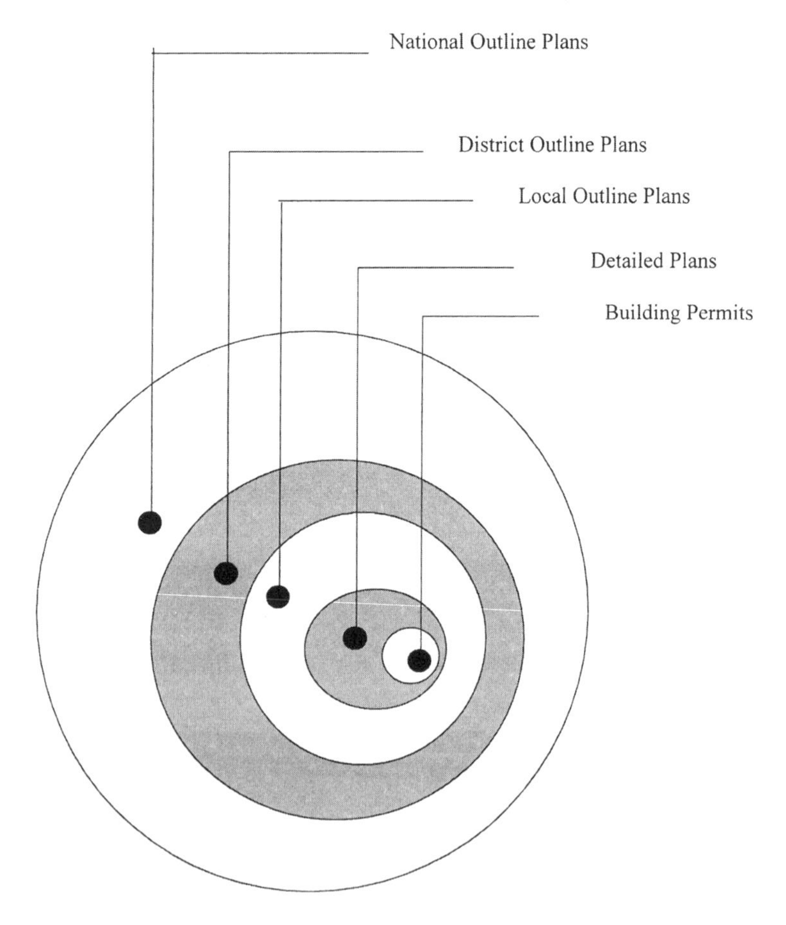

Figure 3 *The hierarchy of statutory plans to which a building permit must conform*

or impact. However, in the 1990s, when development encroached on green areas, regional structures changed, and national plans created new concepts, district plans came under the planning spotlight, and a new wave of preparation and updating is currently under way.

At the lowest level are mandatory local outline plans and optional detailed plans. These are the main instruments for regulating development, and any new construction must usually be anchored in one or both. American readers can view outline plans as a cross between zoning and subdivision regulations, and detailed plans can be viewed as parallel to site plans or planned unit development plans. Readers from most Continental European countries will recognise outline and detailed plans as similar to local plans prevalent in their respective countries. British readers can consider them as similar to their planning schemes before 1947.

The wording of the law leads one to assume that a local outline plan would cover the entire municipal area, but, in fact, very few local authorities have up-to-date comprehensive plans. Instead, most local authorities have a quilt of countless amendments to some older outline plan, and it can be said that 'the dog wags the tail'.[27] The hierarchy in Figure 3 has thus partly collapsed. Otherwise, the notoriously slow plan-approval procedures would have been even worse.

National statutory plans

The law authorises the National Board to prepare national outline plans which are approved by the Cabinet. Preparation of these plans is not mandatory, but, once prepared, they have to be adhered to by all lower plans, down to building permits. The importance attributed to these plans by the 1965 legislators is indicated by the fact that, unlike the lower types of plans, the public has no right to preview or submit objections to national plans and there is no requirement for direct public consultation or a public hearing. In recent years, the Board has developed its own discretionary modes of public input. The rationale for the absence of participation rights is that in the case of national plans (many of which deal with LULUs[28]), the public interest should override local or private opposition. Another part of the rationale is that national-level projects sometimes require speedy approval. Neither of these rationales have proved watertight.

Media coverage of national plans—including sectoral plans—has grown greatly in recent years. Until the early 1990s, most developers and planners knew little about these plans, with the possible exception of the plans for roads and for population distribution. This is not surprising if one considers that, in the absence of any requirements for citizen review, these plans were literally inaccessible even to local government planners. Most national plans could only be perused at the Ministry of the Interior's central office! Recent years have seen a revolution in this unsavoury situation. Today, copies of these plans are much more easily available. Some are well known and their importance is widely recognised. The Hebrew acronym for a national plan—*tama*—which only a few years ago was still planners' exclusive jargon, is today almost a household word.

SECTORAL PLANS

The language of the law speaks about '*the* national plan', indicating that the legislators probably expected that the entire country would be covered by a single, comprehensive national plan. In practice, no such comprehensive plan was prepared until 1990, and it is my assessment that no such plan would have been prepared for many years—or generations—were it not for the mass immigration crisis (the well-known exception—the Population Distribution Plan—was comprehensive only in the geographic sense). Instead, the National Board concentrated on sectoral plans—those initiated by agencies such as those surveyed above. Most of these plans have proved to be extremely important instruments for shaping development according to national policy. I shall turn to comprehensive national plans shortly, but first, let us take a closer look at sectoral plans.

Each of the 30 or so sectoral plans prepared to date (some in the consecutive list of 38 have been aborted) deals with a subject area which is authorised in the law and which members of the Board regard as having national importance. These plans tend to be quite detailed—sometimes too detailed—traditional land-use plans. The range of sectoral national land-use plans is probably unique to Israel. In the other countries covered in this volume, some of these topics are dealt with by separate national legislation, while others are not handled at the national level at all, but at the regional or local levels.

The list of sectoral plans includes roads, airports, railways, parks and nature reserves, surface and underground water reservoirs, tourism, coastlines, mining and forests. It also includes a whole range of LULUs which, in a small country, are viewed as a national-level interest: power plants, cemeteries, rubbish disposal sites, prisons, sewage purification plants, oil and gas pipelines, telecommunication sites, and even petrol stations (happily, the anachronistic latter has been phased out). These plans, though 'only' sectoral, have played a major and essential role in shaping Israeli land use, in supplying services, and in protecting its environment. For example, were it not for the plan for parks and nature reserves prepared in the 1970s, open space protection would have had to rely solely on the Commission for the Preservation of Agricultural Land, which has proved only partially effective. The few contiguous high-quality open landscapes still available—such as the Carmel mountain—would have probably become real-estate sites. Were it not for the plan for roads, the rights of way would probably have been gnawed up, and in land-tight Israel, adequate alternatives would not have been forthcoming. Similarly, water reservoirs and coastline areas would have become built up. The recent plan for afforestation has managed to preserve, at the last moment, some of the extant open spaces not declared as parks and not used for agricultural purposes. A plan currently in the spotlight of professional and public debate is the coastline preservation plan which has tried to protect the last patches of open coast (a gold-mine in real estate), but it is criticised for not being stringent enough.

THE NATIONAL PLAN FOR POPULATION DISTRIBUTION (NUMBER 6)

This plan used to be mistakenly regarded by some analysts as *the* national plan.[29] The population distribution policy has been the country's most long-standing and consistent urban and regional policy, a legacy from Israel's formative years. While the first versions of this plan were issued in the 1950s (Dash and Ephrat, 1964), the first statutory version was approved in 1975 and (unofficially) updated in 1985; it is still in effect, though overtaken by Plan 31 and, soon, Plan 35.

The National Plan for Population Distribution sets quantitative population caps or goals for each town and village. The rationale for the caps is not American-style 'growth management', but rather the desire to direct inhabitants and investment to peripheral areas, against market forces. A cap lower than estimated demand is placed on towns and cities in the central area, while an overly optimistic growth goal is placed on towns on the peripheries (Reichman, 1973). In practice, however, one does not see major cities in the central districts halting their population intake because of the plan. The plan has probably had a more significant effect on policies for new towns in the peripheral areas. That is not to say that the population distribution policy itself has not been effective, but that it has been implemented more through economic incentive tools than through the statutory plan.

THE NEW GENERATION OF COMPREHENSIVE PLANS

The mass immigration crisis of the early 1990s should be credited with what can only be called a revolution in national land-use planning in Israel. It is my assessment (reasoned in detail in Alterman, 1995) that were it not for the crisis, this revolution would not have happened at this time, if at all, and certainly not on the same scale. The story of two of the new-wave national plans completed to date deserves to be told in some detail. The latest such plan currently under preparation (national plan 35) awaits a future analysis.

The Combined National Plan for Construction, Development, and Immigrant Absorption (Number 31)

During the mass immigration crisis, senior planners from the Ministry of the Interior were dismayed to see how the physical and social landscape of the country was being permanently altered through short-range planning and action. They recognised the opportunity that the crisis offered for initiating long- and middle-range national planning. By then, the budgets and person-power allocated to the Ministry of the Interior for planning had withered to a ludicrously low level. Not since the 1950s had there been a serious attempt at countrywide integrated planning—until that date, all national plans had been sectoral. A senior planner in the Ministry of the Interior who headed the national and district plans office seized the moment, and wrote to the Minister:

> For the first time since the establishment of the State, the Ministry of the Interior is facing an exceptional challenge and the opportunity to take the lead in comprehensive national planning and land-use alignment for the

absorption of mass immigration. (Internal memo from Ms Dina Rachevsky, director of national and district plans, to the Minister of the Interior, Arie Deri, dated 20 May 1990; translated from the Hebrew by the author)

The planner's first step was to convince the heads of the Ministry of the Interior of the importance of preparing a special statutory national plan for the immigrant absorption challenge. It would be middle-range and would be prepared speedily so as to respond to the new needs created by the crisis. The idea was so well received that the bill of the crisis-time Interim Law mentions a 'national plan for immigrant absorption', and gives it special standing over and above other national plans, even though, at that time, the new plan had not yet been commissioned!

The architect-planners' team[30] that won the tender to prepare the plan quickly assembled a 30-person planning team—the largest since the 1950s—that included not only land-use and infrastructure planners, but also economists, demographers and social planners, but no public participation. The plan's steering committee was also unprecedented in scope, including every relevant government agency, but there was no access to public interest groups (except through media coverage) and no direct public participation. Their task was to prepare a national plan for the coming five to seven years so as to guide the accelerated planning and development decisions country-wide. The plan's terms of reference called for integrating land-use, environmental, transportation, economic development and social policies. Such terms of reference greatly expanded the hitherto accepted conception of the issues addressed by the National Planning Board, but the non-land-use aspects were not viewed as implementable through a national statutory plan under the Planning and Building Law.[31] The team produced the first draft document in record time (six months), at the end of 1991. The plan quickly received clearance from the National Planning Board in 1992, and Cabinet approval in 1993 (recall the *unanimous* story . . .).

One of the central issues that the plan brought to the forefront of heated professional debate was the desirability of continuing the long-entrenched national population distribution policy. The Plan 31 team and steering committee accepted the thesis which the economists from the Ministry of Finance had recommended—that during the plan's first period, economic considerations be maximised, and the development effort be directed to the country's central areas. Not turning its back entirely on the population distribution policy, the plan recommended that in the 'second phase', the outlying areas be targeted for enhanced development. However, since the plan's life was for only five to seven years, the 'second phase' was to be *après le deluge*.

Plan 31 became the most important planning document to guide land-use planning and development in Israel for the next five years. Its most important effect was that it successfully placed some restraint on the development appetite of the Ministry of Housing and private band-wagon riders. During this crisis, such an appetite might have endangered most of Israel's open space reserves—modest enough even before the crisis in this small and densely populated country.[32]

The planning process of Plan 31 brought to national planning a degree of public–professional exposure they had never received before. At the time, however, that was not a very high level to reach, given the almost unknown and inaccessible status of most of the sectoral plans. Plan 31 established a new norm for public exposure of land-use planning in Israel, which was to be amplified even more in the two subsequent comprehensive plans, described below.

An integrated long-range, non-statutory plan: 'Israel 2020'

The most ambitious planning initiative in Israel's history was actually launched a year before the crisis. The idea was conceived by one of the country's best-known planner-architects (Adam Mazor), and was modestly launched as the initiative of the Israel Association of Architects and Planners and the Technion—Israel Institute of Technology, where the planner served as a part-time professor. But the project—called 'Israel 2020: A Master Plan for Israel in the 21st Century'—found no financial support. Were it not for the crisis, the project would probably have remained a modest voluntary, rather 'academic' effort by a small group of professionals and academics outside government, willing to volunteer their time.

But as thousands of immigrants poured into the country month after month, Mazor sensed the opportunity to convince all relevant government and quasi-government bureaux of the utility of long-range national-level planning. He managed to raise funding from almost every conceivable government and quasi-government agency interested in some aspect of spatial planning, all of which served on the plan's steering committee. Mazor stressed the fact that the land resources of Israel were limited, and that the country was already one of the more densely inhabited countries in the world. This combination of factors, along with the mass immigration wave, he argued, would lead to unbearable environmental, economic and social conditions if land resources were not planned within a long-range, comprehensive view.

The National Planning Board acceded to the request of senior planners at the Ministry of the Interior to give its blessing to this long-term planning enterprise. It commenced in parallel with the statutory middle-range Plan 31, but it would be much broader in scope, would have Israel's long-range future as its horizon, and would take much longer to prepare. The involvement of the National Board—though symbolic—was unprecedented. This was not to be a statutory plan, not even a government plan, so this decision reflected an unusual recognition of the 'net' importance of long-range integrative planning.

The subject span of this ambitious strategic plan included almost every sphere of public policy that directly or indirectly impinges on spatial development: land use, environmental issues, the economy, social and demographic issues, water, agriculture, transportation, infrastructure, education, institutional and legal structures, relations with the world's Jewish communities, and even security issues. The team of professionals from many areas of professional and academic specialisation numbered 250 at the height of the project. The plan, which had begun as a conventional master plan, took on a wide strategic, multi-sectoral view and developed innovative planning methods, but it, too, remained an expert-led plan and did not make an effort to involve interest groups or the broader public directly.

Perhaps the greatest 'political' achievement of the Israel 2020 planning enterprise was the fact that, during its seven years of preparation, it managed to survive three changes of government and receive the continuing financial and steering support of each consecutive regime. In Israel, in the volatile Middle East, a planning enterprise such as 'Israel 2020' deals with issues that are, potentially, very sensitive politically. Demographic projections of Jews and Arabs within Israel, the spread of development within the country's various regions, and even social relations among various national and ethnic groups are issues that might have easily activated the raw nerves on both sides of the party-political rift, between doves and hawks in turbulent Israel. This remarkable immunity to the stormy political situation is due not only to the professional prestige of the project's leader and team, but also to the commitment and professional-political skills of the senior planners in the various government bureaux who served on the steering committee, each of whom had to continuously 'sell' the plan to their (changing) ministers, and to changing governments.

The set of over 30 volumes produced by this project during its six years of operation—the final reports were published in 1997 in printed form and on a widely distributed CD-ROM—functions in some ways like a national think-tank covering, in a systematic manner, public policy issues never before placed back-to-back. Although the plan was not intended to become statutorily-binding, the plan was presented before the National Planning Board and was symbolically presented to the President of Israel in a special ceremony held at the Technion in January 1998 (Alterman, 1997b).

Even before its official completion, the Israel 2020 plan had already had considerable influence as a policy document, and this without its being a statutory plan and without its having any official status. The project has raised the planning and public policy discourse in Israel to a new plateau, providing a set of concepts and a new language spoken today as a matter of course not only by professional planners, but also by other policy-makers and politicians. The Israel 2020 plan has also granted potential users a sophisticated database and new planning methodologies on which to base future national or regional plans.

A long-range, comprehensive statutory plan (Plan 35)

As Plan 31 approached the end of its term in the late 1990s, the Ministry of Housing, the Lands Administration and the Ministry of Finance argued against extending it. They contended that Plan 31 was overly detailed and restrictive in its population caps, that it preserved too much open space, and therefore did not allow enough land reserves for development for Israel's medium- and long-range future.

By this time too, the norms of what should be expected of national planning had thoroughly changed as a result of Plan 31 and of the Israel 2020 project. So, in 1997, the National Planning Board decided to commission Israel's first comprehensive, long-range statutory national plan—Plan Number 35, targeted for the year 2020. The new plan covers the entire country and deals with all major land uses, infrastructure and environmental issues and is to be more flexible, partially taking on the character of a policies plan rather than a traditional land-use plan (that remains to be seen). It also has chapters on economic development, social issues, etc., but these are not adoptable through the statutory planning channels and are therefore virtually ignored. The plan is intended to provide a long-term continuation for National Plan 31 and its terms of reference stipulate that, where appropriate, it should also incorporate the ideas of the 'Israel 2020' plan. It is the first national plan where some degree of public consultation was included—though in the absence of a public hearing or compulsory consultation process, this did not become much more than a lip service or a 'planningly correct' act.

Whereas, ten years earlier, architects used to condition taking on planning tasks on a promise of getting architectural design jobs, this time there was hot competition among the country's leading private architect-planners' offices. Each one put forward an impressively large and multi-disciplinary team of specialists in various planning and related fields. By 1998, the public profile of land-use planning had risen so much that even the competition over the tender made the daily press, and each interim presentation was visibly covered—an exposure that Plan 31 had not achieved.

National Plan 35 was completed in 1999, but it is my assessment that the plan will be hotly debated and will not receive the National Board's and the Cabinet's approval for a long time. Its degree of usefulness may therefore not be too different from the non-governmental Project 2020 plan. Will the new plan be able to bridge Israel's growing socio-political schisms? Will it be effective, and for how long? Given the growing rifts in Israeli society and politics, I may be permitted some scepticism.

The future of national-level planning in Israel: Will the momentum continue?

Let us turn the movie back, and attempt to recreate Israel's recent planning history in a scenario without the mass immigration crisis of the early 1990s. What would national-level planning have looked like? Would it have been strengthened, or would it have withered away? Israel had been undergoing general trends of decentralisation, deregulation and privatisation, similar to most other Western countries. Planning institutions, laws and professional norms were caught between these conflicting trends of privatisation and national regulation and control.

I would conjecture that, were it not for the mass immigration crisis of the early 1990s which tended to recentralise decision-making powers, integrated national-level planning in Israel would likely have continued its slow decline for some years more, in favour of more market-led development approved 'bottom-up'. At the same time, however, the instruments of national planning would probably not have been officially dismantled, and would have remained in a reduced format—mostly on the sectoral level—to reflect Israel's special needs. These stem from Israel's high population density, high natural growth rate (relative to other advanced-economy countries), commitment to immigrant intake, and economic growth. Mass immigration provided a great service: it advanced the clock of the next crisis that would have come seven to ten years later—a crisis caused by sprawling development, collapsing infrastructure, disappearing green spaces and threatened environment and water. At that time, without effective, integrated national-level planning, everyone would have paid the price.

The mass immigration crisis contributed to changing this trend, bringing national-level planning to a prominence it had never had before, not even during the country's formative years, the heydays of centralised, consensus-generated national planning. The crisis, which brought about highly accelerated development and pressured land and other resources, brought home to decision-makers the usefulness of planning. The crisis provided astute planners with the opportunity to argue their case to willing ears, and to get decent budgets for planning enterprises after decades of decline.

At first glance, the new planning defies the major tenets of planning theory in recent decades: it is centralised, it is comprehensive, and it is long-range. But a closer look shows that the new planning is of a new mode: it is more participatory and more transparent than was customary in Israel before (though it is less participatory than in most other Western countries). It can no longer rely on a socio-political consensus but rather reflects the outcome of a conflict-mediation process, often a litigious one. The new national planning is not a playback of the planning of the 1950s. It is much reformed and reflects the changes that have occurred in planning thought internationally as well as the enormous changes that have occurred in Israeli society and polity.

As land-use planning issues have grown in saliency and prominence, so have the conflicts surrounding them. Will the new plans be implementable as Israel's social and political fabric becomes more and more strained, and as land-use and development issues are becoming increasingly enmeshed with the country's deep-seated social and political controversies? Will the new plans be effective despite the fact that the country is consistently extending its trends of privatisation and decentralisation? Plan 35—Israel's first statutory national plan that is both comprehensive and long-range—could also turn out to be the last of the large-scale planning enterprises. One thing is certain: land-use planning is unlikely to return to the dusty corner of neglect that it occupied not very long ago. In the future, land-use planning at all levels will continue to be prominent in the public eye, will draw upon high-level professionals, and will have to deal with more and more controversial questions, engendering hot public debates.

REFERENCES

ALEXANDER, ERNEST R. (1998), 'Planning Theory in Practice: The Case of Planning Highway 6 in Israel', *Environment and Planning B: Planning and Design*, **25**, 435–45.

ALEXANDER, ERNEST R., RACHELLE ALTERMAN and HUBERT LAW YONE

(1983), *Evaluating Plan Implementation: The National Statutory Planning System in Israel* (*Progress in Planning* monograph series, Vol. 20, Part 2), Oxford, Pergamon.

ALTERMAN, RACHELLE (1981), 'The Planning and Building Law and Local Plans: Rigid Regulations or a Flexible Framework', *Mishpatim* (*Laws*) (Law Journal of the Faculty of Law, Hebrew University, Jerusalem), **11**, 179–220 (Hebrew).

ALTERMAN, RACHELLE (1991), 'Planning and Implementation of Israel's "Project Renewal": A Retrospective View', in Alterman and Cars (eds), 147–69.

ALTERMAN, RACHELLE (1995), 'Can Planning Help in Time of Crisis? Public Policy Responses to Israel's Recent Wave of Imigration', *Journal of the American Planning Association*, Spring, 156–77.

ALTERMAN, RACHELLE (1997a), 'The Challenge of Farmland Preservation: Lessons from a Six-Country Comparison', *Journal of the American Planning Association*, **63**, 220–43.

ALTERMAN, RACHELLE (1997b), *From Long-Range Planning to Implementation: Institutional Changes and Decision-Making* (policy report in the set of final reports of *Israel 2020: A National Plan for Israel in the 21st Century*) Haifa, Technion Research and Development Foundation (Hebrew).

ALTERMAN, RACHELLE (1999), *Israel's Future Land Policy*, Floresheimer Research Institute, Jerusalem (Hebrew).

ALTERMAN, RACHELLE (2000), 'Land-use Law in the Face of a Rapid Growth Crisis: The Case of the Mass Immigration to Israel in the 1990s', *Washington University Journal of Law and Policy*, **3**, 773–840.

ALTERMAN, RACHELLE (2001), 'The Land of Leaseholds: Israel's Extensive Public Land Ownership in the Era of Privatization', in Yu Hung Hong (ed.), *International Experience in Leasing Public Land*, Cambridge, MA, Lincoln Institute of Land Planning.

ALTERMAN, RACHELLE (forthcoming), *Planning in the Face of Crisis: Land and Housing Policy in Israel.*

ALTERMAN, RACHELLE and GORAN CARS (eds) (1991), *Neighborhood Regeneration: An International Evaluation*, London, Mansell.

ALTERMAN, RACHELLE and MORRIS HILL (1986), 'Land Use Planning in Israel', in Nicholas N. Patricios (ed.), *International Handbook on Land Use Planning*, Greenwood, 119–50.

ALTERMAN, RACHELLE and MICHELLE SOFER (1994), *Streamlining Procedures? The Degree of Success of Recent Legislative Attempts to Shorten the Time for Approval of Statutory Plans*, Haifa, Center for Urban and Regional Studies, Technion—Israel Institute of Technology (Hebrew).

BRUTZKUS, ELIEZER (1988), 'The Development of Planning Thought in Israel', *City and Region – Ir Ve'ezor*, **18**, 188–98 (Hebrew).

COHEN, ERIC (1970), *The City in Zionist History* (Urban Studies Series), Jerusalem, Hebrew University.

DASH, J. and E. EPHRAT (1964), *The Israel Physical Master Plan*, Jerusalem, Israel Ministry of the Interior.

DROR, YEHEZKEL (1989), *A Grand Strategy for Israel*, Jerusalem, Academon—the Hebrew University Students' Printing and Publishing House (Hebrew).

DROR, YEHEZKEL (1998), *Grand-Strategic Thinking for Israel* (Policy Papers No. 23), Ariel, Ariel Center for Policy Research, The College of Judea and Samaria.

LAW YONE, HUBERT and RACHEL WILKANSKY (1984), 'From Consensus to Fragmentation: The Dynamics of Paradigm Change in Israel, *Socio-Economic Planning Sciences*, **18**, 367–73.

MARGALIT, LEERIT and RACHELLE ALTERMAN (1998), *From Fees to Contracts: Methods for Involving Developers in the Supply of Public Services*, Haifa, Center for Urban and Regional Studies, Technion—Israel Institute of Technology (Hebrew).

MAZOR, ADAM (1993), 'The Land Resource in Spatial Planning', in *Israel 2020: A Master Plan for Israel in the 21st Century, Stage 1, Vol. 2*, Haifa, Israel Engineers' and Architects' Association/Faculty of Architecture and Town Planning, Technion—Israel Institute of Technology (Hebrew).

RAZIN, ERAN (1996), 'Municipal Reform in the Tel Aviv Metropolis: Metropolitan Government or Metropolitan Cooperation?', *Environment and Planning C: Government and Policy*, **14**, 39–54.

RAZIN, ERAN (1997), 'The Local Government Map in Israel for the Year 2020', in *Spatial Economic and Municipal Aspects* (policy report in the set of final reports of *Israel 2020: A National Plan for the Israel in the 21st Century*), Haifa, The Technion Research and Development Foundation (Hebrew).

RAZIN, E. and A. HAZAN (1995), 'Industrial Development and Municipal Organization: Conflict, Cooperation and Regional Effects', *Environment and Planning C: Government and Policty*, **13**, 297–314.

REICHMAN, SHALOM (1973), 'On the Relevance of Israel's Geographic Population Distribution Plans', *Ir Ve'ezor*, **1**, 26–43 (Hebrew).

SHACHAR, ARIE (1993), 'The National Planning Doctrine in Israel', in *Israel 2020: A Master Plan for Israel in the 21st Century, Stage 1, Vol. 2*, Haifa, Israel Engineers' and Architects' Association/Faculty of Architecture and Town Planning, Technion—Israel Institute of Technology (Hebrew).

SHEFER, GABRIEL (1996), 'Society, Politics, Government and National-Level Planning in Israel', in Rachelle Alterman (ed.), *Towards the Implementation of the 'Israel 2020' Plan: National-level Planning Institutions and Decisions in Ten Countries*, in *Israel 2020: A Master Plan for Israel in the 21st Century, Phase 3, Report no. 9*, Haifa, Israel Engineers' and Architects' Association/Faculty of Architecture and Town Planning, Technion—Israel Institute of Technology.

UNITED NATIONS (1996), *Demographic Yearbook*, New York, United Nations.

YIFTACHEL, OREN (1992), *Planning a Mixed Region in Israel: The Political Geography of Arab–Jewish Relations in the Galilee*, Aldershot, Avebury.

NOTES

1 The West Bank and the Gaza Strip have never been annexed to Israel *de jure*. Most of the Gaza Strip and some areas of the West Bank (the most densely inhabited by Palestinians) have become part of the Palestinian Authority created through the Oslo Peace Accords. The future of the areas still occupied by Israel is a highly charged political issue within Israel, and according to the Oslo Accords, it is to be negotiated with the Palestinians. The Golan Heights, a small area in the north-east, has undergone quasi-annexation by Israel from Syria through Israeli domestic legislation. Its international status is, however, in contention, and the future of this area, too, is often brought up as a topic for peace negotiations.

2 For a detailed analysis of Israel's immigration policy and its impact on urban and regional policy, see Alterman (forthcoming). Israel's immigrant intake relative to population size is considerably higher than Germany's—another country within our sample, which has had a commitment towards taking in a particular group of people (ethnic Germans).

3 To present the full picture one should remember, though, that cumulatively, the vast majority of Israelis were (and still are) living in apartment buildings. Furthermore, the new ground-attached housing, though regarded as very low-density for Israeli urban areas, is typically planned at 12 units to the net acre, which in the USA and Canada would be regarded as rather high density.

4 A reminder: I am referring to the international law status of the borders of Israel proper, without the areas occupied in the 1967 war.

5 In the past, both Labour and Likud supported the population distribution policy. Today, it is partly a partisan issue. The Likud government, back in power in 1996–99 after four years of a Labour-led government, reinstated some of the elements of the population distribution doctrine and its underlying geopolitical goals. Likud is likely to resume that policy in its new term after the 2001 elections.

6 As a *pro bono* adviser to the Knesset Committee for Interior and Environmental Affairs on major amendments to planning legislation, I can testify to this first-hand.

7 Before that law was revised in 1995, there was no numeric cap and Cabinets at times comprised more than 20 ministers.

8 For example, when Israel's exemplary Project Renewal was instituted in 1980, no specific legislation was necessary. The Cabinet allocated budgets and ministerial responsibility for that project, and the minister responsible set up goals and the institutional structure for decision-making (Alterman, 1988).

9 For example, Government Decision 4162, 12 August 1998.

10 Following one of the Cabinet's decisions on this matter, a colleague and I were commissioned to develop guidelines for higher-density housing.

11 The Minister of the Interior may have indirect influence at the local level as well. Although in most local jurisdictions the Minister of the Interior does not appoint any of the members of local statutory planning commissions (these are composed in most cases of the elected councillors), he potentially does have some influence over local planning, above the statutory responsibilities for overseeing local planning decisions that the law assigns to national government bodies.

12 But at the same time, the Minister and his staff are members of the statutory bodies, alongside other members who are appointed by other ministers or agencies, and are legally expected to use their independent discretion in their decisions while taking their respective agency's interests into account.

13 I cannot resist noting that, to the best of my knowledge, all this hefty involvement in urban and regional matters has been carried out without a single urban and regional planner being hired; only economists have been involved. My belief is that the trend of involvement of the Ministry of Finance in urban and regional planning will continue, and even intensify. It is therefore high time to consider an appropriate professional mix of 'referees'—as they are called—assigned by the Ministry of Finance to initiate and oversee planning matters.

14 The Ministry of Finance's aims are to enhance market-led policies; to watch over the national budget; to be concerned with tax implications, etc.

15 The Ministry of Finance has even encroached into the territory of the Ministry of Housing—a stronger ministry than the Interior. This may have to do with the failure of the Ministry of Housing to assuage the Cabinet's concerns regarding housing supply.

16 History fans will remind us that, during the state's formative years, the planning function was, for a short time, under the auspices of the Prime Minister's Office, before the national statutory planning institutions were set up. But the current modest activity is in no way a citation of that long-forgotten role.

17 In the early 1990s I was appointed as a member of one of the sub-committees of that Council. I was never invited to any meeting, nor saw any output of the sub-committee or the Council as a whole.

18 As a member of the Israel 2020 management team who was charged with investigating alternative institutional structures, I twice joined the project's head, Professor Adam Mazor, in meeting Dr Beilin for that purpose.

19 Dr Beilin remained supportive of the Israel 2020 project and helped promote it among key Cabinet members; but in June 1996, Labour lost the elections.

20 For a more detailed description of the Israel state-lands system, see Alterman (2001).

21 For an explanation of this seemingly paradoxical comment, see Alterman (2001).

22 I was not aware of the shift in focus to West Bank planning of that particular agency until the former director general of the Prime Minister's Office, Mr Avigdor Lieberman, explained on a radio broadcast (Network 2, on 18 October 1998) that this is the reason why he has lobbied for the appointment of a non-professional person to head that department—a person linked to Mr Lieberman's ultra-right-wing views.

23 The importance of this shift nationally and internationally is analysed in Alterman (1997).

24 Laws of the State of Israel, 1965 (available in English). Not much has recently been written on the Israeli planning system in the English language. For some more detail on the system and its operation, see Alterman and Hill (1986), Alexander et al. (1983) and Alterman (2000).

25 My guess is that, if such a bill were proposed, the Knesset Committee for Internal and Environmental Affairs would encourage a compromise.

26 In rural areas or regions with a number of small towns, the local commissions are composed of one representative only from each town as well as central government representatives.

27 In older cities, the original plan is usually one prepared under the British before 1948, while in new towns it is one prepared by the Ministry of Housing in the 1950s or 1960s when the town was established.

28 'Locally unwanted land uses'.

29 See, for example, Yiftachel (1992), 95–98.

30 Led by architect Raphael Lehrman of Tel Aviv, who hired a large team of planners and consultants in the areas covered.

31 A conception accepted by many, but not by myself. I had written a law review article in the 1980s showing that, even without amending the Planning and Building Law, plans prepared under it do not have to be restricted to narrow, physical, land-use planning and can deal with broader policies as well as being flexible (Alterman, 1981).

32 The plan's major instruments for achieving the goal of open space preservation were the population caps it placed on ex-urban development and the priority it gave to the development of the major cities and to urban landfill. Although these caps were a somewhat 'old hat' repeat of the instruments of the long-standing National Plan for Population Distribution, they were more flexible and went together with an innovative policy of open space preservation in the 'rural spaces'. This was a new term coined by Plan 31 to indicate a more realistic and up-to-date policy for open space preservation than the rigid instrument of 'agricultural land preservation' which had partly become a misnomer. The plan had some immediate influence in helping to stop several environmentally and socially controversial new towns proposed by the Ministry of Housing.